Interactions of Biochar and Herbicides in the Environment

Interactions of Biochar and Herbicides in the Environment

Edited by
Kassio Ferreira Mendes

CRC Press is an imprint of the
Taylor & Francis Group, an **informa** business

First edition published 2022
by CRC Press
6000 Broken Sound Parkway NW, Suite 300, Boca Raton, FL 33487–2742

and by CRC Press
4 Park Square, Milton Park, Abingdon, Oxon, OX14 4RN

CRC Press is an imprint of Taylor & Francis Group, LLC

ISBN: 978-1-032-06397-3 (hbk)
ISBN: 978-1-032-06375-1 (pbk)
ISBN: 978-1-003-20207-3 (ebk)

DOI: 10.1201/9781003202073

Typeset in Times
by Apex CoVantage, LLC

Contents

Preface

The use of herbicides is a fundamental part of the current agricultural model, and without the use of these products it would likely be unfeasible to guarantee food security worldwide. On the other hand, these products have a series of environmental and social impacts, which demand a constant concern on the part of society. Soil is the final fate of herbicides applied in agriculture in the soil–plant system. When they come into contact with the soil, herbicides are subject to a series of physical–chemical processes that regulate their fate and behavior in the environment, which is extremely complex. The use of biochar in agricultural soils as fertilizer and soil conditioner has been further explored; however, little is known about the effect of this material on soils contaminated with herbicides, and little is known about its effect on weed control as well. The fate and environmental behavior of certain herbicides, as well as their efficacy when applied directly to the soil, such as pre-emergent herbicides with residual action, are strongly influenced by the retention bond with soil particles and organic carbon content. Therefore, the addition of biochar to the soil can easily potentiate the herbicide retention process, which, in addition to contributing positively to the reduction of chemical contaminants in the environment, can exert negative effects on herbicide behavior and effectiveness of these products in weed control. Thus, this book highlights the physical and chemical characteristics of biochar, which can impact on the sorption–desorption, runoff, leaching, and hence efficacy and degradation of herbicides. Furthermore, the removal of herbicides in water is also evidenced with the addition of biochar.

Editor

Dr. Kassio Ferreira Mendes is Professor of Biology and Integrated Management of Weed, Department of Agronomy, Federal University of Viçosa, Brazil; Post-Doctor (2019) and Doctor (2017) in Science – Nuclear Energy in Agriculture (Chemistry in Agriculture and Environment) by Center of Nuclear Energy in Agriculture, University of São Paulo, Brazil with interuniversity exchange doctorate University of Minnesota, USA (2016), Twin Cities Campus – College of Food and Agricultural Sciences in the Department of Soil, Water, and Climate. Master in Agronomy (Crop Science) by Federal University of Viçosa (2013); Agronomist by University of State Mato Grosso (2011); Member of Brazilian Society of Weed Science.

Contributors

Abdelrani Yaacoubi: Ph.D. in Water Sciences (1988) from the University of Poitiers, France. He is currently Full Professor at the Faculty of Sciences Semlalia, Laboratory of Applied Chemistry and Biomass, Cadi Ayyad University, Marrakech, Morocco. His research interests include water treatment and carbon materials.

Bruno Caio Chaves Fernandes: Chemistry laboratory technician at the Universidade Federal Rural do Semi-Árido, working on the behavior of herbicides in the environment. He holds a Ph.D. degree in Phytotechnics from the Federal Rural University of the Semi-Arid (2020). Master's degree in Plant Science from the Federal Rural University of the Semi-Arid (2017). Specialization in Auditing in Health Services by the Faculty of Medical Sciences of Campina Grande (2009). Degree in Pharmacy from the Federal University of Rio Grande do Norte (2008).

Camille Nunes Leite: Industrial Chemist (2016) from Federal University of Uberlândia. She is currently a Master's degree student in the Graduate Program in Soils and Plant Nutrition at the Luiz de Queiroz College of Agriculture, at the University of São Paulo, Brazil.

Candelario Palma-Bautista: Ph.D. student in Biosciences and Agri-food Sciences at the University of Cordoba (UCO) in Spain. Active member of the research group of Professor Rafael De Prado developing research on herbicide resistance in dicotyledonous weeds in the Department of Agricultural Chemistry, Edaphology and Microbiology of the UCO (2022). Research internship at the Department of Genetic Transformation and Functional Genomics, Institute of Sustainable Agriculture (IAS-CSIC) (2021). Master of Science in Plant Production, Protection, and Improvement by UCO (2019). Agronomist engineer specialized in agricultural parasitology by Universidad Autonóma Chapingo in Mexico (2017).

Daniel Valadão Silva: Professor of Agronomy at the Federal Rural University of the Semi-Arid Region – UFERSA, a CNPq Productivity Scholar and coordinator of the Postgraduate Program in Soil and Water Management at UFERSA. Post-doctorate (2015) in Agronomy and Doctor (2014) in Plant Science from the Federal University of Viçosa, Viçosa, Minas Gerais, Brazil. Master in Plant Production from the Federal University of the Jequitinhonha and Mucuri Valleys – UFVJM (2011). Graduated in agronomy from UFVJM in 2009. Conducts research related to Weed Management and Impact of Herbicides on the Environment.

Felipe Hipólito dos Santos: He holds a B.S. degree in Agronomy from the State University of Midwestern Paraná (2018) and an M.Sc. in Soil and Plant Nutrition from the University of São Paulo (2021). He is currently a Ph.D. candidate in the Graduate Program of Soils and Plant Nutrition at the Luiz de Queiroz College of Agriculture, University of São Paulo, Brazil.

Fernando Sarmento de Oliveira: Agronomic Engineer (2015) from the Federal University of Campina Grande, Brazil. Master (2017) and Ph.D. (2021) in Agronomy from the Federal Rural University of Semi-Árido, Brazil. He is currently a professor of Institute Center for Technological Education (CENTEC), Brazil.

Francisca Daniele da Silva: Master's student in Environment, Technology and Society at the Federal University of the Semi-Arid – UFERSA. Graduated in Environmental Management from Universidade Potiguar (2011). She has experience in the field of Environmental Sciences, with an emphasis on environmental management and conservation.

Gabriel da Silva Amaral: Graduated in Chemistry and Master in Organic Chemistry from the *Universidade Estadual de Santa Cruz*, Ilhéus, BA, focused on the isolation and characterization of secondary metabolites. He is currently a Ph.D. candidate in Chemistry at the Federal University of São Carlos (UFSCar), carrying out his research focused on organic chemistry, weed management, and herbicide resistance.

Ghizlane Enaime: Specialized Master Degree (2013) in Engineering and Management of Industrial Environment from the Cadi Ayyad University Marrakech, Morocco. Doctoral degree (2019) in Physical Chemistry and Environment from the Cadi Ayyad University Marrakech, Morocco. She is currently a postdoctoral researcher at the Institute of Urban Water Management and Environmental Engineering of the Ruhr University Bochum, Germany.

Gustavo Vinícios Munhoz-Garcia: Agronomic Engineer (2019) from the State University of Northern Paraná, Brazil. He is currently a Master student from the Center of Nuclear Energy in the Agriculture, University of São Paulo, Brazil, researching the behavior of herbicides in the soil, using radiometric techniques.

José Guadalupe Vázquez-García: Post-doctoral researcher at Department of Agricultural Chemistry, Soil Science and Microbiology, University of Cordoba (UCO). Ph.D. in Bioscience and Agri-food Sciences (2021) and Master of Science in Plant Production, Protection, and Improvement by UCO (2019). Agronomist engineer specialized in agricultural parasitology by Universidad Autónoma Chapingo (UACh), México (2018). With more than 30 papers published in congresses and international journals of high prestige.

Kamila Cabral Mielke: Agronomic Engineer graduated from the Federal Institute of Education, Science and Technology of Rondônia (2016). Master in Tropical Agriculture from the Federal University of Espírito Santo (2019) in the area of phytoremediation of soils contaminated with herbicides. Doctoral student in Crop Science from the Federal University of Viçosa (UFV) in the area of weeds. She is currently a member of the research Integrated Weed Management Group (MIPD-UFV) with activities in the areas of behavior and fate of herbicides in the environment, impact on soil microbiological activity, remediation of environments contaminated with herbicides, using potentially phytoremediation plants and carbonaceous materials (biochar and bonechar).

Kassio Ferreira Mendes: Agronomic Engineer (2011) from the State University of Mato Grosso. Master in Agronomy (2013) from the Federal University of Viçosa and Ph.D. in Nuclear Energy (2017) from the University of São Paulo, Brazil. He is currently an adjunct professor at the Agronomy Department of the Federal University of Viçosa, Brazil.

Manfred Lübken: Diploma in Civil Engineering (2002) from the Leibniz University Hanover, Germany. Doctoral degree (2009) from the Technical University Munich, Germany. He is currently a senior researcher at the Institute of Urban Water Management and Environmental Engineering of the Ruhr University Bochum, Germany. His research focuses on optimization of (waste) water treatment processes and mathematical modeling.

Manuel Alejandro Ix-Balam: Biologist with orientation in agricultural parasitology by the National Technological Institute of Mexico. Master and Doctor of Science in Entomology from the Federal University of Viçosa (UFV), Brazil. He has experience in the biology, ecology, biodiversity, behavior, and chemical ecology of insects of ecological and agricultural importance. He is currently an associate researcher at the Research Institute for the Sustainable Development of Ceja de Selva (INDES-CES), Peru, developing research on diversity and sexual behavior of insects for cocoa pest management.

Maria Fátima das Graças Fernandes da Silva: Graduated in Chemistry (1973) from the Faculty of Philosophy, Science and Letters of Ribeirão Preto, SP, Brazil, Doctor of Organic Chemistry (1978) from the University of São Paulo. She did post-doctorate at the University of Strathclyde, Glasgow, Scotland (1986–1988), and is a full professor in the Department of Chemistry at the Federal University of São Carlos – UFSCar, where she coordinates the research groups on Natural Products and Biorational Control of Insects and Phytopathogens. She has been a full member of IUPAC and the Brazilian Chemical Society (SBQ) and has published more than 250 scientific articles.

Matheus Bortolanza Soares: Agronomic Engineer (2015) and Master in Agronomy (2017) from the Federal University of Mato Grosso. He is currently a Ph.D. candidate in the Graduate Program in Soils and Plant Nutrition at the Luiz de Queiroz College of Agriculture, at the University of São Paulo.

Mingxin Guo: B.S., Environmental Soil Science (1990), Beijing Agricultural University, Beijing, China. M.S., Environmental Chemistry (1995), Chinese Academy of Sciences, Beijing, China. Ph.D., Soil Science (2001), The Pennsylvania State University, University Park, PA. Professor of Soil & Environmental Chemistry, Department of Agriculture & Natural Resources, Delaware State University, Dover, DE, USA.

Mohammed Loudiki: Ph.D. in Hydrobiology and Aquatic Ecology in 1990 from the University AIX Marseille III, Marseille, France. He is currently Full Professor at the Faculty of Sciences Semlalia, Cadi Ayyad University, Marrakesh, Morocco. His research focuses on water and soil quality biomonitoring and ecosystems rehabilitation as well as the research of green bioprocesses for the control, mitigation, and treatment of water contaminated by harmful algal blooms, especially cyanoHABs.

Mountasser Douma: Ph.D. in Microbial Biodiversity and Ecotoxicology from Faculty of Sciences Semlalia Cadi Ayyad University Marrakech, Morocco (2010). His research interests focus on biodiversity, biotechnology, and ecotoxicology of cyanobacteria and microalgae. He is currently Assistant Professor at the Polydisciplinary Faculty of Khouribga, Sultan Moulay Slimane University, Morocco.

Rafael De Prado: Emeritus Professor at the University of Cordoba (UCO), head of the Department of Agricultural Chemistry, Soil Science and Microbiology. He carries out studies of resistance confirmation and mechanisms of resistant weeds to herbicides. With more than 400 papers published in congresses and international journals of high prestige with more than 4,000 citations.

Renata Pereira Lopes Moreira: She is an assistant professor at the Department of Chemistry of the Universidade Federal de Viçosa, Brazil, since 2013, when she started to coordinate the Laboratory of Nanomaterials and Environmental Chemistry. Her current research is based on the synthesis of materials by sustainable processes, based on green chemistry principles, with the use of nanotechnology applied in the development of new products and processes, involved in the generation of income with the generation of zero waste.

Ricardo Alcántara-de la Cruz: Agricultural Engineer specialized in Agricultural Parasitology from the *Universidad Autonoma Chapingo*, Mexico (2008-2013), Master in Plant Production, Protection and Improvement (2013-2014) and Doctor in Biosciences and Agri-food Sciences (2014-2016) from the *Universidad de Córdoba*, Spain. He has extensive experience in herbicide resistance and integrated weed management. He has done postdoctoral studies at the *Universidade Federal de Viçosa* (2016-2016) and the *Universidade Federal University de São Carlos* – UFSCar (2018-2021) in Brazil, both with an emphasis on Plant Protection (biological pest control and weed management). He currently acts as a visiting professor at UFSCar's Lagoa do Sino campus.

Rodrigo Nogueira de Sousa: Agronomic Engineering (2016) and a Master's in Soil Science and Plant Nutrition (2018) from the Federal University of Viçosa, Brazil. He is currently a Ph.D. candidate in Soil Science and Plant Nutrition from the Luiz de Queiroz College of Agriculture, University of São Paulo. He had an exchange experience during his undergraduate studies at North Carolina Agricultural and Technical State University, USA, from 2014 to 2015. He also had an internship at the Department of Crop and Soil Sciences at North Carolina State University, USA, in 2015, for which he studied management of nitrogen fertilization in corn crops.

Taliane Maria da Silva Teófilo: Has a doctorate (2021) and master's degree (2009) in Phytotechnics from the Federal Rural University of the Semi-Arid, Brazil. Degree in

Agronomy (2007) from the same University. She has experience in weed management in agricultural crops and herbicide behavior in soil.

Tatiane Severo Silva: Is a Ph.D. graduate student at Federal Rural University of the Semi-Arid (2019-2023), Brazil, and a visiting PhD graduate student at UW-Madison, United States (2021-2022). Master in Agronomy from the Federal Rural University of the Semi-Arid (2019). Agronomist, graduated from the State University of Maranhao (2014). She has worked on the soil residual activity single and multiple active ingredient PRE-emergence corn herbicides and their effect on cover crops species and investigated the behavior of herbicides applied PRE-emergence and their potential impact on subsequent sensitive crops and the environment.

Tiago Guimarães: He is an chemist graduated from the Federal University of Espírito Santo (2016). Master in Agrochemistry from the Federal University of Espírito Santo (2018) in the area of adsorption of heavy metals by ornamental rock residues. Doctoral student in Agrochemistry at the Federal University of Viçosa (UFV) in the area of Environmental Chemistry, with emphasis on production and characterization of biochar, and its use in the removal of contaminants emerging from aqueous systems. He is currently dedicated to the research group of the Laboratory of Nanomaterials and Environmental Chemistry (LANAQUA-UFV), to the Cellulose and Paper Laboratory (UFV), and to the Integrated Weed Management Group (MIPD-UFV) in developing research in the areas of lignocellulosic biocomposites, dye adsorption, and herbicide degradation.

Valdemar Luiz Tornisielo: Bachelor in Ecology from São Paulo State University "Júlio de Mesquita Filho" (1979), Master in Entomology from University of São Paulo (1986) and Ph.D. in Nuclear Technology from University of São Paulo (1990). He is currently a researcher at the Center of Nuclear Energy in the Agriculture, University of São Paulo, Brazil.

Vanessa Takeshita: Agronomic Engineer (2017) from the State University of Mato Grosso, Brazil. Master in Science (2019) from the Center of Nuclear Energy in the Agriculture, University of São Paulo, Brazil. She is currently a Ph.D. student in the same institution, researching the behavior of nanoherbicides in the soil and plants, using radiometric techniques.

Weiping Song: B.S. Chemistry, Beijing Normal University (1994). PQ M.S. Quantum Chemistry, Beijing Normal University (1997). Ph.D. Physical Chemistry, Pennsylvania State University (2005). Lecturer at Department of Chemistry, Delaware State University, Dover, DE, USA.

Widad El Bouaidi: Master in Ecology and Management of Continentals Aquatics Ecosystems from Faculty of Sciences Semlalia Cadi Ayyad University Marrakech, Morocco (2018). She is currently a Ph.D. student at Phycology, Biotechnology and Environmental Toxicology research unit of the Faculty of Sciences Semlalia, Cadi Ayyad University Marrakech, Morocco.

Abbreviations

AWC	available water capacity
CEC	cation exchange capacity
EBC	European Biochar Certificate
EC	electrical conductivity
IBI	International Biochar Initiative
OC	organic carbon
SHC	saturated hydraulic conductivity
SSA	specific surface area
WHC	water holding capacity

1 Agricultural Applications of Biochar

Mingxin Guo[1*] *and Weiping Song*[2]
[1] Department of Agriculture and Natural Resources, Delaware State University, Dover, DE 19901, USA
[2] Department of Chemistry, Delaware State University, Dover, DE 19901, USA
*Corresponding author: mguo@desu.edu.

CONTENTS

1.1 INTRODUCTION

Biochar, a charcoal-like solid residual product from thermochemical processing of biomass materials, has been intensively explored in the past two decades as a soil amendment for promoting agricultural production. The idea of using biochar to enhance soil health and improve the crop productivity originated from the discovery of *Terra Preta de Índio* (meaning Amazonian Dark Earth in Portuguese) in the Amazon Basin in the late nineteenth century (Guo et al. 2016a). Subsequent research suggested that *Terra Preta de Índio* is a legacy of ancient Amazonian inhabitants who added charred wood, plant ash, manure, and fish bones to local

DOI: 10.1201/9781003202073-1

land soil (Sombroke 1966; Woods and McCann 1999; Glaser et al. 2000). Even after 500–2000 years of natural weathering and leaching, *Terra Preta de Índio* soils demonstrate significantly higher organic carbon (OC) contents and remarkably greater crop yield relative to the adjacent control soils (Sombroke 1966; Lima et al. 2002), implicating the capability of biochar to persistently sustain soil fertility and health.

Biochar is commonly defined as "the fine-grained or granular charcoal made from heating vegetative biomass, bones, manure solids, or other plant-derived organic residues in an oxygen-free or oxygen-limited environment and used as a soil amendment for agricultural and environmental purposes" (Guo et al. 2016b). Theoretically, all solid biomass materials can be used to manufacture biochar. Researchers have tested a wide variety of bio-derived organic residues as biochar feedstocks, including hard wood, soft wood, saw dust, scrap wood and shavings, tree barks, tree leaves, pine needles, greenwaste, switchgrass, Miscanthus, fescue straw, corn stover, corn stalk, corn cobs, cotton stalk, wheat straw, rice straw, rice husk, soybean straw, rapeseed straw, pepper straw, peanut hull, peacan shell, palm shell, coconut shell, orange peel, olive pomace, sugarcane bagasse, cottonseed meal, poultry litter, cow manure, swine manure solids, sewage sludge, cattle carcass, bone meal, and solid organic municipal waste (Guo, Song et al. 2020). The feedstock materials were generally dried; processed into small pieces, grains, or pellets; and then converted to biochar through pyrolysis (heating at 300–700°C without air) or gasification (heating at 500–900°C with controlled air supply) (Guo, Xiao et al. 2020). During the thermochemical carbonization processes, 10–65% of the feedstock OC was recovered in biochar, phosphorus (P) was nearly preserved, while the loss of nitrogen (N) was tremendous (Song and Guo 2012; Guo, Song et al. 2020). Considering the accessibility, collection infrastructure, and economic viability, crop residues (e.g., corn stover, wheat straw, and rice husk) and forest debris (e.g., floor litter, tree trimmings, and wood waste) are promising feedstocks for field-scale production of agricultural-use biochar. Specialty food-processing wastes like nutshells, olive pomace, sugarcane bagasse, and cottonseed meal are limited in supply and, therefore, may serve a localized biochar feedstock or be simply added to the main feedstock stream. Organic refuses in municipal solid waste may contain toxic elements (e.g., heavy metals) that contaminate the resulting biochar products. Conversion to biochar is an effective method to sterilize and stabilize animal wastes such as carcass, bones, manures, and biosolids, yet the significant feedstock N losses incurred by high-temperature thermochemical treatment (e.g., at more than 400°C) are a major concern. Biochar is also a byproduct of wood pyrolysis to harvest bio-oil (a liquid biofuel) and wood gasification to produce syngas (a gaseous biofuel) (Guo, Xiao et al. 2020). Utilization of such biochar byproducts in agricultural production is a value-added approach to promote the bioenergy industry.

Numerous research and trials have demonstrated that biochar, when applied to cropland in methods similar to applying other organic amendments like compost and solid manure, is promising to ameliorate soil health and promote crop productivity. In agricultural applications, biochar functions primarily as a soil conditioner and a fertility enhancer. Biochar is environmentally recalcitrant and can remain in treated soils for a duration of hundred to thousand years, providing consistent effects of improving soil porosity, aggregate stability, water conservation, nutrient availability and retention, microbial activities, and crop yield (Guo et al. 2016b). Biochar also contains more or less plant nutrients such as N, P, K, Ca, Mg, and S. These inherent nutrients supplement the crop nutrient supply and may act as important sources in the initial growing seasons after application to nutrient-poor soils (Ding et al. 2016). Nevertheless, biochar products prepared from various feedstock materials under different carbonization conditions vary greatly in OC content, carbon stability, nutrient contents, pH, porosity, and other quality characteristics (Guo, Song et al. 2020). When used in agricultural production as a soil conditioner and a fertility enhancer, differently sourced biochars demonstrate remarkably varied capacities for improving soil health and crop productivity. The efficacy of biochar amendment is further affected by the application rate and biochar placement in soil (Guo 2020). To maximize the benefits of biochar amendment in agricultural production, scientific biochar application programs, considering the soil type, crop species, biochar source, application rate, and biochar soil placement, are absolutely necessary. This chapter aims to elucidate the quality variations of various biochar products; summarize the ameliorating effects of biochar amendment on the soil's physical, chemical, and biological health; and recommend appropriate biochar application programs in agricultural production.

1.2 QUALITY VARIATIONS OF VARIOUS BIOCHAR PRODUCTS

Not all biochars are the same: products generated from different feedstock materials and carbonization conditions vary drastically in physical and chemical characteristics. Table 1.1 presents the general composition and basic properties of selected biochars reported in the literature. The feedstocks include wood, plant debris, bioenergy grasses, crop residues, food-processing wastes, manures, sewage sludge, and animal bones. The carbonization conditions extend to slow pyrolysis, fast pyrolysis, and gasification at 300–750°C. The typical biochar yields are 25–60% of dry feed mass for slow pyrolysis, 10–42% for fast pyrolysis, and 5–33% for gasification, depending on the feedstock type, feedstock pretreatment (e.g., moisture content, pelletization, and particle size), and carbonization operations (Guo, Li et al. 2020). During the thermochemical treatment, 10–65% of the feedstock OC may be recovered in biochar, with the rest lost in the co-products pyrolysis bio-oil and syngas (Guo, Li et al. 2020). These biochar products demonstrated an OC content ranging from 5.9% to 93.9% and a mineral ash content from 1.4% to 86.0%. Corresponding to the relatively high ash contents, biochars were typically alkaline, with the measured pH levels at 7.2–12.4. The cation exchange capacity (CEC) and specific surface area (SSA) values were in the ranges of 7.3 to 161.6 $cmol_c$ kg^{-1} and 0.5 to 222.5 m^2 g^{-1}, respectively. The total N, P, and K contents of the biochars were 0.0–45.0, 0.2–157.5, and 1.4–91.5 g kg^{-1}, respectively (Table 1.1). In general, biochars derived from wood and plant residues are greater in OC and SSA values but lower in ash content, pH, CEC, and nutrient (N, P, and K) contents than those derived from animal manure, sewage sludge, and bones. For the same feedstock materials, gasification and fast pyrolysis treatments yield biochar products with higher ash contents yet lower OC contents relative to the slow pyrolysis treatment. Increasing

TABLE 1.1

Quality Characteristics of Biochar as Influenced by Feedstock and Carbonization Conditions.

Feedstock	Pyrolysis conditions	pH	Ash %	OC %	TN g kg^{-1}	TP g kg^{-1}	TK g kg^{-1}	CEC cmol$_c$ kg^{-1}	SSA m^2 g^{-1}	Reference
Hard wood	400°C slow pyrolysis	7.5	3.2	79.0	2.5	0.18	3.0	7.9	15.4	Tian et al. (2016)
Hard wood	500°C slow pyrolysis	8.2	4.2	84.8	3.0	.34	3.6	7.5	26.6	Tian et al. (2016)
Soft wood	400°C slow pyrolysis	7.3		74.6	2.5		2.5			Gezahegn et al. (2019)
Pine chips	400°C slow pyrolysis	7.6		73.9	2.6	0.15	1.4	7.3		Gaskin et al. (2008)
Pine needle	300°C slow pyrolysis		1.9	68.9	10.8				19.9	Chen et al. (2008)
Oak bark	425°C slow pyrolysis		11.1	1.2	4.6				1.9	Mohan et al. (2011)
Greenwaste	450°C slow pyrolysis		10.8	71.1	11.7			52.2	7.3	Zheng et al. (2010)
Switchgrass	500°C slow pyrolysis	.0	7.8	84.4	]10.7	2.4	11.6	82	62.2	Ippolito et al. (2012)
Corn cobs	500°C fast pyrolysis	7.8	13.3	7.6		.36	43.4		<1.0	Mullen et al. (2010)
Corn stover	500°C fast pyrolysis	7.2	32.8	57.3		12.9	23.5		3.1	Mullen et al. (2010)
Soybean straw	300°C slow pyrolysis	7.3	10.4	68.8	18.0	3.25			5.6	Vithanage et al. (2017)
Pepper straw	600°C slow pyrolysis	9.3	25.0	70.0	9.0	3.3	49.0			Fryda and Visser (2015)
Pepper straw	670°C gasification	11.0	33.5	59.0	8.0	3.1	44.0			Fryda and Visser (2015)
Rice straw	400°C slow pyrolysis		34.1	46.3	12.6			60.7	19.7	Jiang et al. (2015)
Wheat straw	400°C slow pyrolysis	9.1	9.7	65.7	10.5			161.6	4.8	Kloss et al. (2012)
Sugarcane bagasse	400°C slow pyrolysis	8.3	11.0	70.7	5.7					Nwajiaku et al. (2018)
Rice husk	500°C slow pyrolysis	.2		47.8	0.0			17.6		Ghorbani et al. (2019)
Rice husk	750°C gasification		82.0	12.0	1.6	2.2	5.7			Fryda and Visser (2015)
Coconut shell	600°C slow pyrolysis	8.5	4.1	93.9	4.0				222.5	Windeatt et al. (2014)
Coconut fiber	600°C slow pyrolysis	9.6	13.5	82.6	24.0				23.2	Windeatt et al. (2014)
Peanut hulls	400°C slow pyrolysis	7.9	8.2	74.8	27.0	2.6		13.6	0.52	Novak et al. (2009a)
Olive Pomace	600°C slow pyrolysis	10.5	18.1	71.8	19.0				1.2	Windeatt et al. (2014)
Poultry litter	300°C slow pyrolysis	9.5	47.9	38.0	41.7	22.7	69.3	51.1	2.7	Song and Guo (2012)
Poultry litter	400°C slow pyrolysis	10.3	56.6	36.1	26.3	26.3	81.2	41.7	3.9	Song and Guo (2012)

Poultry litter	500°C slow pyrolysis	10.7	60.6	34.5	12.1	27.9	87.9	35.8	4.8	Song and Guo (2012)
Poultry litter	600°C slow pyrolysis	11.5	60.8	32.5	12.1	7.9	91.5	35.8	4.8	Song and Guo (2012)
Chicken litter	750°C gasification	12.4	86.0	7.0	2.0	8.9	3.3			Fryda and Visser (2015)
Pig manure solids	700°C slow pyrolysis	9.5	52.9	44.1	26.1				4.1	Cantrell et al. (2012)
Pig manure solids	750°C gasification		77.0	21.0	3.0	23.8	30.5			Fryda and Visser (2015)
Cow manure	500°C slow pyrolysis	.4		41.7	18.9	.2	5.3		8.6	Kiran et al. (2017)
Sewage sludge	450°C slow pyrolysis	8.6		21.3	31.7	15.4	13.8			Liu et al. (2014)
Sewage sludge	487°C fast pyrolysis	9.0	65.9	19.7	45.0					Arazo et al. (2017)
Cattle carcass	450°C slow pyrolysis	9.4			39.5	108.0	30.8			Ma and Matsunaka (2013)
Cattle bone	650°C slow pyrolysis			5.9	19.5	157.5				El-Refaey et al. (2015)
Bone meal	350°C slow pyrolysis	7.5		18.0	33.0	127.1	2.5			Zwetsloot et al. (2016)
Range	300–750°C	7.2–12.4	1.4–86.0	5.9–93.9	0.0–45.0	0.2–157.5	1.4–91.5	7.3–161.6	0.5–222.5	

OC: organic carbon; TN: total nitrogen; TP: total phosphorus; TK: total potassium; CEC: cation exchange capacity; SSA: specific surface area the peak carbonization temperature results, generally, in reductions in yield and feed OC recovery, yet improvements in mineral nutrient content and SSA of the biochar products. Additionally, the preparation of feedstock materials (e.g., moisture content, particle size, and envelope density) and the post-treatment of biochar (e.g., steam cooling, activation, and storage) have notable influences on the yield and properties of the final products.

The composition and physiochemical characteristics of biochar determine its quality for applications in agriculture as a soil amendment to facilitate soil health, crop productivity, and environmental benefits. The International Biochar Initiative (IBI 2015) has recommended the following parameters to assess the biochar quality: pH, lime equivalence, electrical conductivity (EC), mineral ash content, OC content, SSA, H/OC molar ratio, particle size distribution, germination inhibition assay, total and available plant nutrients, and the presence of inorganic and organic pollutants. The European Biochar Certificate (EBC 2012) program suggested similar quality variables for biochar: pH, EC, volatile matter content, total ash content, OC content, H/OC molar ratio, O/C molar ratio, bulk density, SSA, macronutrient (N, P, K, Ca, and Mg) contents, and contents of heavy metals and organic contaminants. The primary advantage of biochar over raw biomass residues is its high environmental recalcitrance: the inherent stable carbon constituents (i.e., condensed aromatic carbon fractions) enable biochar to persist against mineralization in the natural environment over a minimum 100-year time interval (Budai et al. 2013). The mean residence time of biochar in field soils is estimated at 90–1600 years (Singh et al. 2012). Research suggests that more than 65% of the OC in biochars with the H/OC molar ratio less than 0.7 would remain in natural soils 100 years after field application (Joseph et al. 2019). The stable proportion of OC, however, varies with the biochar source. Approximately 5–65% of the OC in biochars produced through 300–600°C slow pyrolysis of various biomass residues was mineralizable to 0.056 M acidic dichromate (Guo 2020). The environmental recalcitrance (stability) of biochar is largely determined by its OC aromaticity that is inversely proportional to the biochar H/OC molar ratio. Research has evidenced that biochar products with the H/OC ratio less than 0.7 have an expected life span greater than 100 years in natural soils (Camps-Arbestain et al. 2015). In agricultural applications, more recalcitrant biochar products are preferable to achieve the long-term benefits for promoting soil health. To be entitled as biochar, biomass-derived "char" materials should contain significant contents of stable OC. The stable OC content can be estimated by multiplying the OC content of biochar with its stable (i.e., >100-year soil life) OC proportion (Guo 2020). The IBI biochar standards require all biochar products to possess ≥10% OC and a ≤0.7 H/OC molar ratio. The maximum allowable As, Cd, Cr, Co, Cu, Pb, Hg, Mo, Ni, Se, and Zn concentrations of biochar are also specified (IBI 2015). The EBC program regulates that certified biochar products have a minimum OC content of 50%, a maximum H/OC molar ratio of 0.7, and a maximum O/C molar ratio of 0.4. The maximum concentrations of the heavy metals Cd, Cr, Cu, Hg, Ni, Pb, and Zn in biochar are additionally regulated (EBC 2012). If the EBC criteria were applied, nearly all solid gasification residues and the solid residues from pyrolysis of high-ash plant residues (e.g., rice husks), manures, biosolids, and animal bones were not biochar but "bio carbon minerals" due to the lower OC contents (Table 1.1). It may be more appropriate to adopt "OC content ≥30%" and "H/OC molar ratio ≤0.7" as the quality standards for regulating biochar's stable OC requirement. Clearly, the dark solids "biocoal" and "hydrochar" resulted respectively from torrefaction and hydrothermal liquefaction treatments of biomass materials are not qualified as biochar because of the low environmental recalcitrance of the products (Tekin et al. 2014; Barskov et al. 2019).

The quality of biochar as a soil amendment in agricultural applications should be further assessed from its CEC and water holding capacity (WHC) in addition to the indicators recommended by IBI (2015) and EBC (2012). The WHC and nutrient retention capability of biochar are related to its SSA (or porosity) and CEC, respectively (Guo 2020). Elevating the pyrolysis temperature results typically in biochar products with increased SSA yet decreased CEC (Song and Guo 2012). Activation treatments engender similar effects (Guo, Song et al. 2020). Thus, a moderate pyrolysis temperature (e.g., 400–450°C) may be employed to prepare agricultural-use biochar with balanced SSA and CEC features. At such a treatment temperature, slow pyrolysis of waste wood (OC 53.6%), wheat straw (OC 51.3%), and poultry litter (OC 35.5%) also achieved satisfactory transformation (i.e., 23.7–45.9%) of feedstock OC to highly recalcitrant forms (resistant to laboratory acidic chromate oxidation) in the biochar products (Figure 1.1) and significant feedstock N retention (i.e., 35.3–44.2%) (Song and Guo 2012).

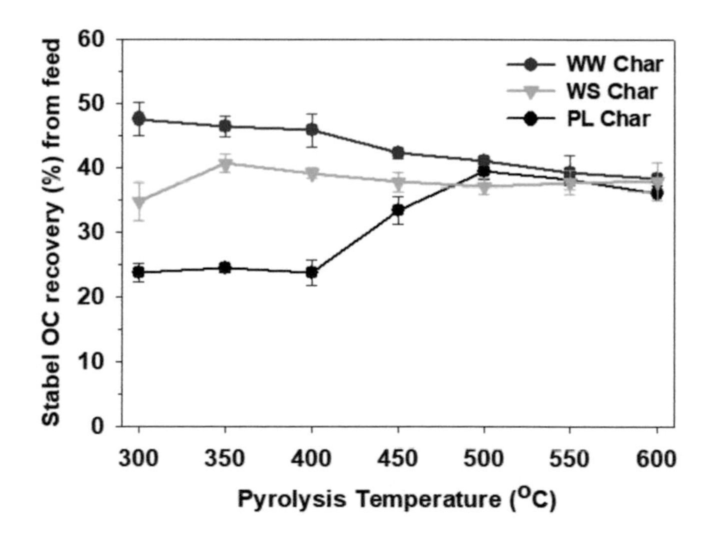

FIGURE 1.1 The recovery (%) of feedstock organic carbon (OC) as stable OC (resistant to 0.056 M acidic dichromate oxidation) in biochars produced from waste wood (WW), wheat straw (WS), and poultry litter (PL) through complete slow pyrolysis at 300–600°C (Guo and Shen 2011, Song and Guo 2012).

1.3 POTENTIAL BENEFITS OF BIOCHAR AMENDMENT IN AGRICULTURAL APPLICATIONS

Biochar can be applied as a soil amendment in nearly all cropping systems. Reported field and greenhouse trials cover the production of corn (*Zea mays*), sorghum (*Sorghum bicolor*), wheat (*Triticum aestivum*), barley (*Hordeum vulgare*), rice (*Oryza sativa*), soybean (*Glycine max*), Faba bean (*Vicia faba*), green bean (*Phaseolus vulgaris*), mung bean (*Vigna radiata*), pea (*Pisum sativum*), peanut (*Arachis hypogaea*), cauliflower (*Brassica oleracea*), mustard (*Brassica nigra*), rapeseed (*Brassica napus*), spinach (*Spinacia oleracea*), lettuce (*Lactuca sativa*), amaranth (*Amaranthus cruentus*), pepper (*Capsicum annuum*), tomato (*Solanum lycopersicum*), carrot (*Daucus carota*), asparagus (*Asparagus officinalis*), pumpkin (*Cucurbita pepo*), watermelon (*Citrullus lanatus*), sweet potato (*Ipomoea batatas*), orchard apple trees (*Malus domestica*), and other crop species (Guo et al. 2016b; Vigay et al. 2021). The benefits of biochar amendment to enhance soil health and promote crop productivity are more evident in highly weathered, acidic, low-fertility soils (Guo 2020). Additional benefits extend to the enlargement of soil carbon sequestration and mitigation of soil greenhouse gas emissions (Vigay et al. 2021). Biochar may also be used as a feed additive or as a bedding material in livestock and poultry production systems to improve the animal health and performance (Toth and Dou 2016; Flores et al. 2021). These atypical agricultural applications of biochar are not discussed in the present chapter.

Soil health is "the continued capacity of soil to function as a vital living ecosystem that sustains plants, animals, and humans" (NRCS 2019). The health of a soil is a comprehensive reflection of its physical, chemical, and biological properties. Soil health, in turn, determines the capability of soil to support plant growth and provide other ecosystem services (Guo 2020). In agricultural applications, biochar amendment improves the crop productivity through enhancing the health of cropland soil in its physical, chemical, and biological attributes.

1.3.1 Soil Physical Health Amelioration by Biochar Amendment

Biochar is more porous, lower in particle density, higher in surface area, and coarser in particle size as compared with common agricultural soils. If appropriately implemented, biochar amendment improves soil physical health via reducing soil bulk density and compaction and increasing soil porosity, water retention, hydraulic conductivity, and aggregate stability (Figure 1.2). The efficacy, however, varies with the soil type and the biochar quality and amendment rate. Soil incorporation of an oil mallee-derived biochar at 5% w/w, for example, remarkably increased the microtomographic porosity (>70 μm) and macropore connectivity of three different soils (Vertisol, Ferralsol, and Arenosol) and resulted in notable changes in soil water movement and retention (Quin et al. 2014). Hardie et al. (2014) observed significant decreases in soil bulk density and increases in WHC and saturated hydraulic conductivity (SHC) 30 months after amending a sandy loam with greenwaste-derived biochar at 47 t ha^{-1}. Incorporation of a wood-derived biochar at 5–25 t ha^{-1} into the top 12 cm of silt loam soil of a corn field reduced evidently the soil bulk density and increased soil macropore volume (>1500 μm), SHC, and available water capacity (AWC) (Eastman 2011). In a 295-day laboratory incubation study, Herath et al. (2013) noticed that the aggregate stability, macroporosity, WHC, and SHC of two New Zealand silt loam soils were markedly enhanced by the addition of a corn-stover-derived biochar at 7.2 t C ha^{-1}. By reviewing the related studies in the literature, Blanco-Canqui (2017) concluded that biochar amendment could reduce soil bulk density by 3% to 31%, increase soil porosity by 14% to 64%, and enhance soil wet aggregate stability by 3% to 226%. Soil AWC could be promoted by 4% to 130%, yet SHC would decrease in coarse-textured soils and increase in fine-textured soils. The effectiveness of biochar amendment increased with the application rate in the range of 0.5% to 15% of soil. Amendments with finer particle-sized biochars may increase soil WHC but decrease SHC (Blanco-Canqui 2017).

1.3.2 Soil Chemical Health Amelioration by Biochar Amendment

Biochar possesses a variety of surface functional groups that are sorptive to nutrient ions (Song and Guo 2012). Its high CEC value implicates the great potential as a soil amendment for nutrient retention. Biochar also contains plant nutrients (e.g., Ca, Mg, K, S, and P) and is generally alkaline (Table 1.1). Biochar amendment typically supplements the soil nutrient supply, increases soil CEC and nutrient availability, reduces soil acidity and nutrient losses, and facilitates soil fertility (Figure 1.2).

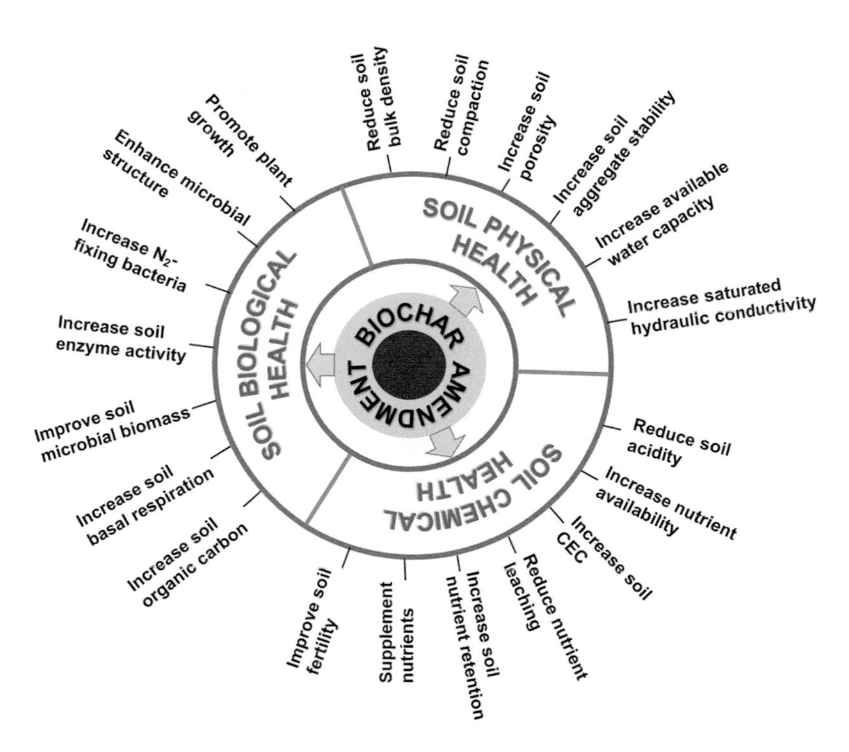

FIGURE 1.2 Potential benefits of biochar amendment for ameliorating physical, chemical, and biological health of soil and promoting plant growth. CEC: cation exchange capacity.

Novak et al. (2009b) reported that amending a strongly acidic sandy soil with pecan shell-derived biochar at 2.0% w/w reduced the soil exchangeable acidity by 0.9 $cmol_c$ kg^{-1} and raised its pH from 4.8 to 6.4. The soil contents of OC, P, Ca, K, and Mn were also considerably elevated. Chintala et al. (2014) noticed that soil amendment with switchgrass- and corn-stover-derived biochars at 52 t ha^{-1} increased the CEC, EC, and exchangeable acidity of a strongly acidic (pH 4.8) clayey soil. It was also through acidity reduction and pH adjustment that biochar reduced the Al toxicity and increased the P availability of a subtropic acidic soil (Van Zwieten et al. 2015). A trial conducted in Italy demonstrated that the application of wheat-bran-derived biochar at 14 t ha^{-1} to a tomato field significantly increased the soil CEC, OC content, and plant-available P, K, M, and NH_4^+ (Vaccari et al. 2015). Increases in CEC and OC content were also observed for an Indonesia upland soil in the cassava field that received treatments of farm waste-derived biochars at 15 t ha^{-1} (Islami et al. 2011). Soil amendment with wood-derived biochar at 10% w/w greatly reduced the N leaching from two potted Amazonian soils spiked with $(NH_4)_2SO_4$ fertilizer (Lehmann et al. 2003). Co-application of wood-derived biochar with biosolids to cropland minimized the nitrate leaching as compared with the biosolids-alone treatment (Knowles et al. 2011). Considerably reduced leaching of nutrients such as N, P, K, Ca, and Mg from biochar-amended cropland soil was additionally reported (Agegnehu et al. 2015; Gautam et al. 2017). Biochar amendment generally mitigates ammonia volatilization and soil denitrification and subsequently increases soil N availability (Cayuela et al. 2013). Through an intensive literature review, Ding et al. (2016) suggested the possible mechanisms for biochar amendment to enhance the overall soil fertility: 1) addition of nutrients inherent in biochar, 2) enhancement of pH-dependent nutrient availability, 3) increase of nutrient retention from adsorption and biological immobilization, 4) reduction of nutrient leaching losses, and 5) mitigation of ammonia volatilization and N denitrification losses.

1.3.3 Soil Biological Health Amelioration by Biochar Amendment

Biochar amendment introduces biodegradable OC and additional habitats to soil microorganisms and, subsequently, improves soil biological health (Figure 1.2). A significant portion (e.g., up to 65%) of OC in biochar is readily mineralizable and serves as food substrate to microorganisms after soil application (Singh et al. 2012; Song and Guo 2012). Biochar matrices possess numerous macropores and micropores. Soil-applied biochar provides less acidic, more aerated, favorable dwelling space to microbes. Biochar generally promotes the microbial structure and activity of the treated soil (Figure 1.2). Adding wood-derived biochar to pasture soil pots at 10% v/v, for instance, increased the abundance of phosphate-solubilizing bacteria in the soil (Anderson et al. 2011). Addition of manure/pine shavings-derived biochar substantially increased the microbial biomass and microbial activity of four distinct temperate soils (Kolb et al. 2009). Harter et al. (2014) reported on increases in the abundance of N_2-fixing microorganisms and the rate of the microbial N_2O reduction in a water-saturated soil from biochar amendment. Biochar amendment to a subtropical acidic soil stimulated the N_2-fixation and grain production of faba bean (Van Zwieten et al. 2015). The microbial biomass, microbial metabolic efficiency, and basal respiration of a tropical acidic soil were all increased by 50–150 g kg^{-1} biochar amendments (Steiner et al. 2008). A meta-analysis indicated that biochar amendment increased the abundances of NH_3-oxidizing archaea and denitrification genes by an average of 25.3% and 21.2%, respectively (Xiao et al. 2019). The efficacy of biochar amendment on improving the soil biological health is, however, dependent on the biochar quality, amendment rate, and the soil type (Farkas et al. 2020). The impacts are more evident in highly weathered, acidic, and nutrient-poor soils (Ding et al. 2016).

1.3.4 Crop Growth and Productivity Improvement by Biochar Amendment

Appropriately practiced biochar amendment promotes crop growth and productivity through enhancing the soil health including the soil fertility. It was reported that the corn grain yield was nearly doubled by a 6 t C ha^{-1} wood-derived biochar amendment trial in Kenya (Kimetu et al. 2008). The leap in crop yield was apparently the primary result of mineral nutrients (e.g., K^+, Ca^{2+}, Mg^{2+}, P, and S) inherent in biochar introduced to Kenya's nutrient-deficient tropical soils. The yield leap would probably disappear in the following growing seasons, whereas a modest yield increase may continue into the far future. The boosting effect of biochar amendment on crop yield faded within 3 years following application as observed worldwide in a number of field trials with diverse biochars and crop species, owing potentially to the over-time loss of biochar base cations (e.g., K^+, Na^+, Ca^{2+}, and Mg^{2+}) and the associated alkalinity for improving nutrient availability (Jones et al. 2012; Cornelissen et al. 2018; Jin et al. 2019). Biochar products especially those derived from wood and crop residues contain minor levels of plant nutrients and, therefore, cannot serve as a major nutrient source for crops. Biochar amendment at significant rates (e.g., >20 t ha^{-1}) may initially boost the soil fertility by introducing additional nutrients in the mineral ash, yet the long-term fertility enhancement from biochar improving soil nutrient retention and availability is insufficient to sustain a satisfactory crop yield (Guo 2020). Biochar amendment cannot replace fertilizer application in modern agricultural production. The long-term crop-yield-boosting effect of biochar amendment would be better manifested with regular soil fertilization. Amendment of a normally fertilized Indonesian soil with tree bark-derived biochar significantly increased the yields of corn and peanut (Yamato et al. 2006). Co-application of wood-derived biochar at 10 t ha^{-1} and chemical NPK fertilizers at agronomic rates improved the grain yields of wheat and corn by 6–10% in a trial in Italy (Baronti et al. 2010). The positive effect of biochar amendment on crop yield is generally more prominent in acidic, nutrient-depleted soils. For fertile soils with already high crop productivity, the effect may be insignificant (Jeffery et al. 2011; Ding et al. 2016).

1.4 RIGHT BIOCHAR APPLICATION PROGRAMS IN AGRICULTURAL PRODUCTION

Scientifically designed biochar amendment programs are critical to achieve the maximum benefits of biochar application for promoting soil health and crop productivity. Implementation of agricultural biochar amendment demands scientific considerations on the biochar source, application rate, and biochar placement in soil (Guo 2020). At the current development stage, there are few guidelines to instruct the best practices of biochar application in agriculture. Among the pioneer biochar users and other stakeholders, a common myth exists as "Biochar is a panacea to all soil problems. One time application of biochar to cropland will secure a forever-fertile soil for high crop yield. No fertilizers or pesticides are further needed". The misunderstanding on biochar, the inappropriate practices of biochar application, and the consequently lower-than-expected results of soil health and crop yield have generated widespread negative impacts, discouraging crop growers and policy makers to adopt biochar amendment in agricultural production. The 3R (right source, right rate, and right placement) principles need to be strictly followed in agricultural biochar applications to achieve the long-term soil health and crop-productivity-promoting benefits (Figure 1.3).

1.4.1 Right Biochar Source

Differently sourced biochar products may vary tremendously in the capability to enhance soil health and promote crop productivity. Relative to biochars derived from wood, crop residues, and other plant debris, products prepared from bones, manure, sewage sludge, spent mushroom substrate, tree barks, tree leaves, rice husks and other mineral element-rich biomass materials exhibit a higher mineral ash content, pH, lime equivalence, and salinity values whereas a lower OC content, SSA, and WHC (Table 1.1). When applied as a soil amendment at the same rate, the latter would yield remarkably greater effects on reducing soil acidity and furnishing plant nutrients yet lower effects on decreasing soil bulk density and improving soil WHC than the former. The yield, composition, and physiochemical characteristics of biochar are further influenced by the carbonization conditions such as peak treatment

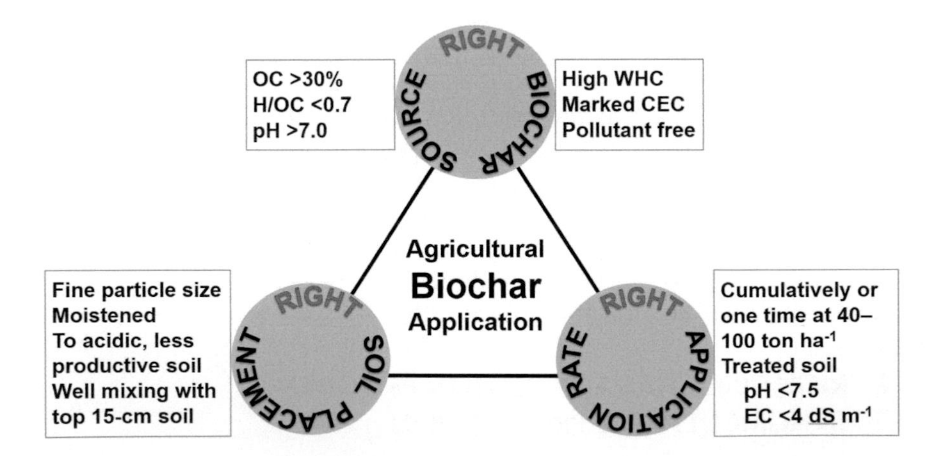

FIGURE 1.3 The 3R (right biochar source, right application rate, and right soil placement) principles for agricultural applications of biochar. CEC: cation exchange capacity; EC: electrical conductivity; OC: organic carbon; WHC: water holding capacity.

temperature, reaction time, heat transfer rate, and O_2 availability (Basu 2010). With elevating the peak treatment temperature in the range of 300–700°C, the resulting biochar products (given adequate reaction time) decreased in yield, total N content, and CEC whereas increased in mineral ash content, pH, EC, SSA, and OC stability (Song and Guo 2012; Zhang et al. 2015).

Commercially available biochar products are typically in 0.05–6 mm particles. To be evenly incorporated in soil, biochar should be processed into less than 2 mm particles prior to land application. Depending on its mineral ash content and the related pH, EC, lime equivalence, and nutrient contents, biochar has a certain capability to rapidly yet transiently promote soil fertility and plant growth by reducing soil acidity and supplying plant nutrients. Products with pH values less than 7.0 may contain substantial organic acid intermediates that inhibit seed germination and seedling development and, therefore, should be avoided (Liu et al. 2013). Biochars generated from manures in particular poultry litter at low pyrolysis temperature (e.g., ≤400°C) contain significant contents of slowly releasable N, P, K, Ca, Mg, and S nutrients and may be applied in organic farming or to strongly acidic, low-fertility soils (e.g., in highly leached cropland and abandoned mine land) as both a soil conditioner and a fertilizer (Song and Guo 2012; Wang et al. 2015). If used in modern, inorganic fertilization-based agriculture, chemical fertilizer application should be reduced accordingly. Wood-derived biochars carry negligible NPK nutrients and should be co-applied with inorganic fertilizers. Manure-derived biochars manufactured at high pyrolysis temperature (e.g., ≥500°C) contain little N (Song and Guo 2012). To pursue the persistent benefits of biochar amendment for conditioning soil and facilitating soil fertility, biochar products possessing greater CEC, SSA (or porosity), and stable OC should be preferentially selected (Figure 1.3). The quality criteria for biochar screening were recommended as: CEC more than 20 $cmol_c\ kg^{-1}$, SSA more than 10 $m^2\ g^{-1}$, and stable OC (against 0.056 M acidic dichromate oxidation) more than 150 $g\ kg^{-1}$ (or OC >300 $g\ kg^{-1}$ if H/OC molar ratio <0.7) (Guo 2020).

1.4.2 Right Biochar Application Rate

Appropriate application rates should be determined in accordance with the biochar source and soil type to achieve the optimal benefits of biochar amendment (Figure 1.3). Too low or too high rates would lead to insignificant or even negative effects on soil health and crop yield.

A rather broad range of application rates from 0.1% to 15% w/w soil (roughly equivalent to 2–300 $t\ ha^{-1}$ assuming top 15-cm field soil incorporation) were employed in biochar agricultural use studies. The more frequent biochar amendment rates were in 1–5% w/w soil (~20–100 $t\ ha^{-1}$) (Guo 2020). Biochar amendment at low rates (e.g., <1% w/w soil) may increase crop yield in the initial growing seasons following application as a result of additionally introduced mineral nutrients, yet the ameliorating effects on soil physical and chemical health such as bulk density, WHC, AWC, SHC, pH, EC, and CEC may become insignificant after months of natural settling and weathering (Guo 2016). For instance, incorporation of wheat-straw-derived biochar at 10 $t\ ha^{-1}$ in the top 10-cm soil of a corn field generated little impact on soil bulk density measured 6 months after biochar amendment (Xiao et al. 2016). At the

1.5 t ha^{-1} application rate, biochars derived from wood and sludge had little influence on soil pH, EC, and WHC and the growth of sunflower (Paneque et al. 2016). In a meta-analysis of biochar amendment impacts on soil physical properties, Blanco-Canqui (2017) noticed that detectable improvements in WHC and AWC occurred mostly in soils receiving biochar amendments at ≥1% w/w soil. Depending on the biochar quality and soil type, the impacts of biochar amendment on soil health may not be evident even at higher rates. Pratiwi and Shinogi (2016) reported, for example, that no significant reductions in soil bulk density were detected 100 days after amending a rice paddy soil pot with rice husk-derived biochar at 2% w/w, soil.

Biochar amendment at over rates, on the other hand, may significantly elevate the levels of soil salinity, alkalinity, sodicity, and C:N ratio and generate inhibiting effects on plant growth. Rondon et al. (2007) noticed that 9% w/w wood-derived biochar amendment significantly suppressed the N uptake and biomass yield of common bean plants in soil pots. Growth inhibition of winter wheat seedlings was visually observed in soil pots amended with poultry litter-derived biochar at 3% w/w (not in 2% w/w soil treatments) (Guo 2020).

The optimal amendment rate is biochar-specific and soil-dependent. For most biochar products and agricultural soils, the positive effects of biochar amendment on soil health and crop productivity escalate with increasing the amendment rate in the range of 0.25% to 2.5% w/w (~5 to 50 t ha^{-1}) (Major 2010). To avoid the potential negative effects from over application, pre-test may be necessary to determine the maximum allowable amendment rates of particular biochar products in single applications to specific soils. For a selected biochar, the application rate should be controlled to ensure a less than 7.5 pH value and a less than 2.7 dS m^{-1} EC value (in 1:1 solid/water slurry; equivalent to 4 dS m^{-1} in saturated soil paste) of the amended soil to minimize the excess alkalinity and salinity risks (Guo 2020). Numerous greenhouse potting and field plot experiments have suggested soil amendment with manure-derived biochars at 2–3% w/w soil (40–60 t ha^{-1}) and with wood or crop residue-derived biochars at 2–5% w/w soil (40–100 t ha^{-1}) to achieve significant, long-term benefits on soil health and crop productivity (Guo 2020). Given the high market prices of commercially available biochar products, however, field application of biochar in agriculture has been mostly implemented at a low rate (e.g., 10 t ha^{-1} or ~0.5% w/w, soil) (Steiner 2016).

1.4.3 Right Biochar Placement to Soil

It is important to consider the type of soil to be treated in designing effective biochar amendment programs. Strongly acidic (i.e., pH <5.5), coarse-textured soils respond generally more in health and crop yield improvements than neutral to alkaline, fine-textured soils to biochar amendment (Steiner 2016). To maximize the potential benefits, biochar amendment should be primarily implemented to cropland with highly weathered, acidic, OC-depleting (e.g., <10 g kg^{-1}) soils (Guo et al. 2016b). The stable OC matrix enables biochar to persist in natural soils over hundreds of years and to continuously function to provide long-term soil conditioning and health enhancement benefits. Therefore, soil amendment with biochar may be practiced in "once-a-life" at the optimal rates or in multiple times at cumulatively the optimal rates (Figure 1.3).

Thoroughly mixing biochar with the top 15–20-cm cropland soil is essential to attain the agricultural benefits of biochar amendment. Surface broadcasting followed by multi-round tillage is commonly practiced to incorporate biochar into the root zone soil. Pre-processing biochar into less than 2 mm particles facilitates the soil incorporation to achieve the relatively uniform distribution of biochar in soil. Dust formation from wind shifting dry, fine biochar particles may be severe during field broadcasting application. The problem can be minimized by applying moistened or compost-ladened biochar under low-wind (i.e., <8 km h^{-1}) weather conditions (Guo 2020). Soil incorporation by tillage may not be practical when biochar is applied to perennial crops, grassland, and no-till systems. Surface application of biochar can be carried out by broadcasting, side dressing, band drilling, trenching, or localized holing. Surface-applied biochar, however, functions primarily as a sterilized physical cover and provides rather limited soil health improvement benefits. Moreover, the material decreases the land albedo and is readily eroded by wind and runoff water (Major 2010).

Biochar in soil may interact with land-applied pesticides and herbicides through stabilization (e.g., adsorption, partition, and pore-filling) and accelerated decomposition (Guo, Song et al. 2020) and, subsequently, decrease the effectiveness of herbicides for weed control and other pesticides. The topic is being extensively discussed in the following chapters.

1.5 STATUS AND CHALLENGES OF AGRICULTURAL APPLICATIONS OF BIOCHAR

It is a tradition in East Asia for farmers to apply wood char and straw ash to cropland as a soil conditioner. Prior to the 1970s when chemical fertilizers were not widely available, crop growers in China, Japan, and Korea intentionally generated char from plant debris smoldering and used the material as an important nutrient supplement in vegetable production. Even today, farmland application of rice-husk-derived char mixed with animal excreta is a common practice in the region. Approximately 40,000 tons of biochar are land-applied in Japan each year (Ogawa and Okimori 2010).

Worldwide commercial production and agricultural application of biochar have been steadily growing. The number of active biochar companies, for example, increased globally from 200 in 2014 to 326 in 2015. The volume of biochar transacted in the market increased from 827 (metric) tons in 2013 to 7,457 tons in 2014 and further to 85,000 tons in 2015 (IBI 2016). In the United States and Canada, there were 112 registered biochar manufacturers and suppliers in 2021 (USBI 2021). The majority of the biochar producers used wood as the feedstock though other biomass materials such as crop residues, animal manures, and food-processing wastes were also used by individual producers to manufacture biochar. Biochars have been applied mainly (~60%) in agriculture as a soil amendment, compost conditioner, potting media substitute, or animal husbandry bedding material. Environmental application (~40%) for soil remediation, water purification, air filtration, and carbon sequestration is another major use of biochar (IBI 2016). The global biochar market reached $1594.5 million in 2020. It is expected to expand at a compound annual growth rate of 16.53% to $3993.7 million by the end of 2026 (Research and Markets 2021).

The major challenges that the biochar industry has been facing extend to lack of market awareness of the biochar application benefits, the evident gap between biochar research and field application, high biochar production costs and market prices, difficulties for biochar transport and shipment, and unavailability of biochar quality regulation and best application programs (Guo et al. 2016c; IBI 2016). The current market price of biochar averages at \$193 – \$234 per dry ton. A more affordable biochar price at approximately \$100 per ton has been desired by agricultural users (USBI 2020). Continuous investments in biochar research, education, and extension are clearly warranted to improve the consumer awareness of biochar, user outreach and training, biochar production technology and business management, and biochar market development.

1.6 SUMMARY AND CONCLUDING REMARKS

Biochar is charcoal generated from the thermochemical carbonization treatment of biomass materials with the pyrolysis or gasification technique and used as a soil amendment for agricultural and environmental purposes. The material, possessing labile and stable OC and mineral ash nutrients, is porous, environmentally recalcitrant, alkaline, and water and nutrient retentive. Research has evidenced that biochar is a superior soil conditioner capable of persistently enhancing soil health and crop productivity. Soil amendment with biochar is a promising solution to sustainable agriculture.

Biochar products manufactured from diverse feedstock materials and carbonization operations, however, demonstrate significantly different physiochemical characteristics and vary greatly in the capability to enhance soil health and crop productivity. The transient capacity of biochar to immediately improve soil fertility and promote plant growth is largely determined by its inherent nutrient contents, whereas the long-term capacity to persistently enhance soil health and crop yield is controlled by its stable OC content and AWC, CEC, and SSA values. High-quality biochar for agricultural uses should contain more than 30% OC and have an H/OC molar ratio less than 0.7, pH value greater than 7.0, and marked levels of SSA and CEC.

Biochar can be directly applied to cropland as an organic amendment or pre-used as a bedding material in animal husbandry and as a conditioner in composting facilities and then applied to cropland in the resulting manure or compost mixtures. To achieve the maximum benefits of biochar amendment in agricultural production, appropriate biochar application programs are necessary with reflecting the 3R design principles: right biochar source, right application rate, and right soil placement. Biochar products showing pH value greater than 7.0, OC greater than 30%, and molar H/OC less than 0.7 should be selected with priority. Biochar should be preferentially applied to cropland with acidic, less-productive soils. The optimal application rate is biochar and soil specific but is recommended at 40–100 t ha^{-1} (equivalent to 2–5% w/w soil) through single or repeated applications with maintaining the pH values of the treated soils at less than 7.5 and EC less than 4 dS m^{-1}. Efficient land application of biochar can be implemented by processing the material into small particles, moistening the material by adding water or mixing with manure and compost, evenly broadcasting biochar to land surface on low wind days, and immediately incorporating the applied biochar into top 15–20-cm soil through multi-round tillage. Regular fertilization at reduced rates is critical to attain a satisfactory productivity of biochar-amended cropland.

Worldwide agricultural application of biochar has been increasingly practiced in field and controlled-environment productions of grains, vegetables, fruits, and animals. The application is expected to expand rapidly in the future with improved awareness of biochar-amendment benefits among agricultural stakeholders and facilitated market and technological development of the biochar industry.

ACKNOWLEDGMENTS

Financial support for compiling the information in this chapter was from the USDA-NIFA Award No. 2021–38821–34702.

REFERENCES

Agegnehu, G., A.M. Bass, P.N. Nelson, B. Muirhead, G. Wright, and M.I. Bird. 2015. Biochar and biochar-compost as soil amendments: Effects on peanut yield, soil properties and greenhouse gas emissions in Tropical North, Queensland, Australia. *Agriculture, Ecosystems & Environment* 213:72–85.

Anderson, C.R., L.M. Condron, T.J. Clough, et al. 2011. Biochar induced soil microbial community change: Implications for biogeochemical cycling of carbon, nitrogen and phosphorous. *Pedobiologia* 54:309–320.

Arazo, R.O., D.A.D. Genuino, M.D.G. de Luna, and S.C. Capareda. 2017. Bio-oil production from dry sewage sludge by fast pyrolysis in an electrically-heated fluidized bed reactor. *Sustainable Environmental Research* 27:7–14.

Baronti, S., G. Alberti, G.D. Vedove, et al. 2010. The biochar option to improve plant yields: First results from some field and pot experiments in Italy. *Italian Journal of Agronomy* 5:3–11.

Barskov, S., M. Zappi, P. Buchireddy, et al. 2019. Torrefaction of biomass: A review of production methods for biocoal from cultured and waste lignocellulosic feedstocks. *Renewable Energy* 142:624–642.

Basu, P. 2010. *Biomass Gasification and Pyrolysis: Practical Design*. Burlington, MA: Academic Press.

Blanco-Canqui, H. 2017. Biochar and soil physical properties. *Soil Science Society of America Journal* 81:687–711.

Budai, A., A.R. Zimmerman, A.L. Cowie, et al. 2013. *Biochar Carbon Stability Test Method: An Assessment of Methods to Determine Biochar Carbon Stability*. Westerville, OH: International Biochar Initiative.

Camps-Arbestain, M., J.E. Amonette, B. Singh, T. Wang, and H.P. Schmidt. 2015. A biochar classification system and associated test methods. In *Biochar for Environmental Management: Science, Technology and Implementation*, ed. J. Lehmann and S. Joseph, 165–194. Abingdon, UK: Routledge.

Cantrell, K.B., P.G. Hunt, M. Uchimiya, J.M. Novak, and K.S. Ro. 2012. Impact of pyrolysis temperature and manure source on physicochemical characteristics of biochar. *Bioresource Technology* 107:419–428.

Cayuela, M.L., M.A. Sánchez-Monedero, A. Roig, K. Hanley, A. Enders, and J. Lehmann. 2013. Biochar and denitrification in soils: When, how much and why does biochar reduce N_2O emissions? *Nature Scientific Reports* 3:17–32.

Chen, B., D. Zhou, and L. Zhu. 2008. Transitional adsorption and partition on nonpolar and polar aromatic contaminants by biochars of pine needles with different pyrolytic temperatures. *Environmental Science and Technology* 42:5137–5143.

Chintala, R., J. Mollinedo, T.E. Schumacher, et al. 2014. Effect of biochar on chemical properties of acidic soil. *Archives of Agronomy and Soil Science* 60:393–404.

Cornelissen, G., Jubaedah, N.L. Nurida, et al. 2018. Fading positive effect of biochar on crop yield and soil acidity during five growth seasons in an Indonesian Ultisol. *Science of the Total Environment* 634:561–568.

Ding, Y., Y. Liu, S. Liu, et al. 2016. Biochar to improve soil fertility. A review. *Agronomy for Sustainable Development* 36:36.

Eastman, C.M. 2011. Soil physical characteristics of an Aeric Ochraqualf amended with biochar. Master degree thesis. Ohio State University, Columbus, OH.

EBC. 2012. *European Biochar Certificate – Guidelines for a Sustainable Production of Biochar*. Arbaz, Switzerland: European Biochar Foundation. www.europeanbiochar.org (accessed September 17, 2021).

El-Refaey, A.A., A.H. Mahmoud, and M.E. Saleh. 2015. Bone biochar as a renewable and efficient P fertilizer: A comparative study. *Alexandria Journal of Agricultural Research* 60(3):127–137.

Farkas, É., V. Feigl, K. Gruiz, et al. 2020. Long-term effects of grain husk and paper fibre sludge biochar on acidic and calcareous sandy soils – A scale-up field experiment applying a complex monitoring toolkit. *Science of the Total Environment* 731:138988.

Flores, K.R., A. Fahrenholz, and J.L. Grimes. 2021. Effect of pellet quality and biochar litter amendment on male turkey performance. *Poultry Science* 100:101002.

Fryda, L. and R. Visser. 2015. Biochar for soil improvement: Evaluation of biochar from gasification and slow pyrolysis. *Agriculture* 5:1076–1115.

Gaskin, J.W., C. Steiner, K. Harris, K.C. Das, and B. Bibens. 2008. Effect of low-temperature pyrolysis conditions on biochar for agricultural use. *Transactions of the ASABE* 51(6):2061–2069.

Gautam, D., R. Bajracharya, and B. Sitaula. 2017. Effects of biochar and farm yard manure on soil properties and crop growth in an agroforestry system in the Himalaya. *Sustainable Agriculture Research* 6:74.

Gezahegn, S., M. Sain, and S.C. Thomas. 2019. Variation in feedstock wood chemistry strongly influences biochar liming potential. *Soil Systems* 3:26.

Ghorbani, M., H. Asadi, and S. Abrishamkesh. 2019. Effects of rice husk biochar on selected soil properties and nitrate leaching in loamy sand and clay soil. *International Soil and Water Conservation Research* 7:258–265.

Glaser, B., E. Balashov, L. Haumaier, et al. 2000. Black carbon in density fractions of anthropogenic soils of the Brazilian Amazon region. *Organic Geochemistry* 31:669–678.

Guo, M. 2016. Application of biochar for soil physical improvement. In *Agricultural and Environmental Applications of Biochar: Advances and Barriers*, ed. M. Guo, Z. He, and S.M. Uchimiya, 101–122. Madison, WI: Soil Science Society of America.

Guo, M. 2020. The 3R principles for applying biochar to improve soil health. *Soil Systems* 4:9.

Guo, M., Z. He, and S.M. Uchimiya. 2016a. Pyrogenic carbon in Terra Preta soils. In *Agricultural and Environmental Applications of Biochar: Advances and Barriers*, ed. M. Guo, Z. He, and S.M. Uchimiya, 15–27. Madison, WI: Soil Science Society of America.

Guo, M., Z. He, and S.M. Uchimiya. 2016b. Introduction to biochar as an agricultural and environmental amendment. In *Agricultural and Environmental Applications of Biochar: Advances and Barriers*, ed. M. Guo, Z. He, and S.M. Uchimiya, 1–14. Madison, WI: Soil Science Society of America.

Guo, M., Z. He, and S.M. Uchimiya. 2016c. Agricultural and environmental applications of biochar: Advances and barriers. In *Agricultural and Environmental Applications of Biochar: Advances and Barriers*, ed. M. Guo, Z. He, and S.M. Uchimiya, 495–504. Madison, WI: Soil Science Society of America.

Guo, M., H. Li, B. Baldwin, and J. Morrison. 2020. Thermochemical processing of animal manure for bioenergy and biochar. In *Animal Manure: Production, Characteristics, Environmental Concerns and Management*, ed. H.M. Waldrip, P.H. Pagliari, and Z. He, 255–274. Madison, WI: Soil Science Society of America.

Guo, M. and Y. Shen. 2011. Transformation and persistence of biochar carbon in soil. Presentation at the 2011 ASA-CSSA-SSSA Annual Meetings, San Antonio, TX, USA, October 18, 2011. Soil Science Society of America, Madison, WI.

Guo, M., W. Song, and J. Tian. 2020. Biochar-facilitated soil remediation: Mechanisms and efficacy variations. *Frontiers in Environmental Science* 8:521512.

Guo, M., P. Xiao, and H. Li. 2020. Valorization of agricultural byproducts through conversion to biochar and bio-oil. In *Byproducts from Agriculture and Fisheries: Adding Value for Food, Feed, Pharma, and Fuels*, ed. B.K. Simpson, A.N.A. Aryee, and F. Toldrá, 501–522. Hoboken, NJ: John Wiley & Sons, Ltd.

Hardie, M., B. Clothier, S. Bound, et al. 2014. Does biochar influence soil physical properties and soil water availability? *Plant and Soil* 376:347–361.

Harter, J., H. Krause, S. Schuettler, et al. 2014. Linking N_2O emissions from biochar-amended soil to the structure and function of the N-cycling microbial community. *ISME Journal* 8:660–674.

Herath, H.M.S.K., M. Camps-Arbestain, and M. Hedley. 2013. Effect of biochar on soil physical properties in two contrasting soils: An Alfisol and an Andisol. *Geoderma* 209–210:188–197.

IBI. 2015. *Standardized Product Definition and Product Testing Guidelines for Biochar that is Used in Soil – Version 2.1*. Vector, NY: International Biochar Initiative.

IBI. 2016. *State of the Biochar Industry 2015*. Vector, NY: International Biochar Initiative.

Ippolito, J.A., J.M. Novak, W.J. Busscher, et al. 2012. Switchgrass biochar affects two aridisols. *Journal of Environmental Quality* 41:1123–1130.

Islami, T., B. Guritno, N. Basuki, and A. Suryanto. 2011. Biochar for sustaining productivity of cassava based cropping systems in the degraded lands of east Java, Indonesia. *Journal of Tropical Agriculture* 49:40–46.

Jeffery, S., F.G.A. Verheijen, M. van der Velde, et al. 2011. A quantitative review of the effects of biochar application to soils on crop productivity using meat-analysis. *Agriculture, Ecosystems & Environment* 144:175–187.

Jiang, J., Y. Peng, M. Yuan, et al. 2015. Rice straw-derived biochar properties and functions as Cu(II) and cyromazine sorbents as influenced by pyrolysis temperature. *Pedosphere* 25:781–789.

Jin, Z., C. Chen, X. Chen, et al. 2019. The crucial factors of soil fertility and rapeseed yield – a five year field trial with biochar addition in upland red soil, China. *Science of the Total Environment* 649:1467–1480.

Jones, D.L., J. Rousk, G. Edwards-Jones, T.H. DeLuca, and D.V. Murphy. 2012. Biochar-mediated changes in soil quality and plant growth in a three year field trial. *Soil Biology and Biochemistry* 45:113–124.

Joseph, S., P. Taylor, F. Rezende, K. Draper, and A. Cowie. 2019. *The Properties of Fresh and Aged Biochar*. Armidale: Biochar for Sustainable Soils, Starfish Initiatives.

Kimetu, J.M., J. Lehmann, S.O. Ngoze, et al. 2008. Reversibility of soil productivity decline with organic matter of differing quality along a degradation gradient. *Ecosystems* 11:726–739.

Kiran, Y.K., A. Barkat, X. Cui, et al. 2017. Cow manure and cow manure-derived biochar application as a soil amendment for reducing cadmium availability and accumulation by Brassicachinensis L. in acidic red soil. *Journal of Integrative Agriculture* 16:725–734.

Kloss, S., F. Zehetner, A. Dellantonio, et al. 2012. Characterization of slow pyrolysis biochars: Effects of feedstocks and pyrolysis temperature on biochar properties. *Journal of Environmental Quality* 41:990–1000.

Knowles, O.A., B.H. Robinson, A. Contangelo, et al. 2011. Biochar for the mitigation of nitrate leaching from soil amended with biosolids. *Science of the Total Environment* 409:3206–3210.

Kolb, S.E., K.J. Fermanich, and M.E. Dornbush. 2009. Effect of charcoal quantity on microbial biomass and activity in temperate soils. *Soil Science Society of America Journal* 73:1173–1181.

Lehmann, J., J.P. da Silva, C. Steiner, et al. 2003. Nutrient availability and leaching in an archaeological Anthrosol and a Ferralsol of the Central Amazon basin: Fertilizer, manure and charcoal amendments. *Plant and Soil* 249:343–357.

Lima, H.N., C.E.R. Schaefer, J.W.V. Mello, R.J. Gilkes, and J.C. Ker. 2002. Pedogenesis and pre-Colombian land use of "Terra Preta Anthrosols" ("Indian black earth") of Western Amazonia. *Geoderma* 110:1–17.

Liu, T., B. Liu, and W. Zhung. 2014. Nutrients and heavy metals in biochar produced by sewage sludge pyrolysis: Its application in soil amendment. *Polish Journal of Environmental Studies* 23:271–275.

Liu, X., A. Zhang, C. Ji, et al. 2013. Biochar's effect on crop productivity and the dependence on experimental conditions – A meta-analysis of literature data. *Plant and Soil* 373:583–594.

Ma, Y.L. and T. Matsunaka. 2013. Biochar derived from dairy cattle carcasses as an alternative source of phosphorus and amendment for soil acidity. *Soil Science and Plant Nutrition* 59:628–641.

Major, J. 2010. *Guidelines on Practical Aspects of Biochar Application to Field Soil in Various Soil Management Systems*. Westerville, OH: International Biochar Initiative.

Mohan, D., S. Rajput, V.K. Singh, P.H. Steele, and C.U. Pittman. 2011. Modeling and evaluation of chromium remediation from water using low cost bio-char, a green adsorbent. *Journal of Hazardous Materials* 188:319–333.

Mullen, C.A., A.A. Boateng, N.M. Goldberg, et al. 2010. Bio-oil and bio-char production from corn cobs and stover by fast pyrolysis. *Biomass and Bioenergy* 34:67–74.

Novak, J.M., W.J. Busscher, D.A. Laird, et al. 2009b. Impact of biochar amendment on fertility of a southeastern Coastal Plain soil. *Soil Science* 174:105–112.

Novak, J.M., I. Lima, B. Xing, et al. 2009a. Characterization of designer biochar produced at different temperatures and their effects on loamy sand. *Annals of Environmental Science* 3:195–206.

NRCS. 2019. *Soil Health*. Washington, DC: Natural Resources Conservation Service, U.S. Department of Agriculture. www.nrcs.usda.gov/wps/portal/nrcs/main/soils/health/ (accessed September 30, 2021).

Nwajiaku, M., J.S. Olanrewaju, K. Sato, et al. 2018. Change in nutrient composition of biochar from rice husk and sugarcane bagasse at varying pyrolytic temperatures. *International Journal of Recycling Organic Waste in Agriculture* 7:269–276.

Ogawa, M. and Y. Okimori. 2010. Pioneering works in biochar research, Japan. *Australian Journal of Soil Research* 48:489–500.

Paneque, M., J.M. Rosa, J.D. Franco-Navarro, J.M. Colmenero-Flores, and H. Knicker. 2016. Effect of biochar amendment on morphology, productivity and water relations of sunflower plants under non-irrigation conditions. *Catena* 147:280–287.

Pratiwi, E.P. and Y. Shinogi. 2016. Rice husk biochar application to paddy soil and its effects on soil physical properties, plant growth, and methane emission. *Paddy and Water Environment* 14:521–532.

Quin, P.R., A.L. Cowie, R.J. Flavel, et al. 2014. Oil mallee biochar improves soil structural properties – A study with x-ray micro-CT. *Agriculture, Ecosystems & Environment* 191:142–149.

Research and Markets. 2021. *Biochar Market Research Report by Row Material, by Technology, by Application, by Region – Global Forecast to 2026 – Cumulative Impact of COVID-19*. Dublin, Ireland: Research and Markets.

Rondon, M.A., J. Lehmann, J. Ramirez, and M. Hurtado. 2007. Biological nitrogen fixation by common beans (*Phaseolus vulgaris* L.) increases with bio-char additions. *Biology and Fertility of Soils* 43:699–708.

Singh, B.P., A.L. Cowie, and R.J. Smernik. 2012. Biochar carbon stability in a clayey soil as a function of feedstock and pyrolysis temperature. *Environmental Science and Technology* 46:11770–11778.

Sombroek, W.G. 1966. *Amazon Soils: A Reconnaissance of the Soils of the Brazilian Amazon Region*. Wageningen, The Netherlands: Centre for Agricultural Publications and Documentation.

Song, W. and M. Guo. 2012. Quality variations of poultry litter biochars generated at different pyrolysis temperatures. *Journal of Analytical and Applied Pyrolysis* 94:138–145.

Steiner, C. 2016. Considerations in biochar characterization. In *Agricultural and Environmental Applications of Biochar: Advances and Barriers*, ed. M. Guo, Z. He, and S.M. Uchimiya, 87–100. Madison, WI: Soil Science Society of America.

Steiner, C., K.C. Das, M. Garcia, et al. 2008. Charcoal and smoke extract stimulate the soil microbial community in a highly weathered xanthic Ferralsol. *Pedobiologia* 51:359–366.

Tekin, K., S. Karagoa, and S. Bektas. 2014. A review of hydrothermal biomass processing. *Renewable and Sustainable Energy Reviews* 40:673–687.

Tian, J., V. Miller, P.C. Chiu, J.A. Maresca, M. Guo, and P.T. Imhoff. 2016. Nutrient release and ammonium sorption of poultry litter and wood biochars in stormwater treatment. *Science of the Total Environment* 553:596–606.

Toth, J. and Z. Dou. 2016. Use and impact of biochar and charcoal in animal production systems. In *Agricultural and Environmental Applications of Biochar: Advances and Barriers*, ed. M. Guo, Z. He, and S.M. Uchimiya, 199–224. Madison, WI: Soil Science Society of America.

USBI. 2020. *USBI Update – Tom Miles, Executive Director. USBI December 2020 Newsletter*. Portland, OR: US Biochar Initiative.

USBI. 2021. *Biochar Suppliers and Manufacturers*. Portland, OR: US Biochar Initiative. https://biochar-us.org/suppliers-and-manufacturers.

Vaccari, F.P., A. Maienza, F. Miglietta, et al. 2015. Biochar stimulates plant growth but not fruit yield of processing tomato in a fertile soil. *Agriculture, Ecosystems & Environment* 207:163–170.

Van Zwieten, L., T. Rose, D. Herridge, et al. 2015. Enhanced biological N_2 fixation and yield of faba bean (*Vicia faba* L.) in an acid soil following biochar addition: Dissection of causal mechanisms. *Plant and Soil* 395:7–20.

Vigay, V., S. Shreedhar, K. Adlak, et al. 2021. Review of large-scale biochar field-trials for soil amendment and the observed influences on crop yield variations. *Frontiers in Energy Research* 9:710766.

Vithanage, M., I. Herath, Y.A. Almaroai, et al. 2017. Effects of carbon nanotube and biochar on bioavailability of Pb, Cu and Sb in multi-metal contaminated soil. *Environmental Geochemistry and Health* 39:1409–1420.

Wang, Y., Y. Lin, P. Chiu, P. Imhoff, and M. Guo. 2015. Phosphorus release behaviors of poultry litter biochar as a soil amendment. *Science of the Total Environment* 512–513:454–463.

Windeatt, J.H., A.B. Ross, P.T. Williams, et al. 2014. Characteristics of biochars from crop residues: Potential for carbon sequestration and soil amendment. *Journal of Environmental Management* 146:189–197.

Woods, W.I. and J.M. McCann. 1999. The Anthropogenic origin and persistence of Amazonian dark earths. *Yearbook of the Conference of Latin American Geography* 25:7–14.

Xiao, Q., L. Zhu, H. Zhang, X. Li, Y. Shen, and S. Li. 2016. Soil amendment with biochar increases maize yields in a semi-arid region by improving soil quality and root growth. *Crop and Pasture Science* 67:495–507.

Xiao, Z., S. Rasmann, L. Yue, F. Lian, H. Zou, and Z. Wang. 2019. The effect of biochar amendment on N-cycling genes in soils: A meta-analysis. *Science of the Total Environment* 696:133984.

Yamato, M., Y. Okimori, I.F. Wibowo, et al. 2006. Effects of the application of charred bark of *Acacia mangium* on the yield of maize, cowpea and peanut, and soil chemical properties in South Sumatra, Indonesia. *Soil Science and Plant Nutrition* 52:489–495.

Zhang, J., J. Liu, and R. Li. 2015. Effects of pyrolysis temperature and heating time on biochar obtained from the pyrolysis of straw and lignosulfonate. *Bioresource Technology* 176:288–291.

Zheng, W., M. Guo, T. Chow, D. Bennett, and N. Rajagopalan. 2010. Sorption properties of greenwaste biochar for two triazine pesticides. *Journal of Hazardous Materials* 181:121–126.

Zwetsloot, M.J., J. Lehmann, T. Bauerle, et al. 2016. Phosphorus availability from bone char in a P-fixing soil influenced by root-mycorrhizae-biochar interactions. *Plant and Soil* 408(1–2):95–105.

2 Biochar Production, Properties, and Its Influencing Factors

Tiago Guimarães[1*] *and Renata Pereira Lopes Moreira*[1]
[1]Department of Chemistry, Federal University of Viçosa, Viçosa, MG, Brazil.
*Corresponding author: tiago.g.guimaraes@ufv.br.

CONTENTS

2.1 INTRODUCTION

Biomass can be defined as the mass of living organisms that are capable of transforming solar energy into chemical energy through photosynthesis (Silva et al. 2018). From a biochemical perspective, it can refer to materials made up of cellulose, lignin, sugars, fats, and proteins (Anukam et al. 2016). Biomass residues are abundant, and, therefore, the development of valorization technologies for these materials is very important and recent. Among the biomass valorization processes are carbonization processes to produce energy (Lopes and Astruc 2021).

DOI: 10.1201/9781003202073-2

The use of biomass to produce clean and sustainable energy has attracted attention on the world stage for being a renewable raw material and for being labeled as carbon neutral, because it sequesters carbon in its life cycle (Arpia et al. 2021). However, its use in nature presents limitations such as low bulk density, high moisture content, hydrophilic nature, and low calorific value (Tumuluru et al. 2011). In addition, it decomposes easily through the action of microorganisms. Therefore, carbonization processes are employed aiming to improve these characteristics and, mainly, to improve the fuel potential of these materials (Leng et al. 2021). Another advantage is the combustion of biomass to produce steam for power generation, as well as gasification to produce a fuel gas, called producer gas (N_2 and CO_2) and synthesis gas (CO and H_2, with low amounts of nitrogen and CO_2) (Tumuluru et al. 2011).

Thermochemical conversion produces several products in a short time (biogas, bio-oil, and biochar), and it is possible to improve the process by using catalysts (Aliyu et al. 2021). Pyrolysis, gasification, hydrothermal carbonization are among the thermochemical processes used, as schematized in Figure 2.1. The amount of biogas, bio-oil, and biochar depends on the carbonization process employed. Thermochemical processes are commercially more attractive because of their higher process efficiency, better flexibility, higher conversion rate, product selectivity, and alternative market for by-products (Sarker et al. 2021).

During biomass carbonization, several reactions occur including hydrolysis, dehydration, decarboxylation, polymer condensation, and aromatization (Kan et al. 2016; Wilk et al. 2021). In Table 2.1 are listed the calorific value of some biomasses and some carbonization products obtained by different processes. As can be seen, bio-oils or biochar with high calorific value can be obtained, depending on the carbonization process.

Biomass carbonization can be carried out using biochemical and thermochemical processes. In biochemical processes, biomass is converted into gas (CO_2/CH_4), waste (compost or fertilizer), and water (water or C_2H_5OH) using microorganisms (Küçük and Demirbaş 1997), while thermochemical processes involve the use of high temperatures (Solarte-Toro et al. 2021).

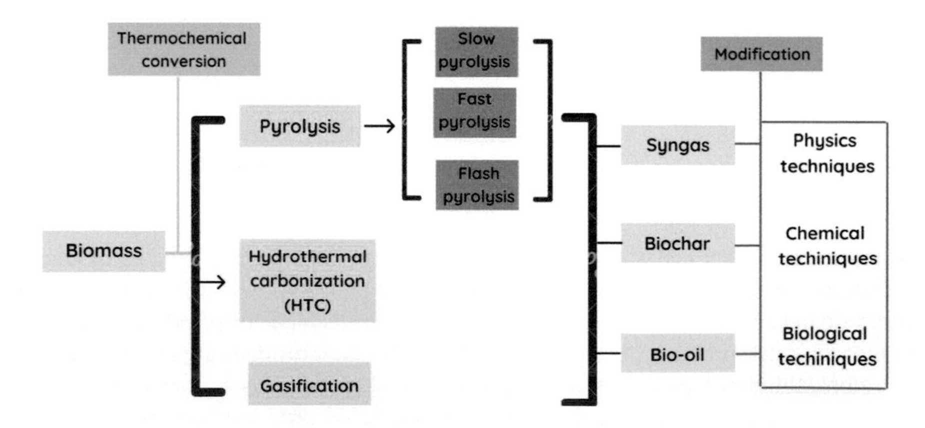

FIGURE 2.1 Thermochemical processes of biomass conversion and products obtained.

TABLE 2.1
Calorific Value (MJ/kg) of Carbonaceous Materials from Different Biomasses.

Biomass	Bio-oil	Bio gas	Biochar
15.30–19.28[a]	18.00–37.20[b]	13.00–14.00[c]	6.00–35.48[d]

[a] Yoder et al. (2011), El Hanandeh et al. (2021), Aliyu et al. (2021), Sarker et al. (2021), Kan et al. (2016); [b] Aliyu et al. (2021), Sarker et al. (2021), Kan et al. (2016); [c] Aliyu et al. (2021); and [d] Yoder et al. (2011), El Hanandeh et al. (2021), Aliyu et al. (2021), X. Zhu et al. (2019), Sarker et al. (2021), Kan et al. (2016).

Biochemical conversion processes produce specific compounds, such as biogas or ethanol, and conversion is a relatively slow process, depending on the type of feedstock (Aliyu et al. 2021). Other disadvantages can be cited: (i) they require complicated pretreatment; (ii) enzymes are expensive; and (iii) conversion efficiency is relatively low (Zhu et al. 2019). In addition, the biological conversion of lignocellulosic biomass is hindered due to the recalcitrance of plant cells and the structural complexity of cellulose, hemicellulose, and lignin (Sarker et al. 2021).

2.2 BIOCHAR CONCEPT

Biochar (BC) can be defined as a carbon-rich, finely divided, porous material produced under limited oxygen conditions and high temperatures (Brassard et al. 2016). It can be considered a hierarchical porous material, because it presents different pore sizes, which make this material highly recommendable for use as adsorbents. These pores are generated in the dehydration process due to water loss during pyrolysis, generating pores of different sizes, which can be categorized into three levels, nano (<0.9 nm), micro (<2 nm), and macro (> 50 nm) (Veiga et al. 2021).

Due to their adsorptive properties, they can be used as precursors for the production of activated carbon due their physicochemical activation (Yogalakshmi et al. 2022). Biochar has a carbonaceous skeleton, a small number of heteroatom functional groups, mineral material (ash), and water (Sajjadi et al. 2019), as schematized in Figure 2.2. BC generally contain carbon (40–70%), oxygen (10–45%), hydrogen (1–5%), nitrogen (0–3%), sulfur (<1%), and other trace elements (Sajjadi et al. 2019).

A lower H/C ratio indicates greater carbonization and therefore a greater amount of aromatic portions in the material, approximating the composition of coal and anthracite. The O/C ratio indicates the gain or loss of oxygenated chemical functions, and the lower these values are, the more deoxygenated the materials can be (Melo et al. 2017). Such properties can best be visualized in an H/C versus O/C plot, van Krevelen diagram, as shown in Figure 2.3.

As shown in Figure 2.3, Ortiz et al. (2020) used the Van Krevelen diagram to compare the elemental compositions of BC obtained from almond husk (almond biomass) at temperatures of 673 K (Δ ASB), 773 K (× ASB), and 873 K (* ASB) and walnut husk (nut biomass) at temperatures of 673 K (•NSB), 773 K (+ NSB), and 873 K (o NSB). A decrease in the H/C ratio was observed, indicating further aromatization of the BC. A decrease in the O/C ratio is also observed with increasing temperature,

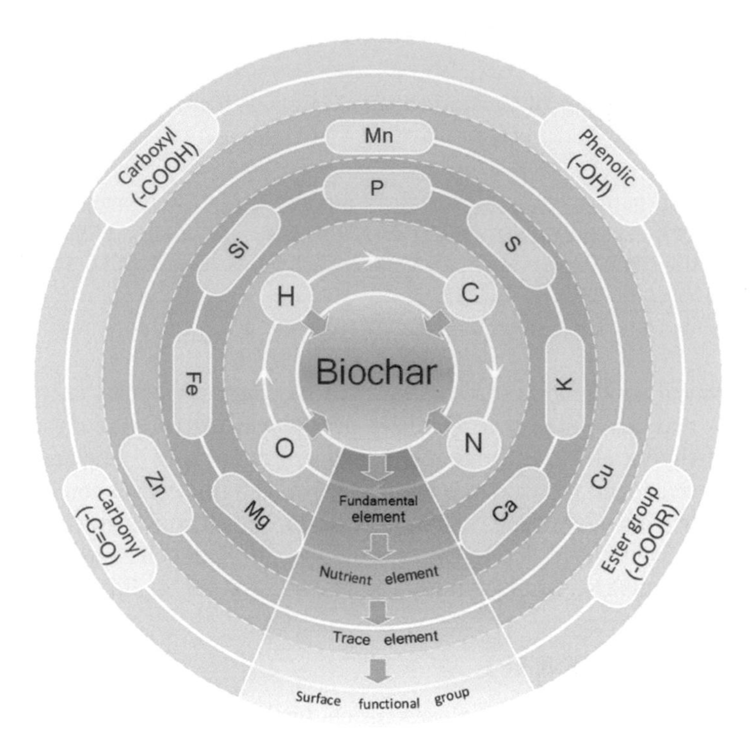

FIGURE 2.2 Main functional groups and ash of a biochar (adapted from Zhang et al. 2021).

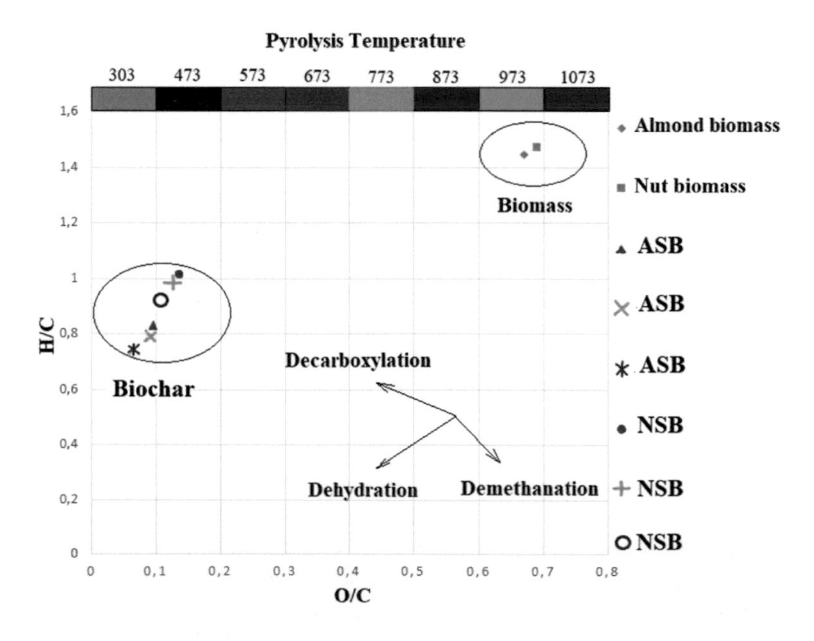

FIGURE 2.3 Example of the Van Krevelen diagram (Rodriguez Ortiz et al. 2020).

suggesting that the oxygenated functional groups on the biochar surface were lost according to the decreasing hydrophilicity of the material (Ortiz et al. 2020).

These carbonaceous materials also have a surface functionalized with different organic groups, which can also bind to metals of interest, such as plant nutrients. Among the oxygen-containing groups are carboxylic, hydroxyl, phenolic, carbonyl, lactone, carboxylic acid anhydride, and cyclic peroxide groups (Sajjadi et al. 2019). The use of BC dates to the Amerindian's time, who used this compost as a soil nutrient. The rich soil organic matter in the Amazon is due to the use of BC in the soil. Interest in biochar has increased recently, thanks to its possible applications in soil improvement, carbon sequestration, and activated carbon production, as well as for more advanced applications for the manufacture of catalysts, composites, and electronic components or as an additive for food and feed (Zheng et al. 2010).

The precursor biomass as a source of feedstock for biochar production directly influences the properties of the final product due to its elemental composition, ash content, among others. Parameters of the carbonization process and physical–chemical properties of BC have been traditionally studied (Rasa et al. 2021). Obtaining BC from waste biomass has become an interesting alternative in the management of industrial waste. This is because this material presents interesting physical–chemical characteristics, as mentioned previously. Thus, such management enables the obtainment of value-added products, being an economic alternative to minimize environmental problems arising from inadequate waste management.

2.3 PRECURSOR RAW MATERIALS

Depending on the precursor biomass, BC will have different physicochemical properties. Biomass can be categorized as forest, marine/aquatic, agricultural, industrial, soil, and animal waste, with a varying composition of carbon, nitrogen, oxygen, ash, volatile matter, and fixed carbon (Shyam et al. 2022), as shown in Table 2.2.

Biomass is considered a carbon-neutral raw material, which means that it does not play any role in any addition to the planet's CO_2 balance (Saxe et al. 2019). In

TABLE 2.2
Chemical Composition of Different Biomasses for Bioenergy Production (Solid, Liquid, and Gas).

Type of biomass*	Carbon (%)*	Nitrogen (%)*	Oxygen (%)*	Ash (%)*	Matter volatile (%)*	Fixed carbon (%)*
Agriculture	27–65	0.5–4.5	17–49	0.8–12.6	73–88	5–20
Aquatic	38–54	1–12	26–53	13–43	37–62	9–29
Food	44–53	0.5–1.3	34–46	0.3–8.1	72–86	14–26
Forest	49–58	2–3	29–38	5–11	78–80	11–15
Industrial	20–48	.5–7.8	15–58	13–57	40–72	3–24

(%) on a dry basis

addition, the use for biochar production of nonedible agricultural waste biomass (Ahmad et al. 2014), particularly herbaceous plants that have short growing seasons, leads to interesting environmental and socioeconomic results compared to the use of conventional biomass such as wood and coconut shells (Lopes et al. 2021).

2.3.1 Lignocellulosic Biomass

As can be seen in Table 2.2, the raw materials used for energy production are completely different in composition, and therefore BC will be as well. For example, plant biomasses, consisting of forest residues, agricultural residues, energy crops, and invasive plant species, are largely made up of lignocellulosic material (Zhu et al. 2019). The three main components of these biomasses are cellulose (30–45% w/w), lignin (15–30% w/w), and hemicellulose (15–35% w/w) (Zhu et al. 2021), as presented in Figure 2.4. Lignocellulosic biomasses also have a minimal amount of extractives (lipids, essential oils, resins, proteins, simple sugars, starch, water), hydrocarbons, and inorganic ash (Yogalakshmi et al. 2022).

Lignocellulosic biomass is difficult to convert with the help of biological processes because of its complex nature, mainly due to the presence of lignin and extractives, requiring the use of acid chemicals and organic solvents (Sarker et al. 2021). Cellulose and hemicellulose consist of carbohydrates, but while cellulose is linear and crystalline, hemicellulose is cross-linked and amorphous (Othmani et al. 2021).

Cellulose is a homopolymer consisting of β-D-glucopyranose units linked by β-glycosidic bonds (Alonso et al. 2012), which comprises 7000–15000 glucose

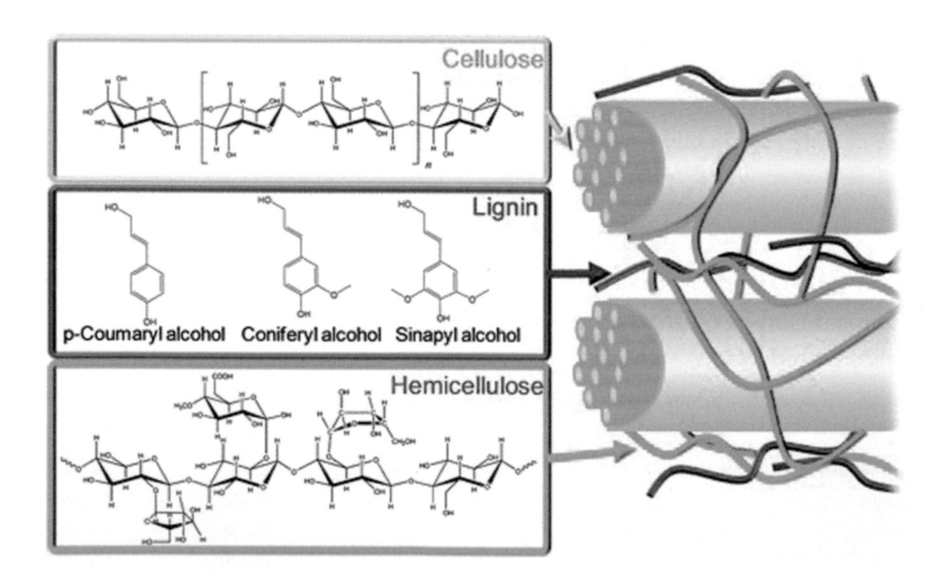

FIGURE 2.4 Structure of lignocellulosic biomass with cellulose, hemicellulose, and lignin depicted. The building blocks of lignin, p-coumaryl, coniferyl, and sinapyl alcohol are also shown (Alonso et al. 2012).

monomer units with a degree of polymerization of 1510–5500 (Sarker et al. 2021). At temperatures below 300°C, decomposition and polymerization reactions occur, leading to the formation of low molecular weight compounds such as H_2O and CO_2. When the temperature approaches 800°C, the biomass begins to produce bio-oil and BC (Yogalakshmi et al. 2022). Holocellulose is a term used for the combination of cellulose and hemicellulose.

Hemicellulose consists of 500–3000 sugar monomers with a degree of polymerization of 50–200 (Sarker et al. 2021). Similar to cellulose, at temperatures below 300°C, decomposition and polymerization reactions occur, leading to the formation of low molecular weight compounds such as H_2O and CO_2, and at temperatures around 800°C, the biomass starts to produce bio-oil and BC (Yogalakshmi et al. 2022).

Lignin, which is an important component of biomass resources, is one of nature's renewable energies that contain aromatic rings. It can be applied in many fields such as chemical engineering, materials, and food. The production of biofuels and high-value chemicals from lignin has become a hotspot, especially in the context of the "lignin first" strategy (Zhu et al. 2020). Ether bonds mainly include β–O–4, α–O–4, γ–O–4, and 4–O–5, and the β–O–4 bond is dominant and consists of more than half of the bonding structure of lignin (Sun et al. 2021). The C bonds mainly include β–β, 5–5, β–5, β–4, and β–1 bonds. Therefore, the complexity of lignin structures makes its depolymerization a challenging problem (Ye et al. 2021).

The carbonization process of cellulose, hemicellulose, and lignin is illustrated in Figure 2.5.

2.3.2 Biomass from Industrial and Domestic Activities' Sludge

The generation of wastewater from domestic and industrial activities is increasing due to rapid population growth and industrial activities. Among the various treatment processes employed for these effluents, the physical–chemical and biological processes stand out but generate large amounts of sludge, called biosolids (Nunes et al. 2021). These biosolids can be disposed of in various ways, and the most common ways include incineration, agricultural application, landfill, and land reclamation (Wilk et al. 2021).

Urban and industrial biosolids have a very varied composition that depends on the efficiency of the treatment processes employed. Besides carbon, macro and micronutrients, toxic elements such as cadmium, chromium, arsenic, mercury, traces of poorly biodegradable organic compounds such as polychlorinated biphenyls (PCBs), polycyclic aromatic hydrocarbons (PAHs), dioxins, pesticides (e.g., herbicides), linear alkyl sulfonates, and potentially pathogenic organisms can be present (Jellali et al. 2021; Kumar et al. 2017; Nunes et al. 2021; Rulkens 2008).

As a bioproduct of biological wastewater treatment, sewage sludge is an important source of secondary pollution or resource recovery for human society (Jiang et al. 2021). Ribeiro et al. (2021) used a sustainable synthesis from carbonization of cosmetics industry waste sludge to produce bio-oil and BC, which can be used for different applications, such as biofuel and bio adsorbent, respectively (Ribeiro et al. 2021).

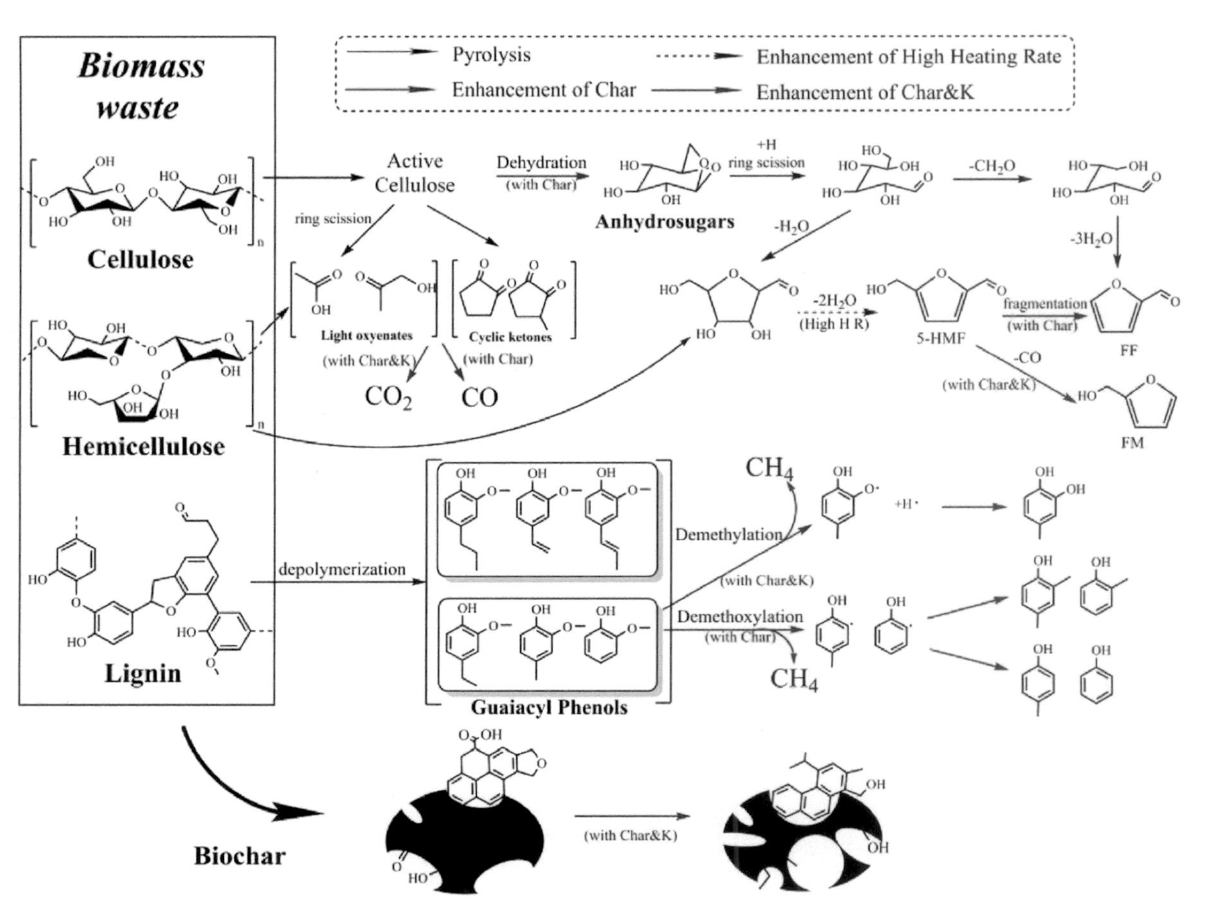

FIGURE 2.5 Cellulose, hemicellulose, and lignin carbonization process (Zhu et al. 2021).

2.3.3 Seaweed

Algae have received considerable attention in recent decades, as BC derived from red algae, green algae, macroalgae, and other types of algae. These materials have been suggested to remove organic pollutants or potentially toxic metal from aqueous system (Liu et al. 2021). For example, Hung et al. (2020) synthesized biochar from red algae at 300–900°C under inert atmosphere and used it in the remediation of 4-nonylphenol contaminants (Hung et al. 2020). Biochars derived from macroalgae and natural seaweed are promising adsorbents for removing organic pollutants (Fazal et al. 2020).

BC produced from the pyrolysis process of blue algae can be applied in the adsorption of Cd(II). The results as found by Liu et al. (2021) showed that the adsorption isotherms can be described by the Langmuir model and that the pseudo-second order model better fits the Cd(II) adsorption kinetics, indicating that the process occurs in a monolayer and is controlled by chemisorption. The best condition found was BC produced at 600°C for 2 h, with a capacity of Cd(II) adsorption of 135.7 mg g^{-1} (Liu et al. 2021).

In turn, Wu et al. (2021) used the seaweed *Enteromorpha prolifera* to produce an algae-based biochar, modified with potassium hydroxide, which was then used for the efficient sorption of a typical antibiotic, sulfamethoxazole (Wu et al. 2021). The study indicated that the structure and properties of algae BC are improved by the modification of potassium hydroxide at high temperature and that the maximum sorption capacity for sulfamethoxazole can reach 744 mg g^{-1}, which is a value greater than that reported for adsorbents in the literature (Wu et al. 2021).

The BCs produced, in general, have a porous structure with abundant functional groups and a high surface area and also contain minerals and trace metals, and the most commonly used production methods include slow or fast pyrolysis, gasification, hydrothermal carbonization, and roasting (Li et al. 2020).

2.4 BC PRODUCTION

There are several carbonization processes, such as pyrolysis (Guimarães et al. 2020), gasification (Brindhadevi et al. 2021), hydrothermal synthesis (Fang et al. 2018), and microwave-assisted hydrothermal carbonization (Wang et al. 2021). However, the most conventional processes for producing the solid product (char), pyrolysis and hydrothermal carbonization (HTC), also called wet pyrolysis, will be described. The product obtained from pyrolysis has been called pyrochar, while the product from HTC is called hydrochar.

2.4.1 Pyrochar Production via Pyrolytic Carbonization

Pyrolysis is a thermochemical process involving the thermolysis of carbon-based materials in the absence of an oxidizing agent and is one of the most efficient processes used in biochar production (Chen et al. 2017; Moreira et al. 2017). However, other gases such as CO_2, steam, hydrogen, helium, carbon monoxide, and methane can also be used as reaction environments, and it has been observed that CO_2, for example, can increase the specific surface area (SSA) of biochar (Kan et al. 2016).

In pyrolysis, biomass is thermally degraded generating three products, bio-oil (liquid), BC (solid), and biogas (Chandra and Bhattacharya 2019). To obtain a greater amount of one of its products, some parameters can be optimized, such as temperature, residence time, and heating rate.

Biomass pyrolysis consists of three main steps: (i) initial moisture evaporation, (ii) primary decomposition followed by (iii) secondary reactions (oil cracking and repolymerization) (Kan et al. 2016). Generally, milder temperatures up to 600°C favor the production of BC (Leng and Huang 2018). On the other hand, longer residence times in the reactor maximize the thermal cracking of tar into gas (Guizani et al. 2014).

This process can be classified based on the rate of heating and its residence time, presenting its own advantages and limitations for each type of pyrolysis. Table 2.3 summarizes the operating parameters for each of these classifications. The largest amount of BC is produced by slow and instant pyrolysis, while condensable gas is the main product of fast pyrolysis. Gasification needs the highest temperature and a limited amount of oxygen (Sajjadi et al. 2019).

Some pretreatments are necessary before pyrolytic carbonization such as:

(1) *Physical pretreatment*: The biomass is processed to decrease its size, allowing the smaller particles to promote heat and mass transfer to form a uniform temperature within the particles during pyrolysis.
(2) *Thermal pretreatment*: Drying and roasting processes (temperatures at 200–300°C) increase the energy efficiency of the pyrolysis process and improve the quality of the products obtained. The presence of water lowers the temperature and prevents the complete cracking of the molecules in the pyrolysis zone (Marzbali et al. 2021).
(3) *Chemical pretreatment*: Aiming to perform the chemical activation of the material, the raw material is impregnated with chemical agents such as KOH, $ZnCl_2$, H_3PO_4, K_2CO_3, among others, and then heated in an inert atmosphere (Miliotti et al. 2020).
(4) *Biological pretreatment*: The use of microorganisms is employed to improve the efficiency of the pyrolytic process (Wu et al. 2021).

TABLE 2.3
Different Types of Pyrolysis and Operating Conditions.

	Slow	Fast	Snapshot	Midway	Vacuum	Hydro
Temperature (°C)	550–950	850–1250	900–1200	500–650	300–600	350–600
Heating rate (°C/s)	0.1–1.0	10–200	≥ 1000	1.0–10	0.1–1.0	10–300
Time of residence (s)	300–550	0.5–10	≤ 1	0.5–20	0.001–1.0	≥ 15
Pressure (MPa)	0.1	0.1	0.1	0.1	0.01–0.02	5.0–20
Main product obtained (%performance)	Biochar	Bio-oil (50–70%)	Bio-oil (75–80%)	-	-	-

Source: Adapted from Tripathi et al. (2016).

In general, pyrochar have undeveloped pore structures and limited functional groups. Therefore, post-treatment steps can be performed, such as physical or chemical activation processes, to improve these characteristics. In physical activation, an activating agent such as steam, CO_2, air or a combination of these gases gasifies at high temperatures part of the carbon structure, creating porosities (Miliotti et al. 2020).

The oxidizing agents used in physical activation diffuse into the porous structure of the biochar and gasify the carbon atoms, resulting in a high surface area (Madhubashani et al. 2021). The accepted mechanism for BC activation by steam is described in Equations (1–8) (Sajjadi et al. 2019).

$$\equiv CC_f(s) + H_2O(g) \rightarrow \equiv CC(O)(s) + H_2(g) \qquad \Delta_r^0 = 131\ \text{kJ mol}^{-1} \qquad \text{(Eq. 1)}$$

$$\equiv CC(O)(s) \rightarrow C(g) + \equiv C_f(s) \qquad \text{(Eq. 2)}$$

$$CO(g) + \equiv CC(O)(s) \rightarrow CO_2(g) + \equiv CC_f(s) \qquad \text{(Eq. 3)}$$

$$CO(g) + H_2O(g) \rightarrow CO_2(g) + H_2(g) \qquad \Delta_r^0 = -41\ \text{kJ mol}^{-1} \qquad \text{(Eq. 4)}$$

$$\equiv CC_f(s) + 2H_2O(g) \rightarrow \equiv C(s) + CO_2(g) + 2H_2(g) \qquad \text{(Eq. 5)}$$

$$\equiv CC_f + CO_2(g) \rightarrow 2CO(g) + \equiv C_f(s) \qquad \Delta_r^0 = 171\ \text{kJ mol}^{-1} \qquad \text{(Eq. 6)}$$

$$\equiv CC_f(s) + 2H_2(g) \rightarrow \equiv C_f + CH_4(g) \qquad \Delta_r^0 = -74{,}5\ \text{kJ mol}^{-1} \qquad \text{(Eq. 7)}$$

$$CH_4(g) + H_2O(g) \rightarrow CO(g) + 3H_2(g) \qquad \Delta_r^0 = 208\ \text{kJ mol}^{-1} \qquad \text{(Eq. 8)}$$

where CC_f consists of the free carbon attached to the pyrochar carbonaceous chain.

The process begins with the exchange of oxygen from the water molecule to the carbon surface to create a surface oxide (Eq. 1), which consists of an endothermic process (Δ_r^0 = 131 kJ mol^{-1}). The surface oxide can be eliminated as CO (Eq. 2). CO can increase the gasification rate by eliminating the surface oxide to produce CO_2 (Eq. 3). CO can also react with water vapor to produce hydrogen gas and CO_2 (Eq. 4), an exothermic process (Δ_r^0 = –41 kJ mol^{-1}). Both gases can activate the surface via reactions described by Eqs. (6) and (7), respectively, the former being an endothermic reaction (Δ_r^0 = 171 kJ mol^{-1}), and the latter being exothermic (Δ_r^0 = –74.5 kJ mol^{-1}). The gasification process by water vapor can also occur according to Eq. 5. On the other hand, H_2 can inhibit the gasification of carbon (Sajjadi et al. 2019).

Activation with CO_2 is a very popular process in which it participates in pyrolysis through the reverse Boudouard reaction, which decreases the carbon yield and leads to the successive removal of carbon from the pores according to Equations (9–11) (Lopes and Astruc 2021).

$$\equiv CC_f(s) + CO_2(g) \rightarrow \equiv CC(O)(s) + CO(g) \qquad \text{(Eq. 9)}$$

$$\equiv CC(O)(s) \rightarrow \equiv C_f + CO(g) \qquad \text{(Eq. 10)}$$

$$\equiv C_f(s) + CO(g) \rightarrow \equiv CC(O)(s) \qquad \text{(Eq. 11)}$$

Water-vapor-activated carbon is more active compared to CO_2-activated carbon, probably due to the smaller size of the water molecule that favors its diffusion in the porous network (Feng et al. 2017). Physical activation can be done in two steps or a single step (Lopes and Astruc 2021). Because it uses high temperatures, it has a high energy expenditure. Other activation processes can also be used such as ultrasonic and microwave activation.

Chemical activation, on the other hand, requires low energy consumption, but the purchase of chemicals and subsequent drying steps must be taken into consideration (Yuan et al. 2021). Chemical activation can be performed in a post-treatment step after carbonization. There are different types of pyrolysis reactors, such as fixed bed reactors, batch or semi-batch reactors, rotary kilns, fluidized bed reactors, microwave-assisted reactors, and some contemporary types such as plasma or solar reactors (Yogalakshmi et al. 2022).

Currently, it is known that the main functional groups of BC are carboxylic and phenolic groups, containing aromatic carbons and heteroatoms; however, the characteristics of these materials vary greatly according to their thermal processing conditions (Fakayode et al. 2020). Another important parameter in biochar production is the biomass characteristics, and biochar produced under the same pyrolysis conditions will have different properties, if coming from different biomasses (El Hanandeh et al. 2021).

Generally, the volatile matter content, ash content, fixed carbon, and thermal stability of the produced BC varied from 10% to 11%, 6% to 28%, 3% to 68%, and 0.03% to 0.87%, respectively, according to pyrolysis temperature, biomass types, and particle sizes (Lu et al. 2021).

Higher temperatures produce alkaline biochar, while lower temperatures produce acidic BC. Such phenomena can be explained by the loss of acidic functional groups during pyrolysis at high temperatures and, second, by the increase in ash content with temperature (Sajjadi et al. 2019).

2.4.2 Hydrochar Production via Hydrothermal Carbonization

The hydrothermal process has been considered as a promising approach for the treatment of wet biomass due to its high tolerance to moisture content and the ability to convert low energy density feedstock into high calorific value solid fuel (Li et al. 2021). The process takes place in an autoclave, as schematized in Figure 2.6.

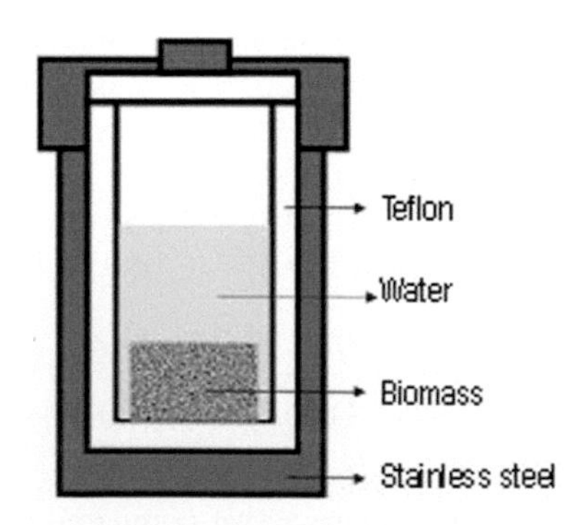

FIGURE 2.6 Schematic of hydrothermal process in autoclave.

An advantage of the hydrothermal process is the low activation energy requirement and flexibility in producing high-quality coal, oil, and gas, and the water itself can act as a reactant medium and assist in starting the reaction via hydrolysis (Marzbali et al. 2021). Other advantages are lower energy consumption compared to pyrolysis and the elimination of biological hazards including pathogens from the waste due to the inherent autoclaving above 121°C (Marzbali et al. 2021). The hydrothermal process can be subcategorized into hydrothermal carbonization (HTC), hydrothermal liquefaction (HTL), and hydrothermal gasification (HTG) (Khan et al. 2019).

Hydrothermal processing operates in both subcritical and supercritical water, as shown in Figure 2.7 (Aliyu et al. 2021). The critical point of water occurs at 374°C and 22.1 MPa. HTL, also called hydrous pyrolysis, occurs under subcritical conditions, that is, over a temperature range of 250 to 374°C and with pressure between 4 and 22 MPa. In the HTL process, the main product obtained is crude bio-oil, being obtained as a by-product of gases, aqueous residue, and solid (Swetha et al. 2021).

HTG takes place under supercritical conditions, that is, temperature and pressure greater than 374°C and 22 MPa, respectively (Sengottian 2020). To obtain more hydrochar by the HTC process, the process temperature can be limited to approximately 180–260°C (Khan et al. 2019). When the residence time is short, the solid hydrocarbon content is high, and when the residence time is long, the solid hydrocarbon content is low (Khan et al. 2019).

The hydrochar formed in HTC is derived from a series of hydrolysis reactions, dehydration and/or elimination of oxygen functional groups, as well as condensation reactions (Melo et al. 2017). In the first step, biomass macromolecules are broken down into their building blocks by hydrolysis, which are then reconstructed to form

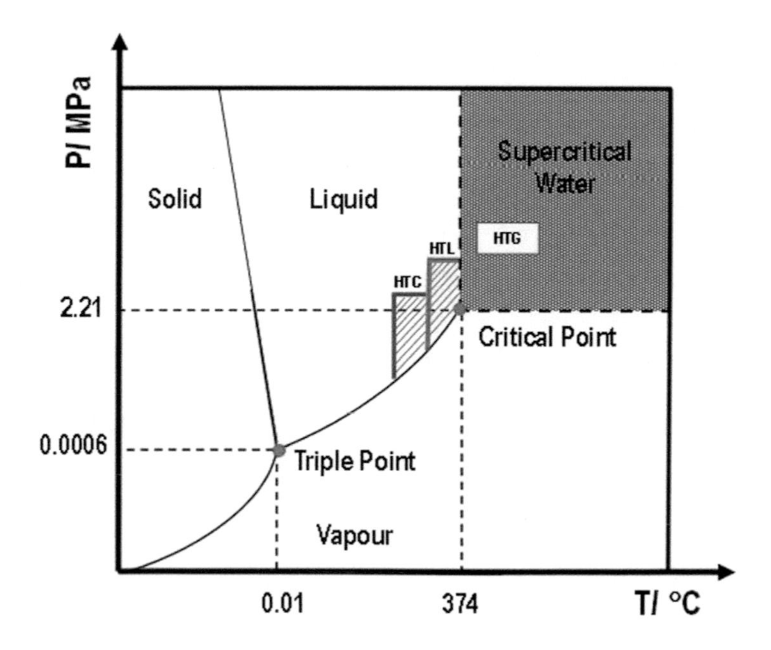

FIGURE 2.7 Hydrothermal processing as a function of temperature.

products (Marzbali et al. 2021). For lignocellulosic biomasses, the hydrothermal carbonization reaction destroys the crystalline structure of cellulose and produces more aromatic cyclization products (Y. Zhang et al. 2021).

The hydrochar obtained by HTC exhibits highly hydrophobic and brittle properties and is easier to separate from the liquid product (Khan et al. 2019). Depending on the temperature employed in the hydrothermal treatment, the hydrochar presents a brown coloration, indicating partial carbonization, or black, indicating total carbonization (Melo et al. 2017).

In the HTC process, water acts as a reactant to reorganize the structure of the biomass (Khan et al. 2019). The decrease in the dielectric constant of water with increasing temperature decreases the polarity of the water, allowing it to dissolve hydrophobic organic compounds, such as free fatty acids, which are poorly soluble at room temperature (Marzbali et al. 2021). Such conditions also weaken the hydrogen bonds of water, producing high ionization constants, increasing the concentration of H^+ and OH^- ions in solution (Tekin et al. 2014). The viscosity of water decreases with increasing temperature, thus increasing the mass transfer and diffusion coefficient (Aliyu et al. 2021). These phenomena allow water to penetrate the biomass more easily (Tekin et al. 2014).

The basic parameters of the hydrothermal carbonization process involve lower temperatures compared to the pyrolytic route, as mentioned before (Wilk et al. 2021). The pressure is often autogenic and can range from 2 to 10 MPa (Wilk et al. 2021). The critical process parameters are temperature, pressure, pH, and residence time.

Higher temperatures can lead to simultaneous dehydration, decarboxylation, and condensation and, with adequate time, can lead to a higher degree of intermediate dissolution and consequent transformation via polymerization and secondary coal formation, which controls the mechanism of hydrocarbon development (Khan et al. 2019), as can be seen in Table 2.4.

TABLE 2.4
Physicochemical Properties of Water as a Function of Temperature and Pressure.

Parameters	Ambient water	Subcritical water	Supercritical water	
Temperature, T (°C)	25	250	400	400
Pressure, p (MPa)	0.1	5	25	50
Density, ρ (g cm^{-3})	0.997	0.80	0.17	0.58
Dielectric constant, (ε)	78.5	27.1	5.9	10.5
pK_{su}	14.0	11.2	19.4	11.9
Heat capacity, c_p (kJ kg^{-1} K^{-1})	4.22	4.86	13	6.8
Viscosity, μ (mPa s)	0.89	0.11	0.03	0.07
Thermal conductivity, λ (mW m^{-1} K^{-1})	608	620	160	438

In addition, water at elevated temperatures promotes the generation of free radicals, whose mechanism occurs in two stages, the first stage is the generation of the free radical pool. The second is the reaction of these free radicals with the biomass.

The main product obtained from this HTC pathway is the solid; the gas occupies a small percentage of the input mass and consists mainly of CO_2; the proportion of liquid and solid product depends on the initial moisture content of the sewage sludge and the process parameters (Wilk et al. 2021). Generally, the mass of the solid phase decreases due to the release of water and carbon dioxide, as well as the leaching of some compounds into the liquid.

Another advantage of the HTC process is that activation and functionalization can be accomplished via one-pot synthesis, where only one reaction vessel is used. Different activation agents can be used, the main ones being H_3PO_4, $ZnCl_2$, and NaOH. These compounds can be used to increase the specific surface area of the materials. Functionalization, on the other hand, consists of introducing some specific properties to the material, such as magnetization.

As the hydrothermal carbonization is initiated by H^+ ions, due to the autoionization of water and organic acids co-generated in the process, several works have reported the introduction of acids to the hydrothermal reaction system, aiming to accelerate the reaction, reduce energy consumption, and improve the quality of the hydrochar (Zhang et al. 2021). Among the acids, H_3PO_4 presents the least corrosiveness. Phosphoric acid can precipitate inorganic compounds, forming phosphatized solids, as shown in Figure 2.8 (Melo et al. 2017).

This acid forms ester bonds with -OH groups on the cellulose, cross-linking the polymer chains and introducing P atoms into the carbon matrix mainly as reduced states (Sajjadi et al. 2019).

Alkaline activation employing NaOH or KOH helps to increase porosity and clear partially blocked pores, resulting in an increase in specific surface area, according to Eqs. (12–13) (Jais et al. 2021). NaOH is a favorable activator due to its high performance, capable of producing improved mesoporous carbon, with high cost-effectiveness and low dosage. In addition, it has low corrosivity and environmental compatibility (Islam et al. 2017).

$$6NaOH + 2C \rightarrow 2Na + 2Na_2CO_3 + 3H_2 \quad \text{(Eq. 12)}$$

$$Na_2CO_3 \rightarrow Na_2O + CO_2 \quad \text{(Eq. 13)}$$

$$2Na + CO_2 \rightarrow Na_2O + CO \quad \text{(Eq. 14)}$$

Hydrochar has less aromatic carbon compared to pyrochar and is chemically similar to lignite (Cao et al. 2013). However, hydrochar contains higher oxygen

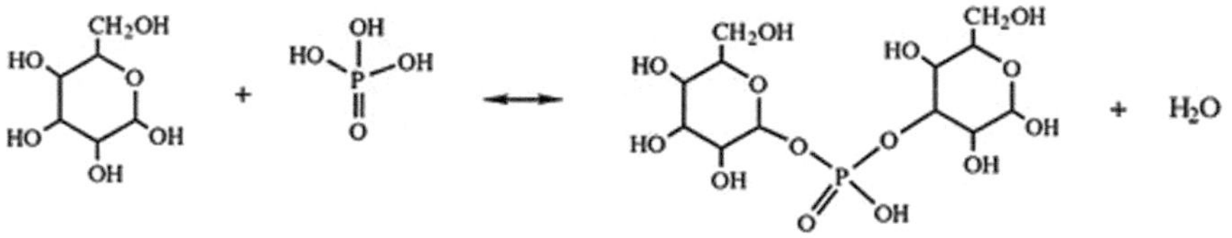

FIGURE 2.8 Precipitation reaction of inorganic compounds with phosphoric acid.

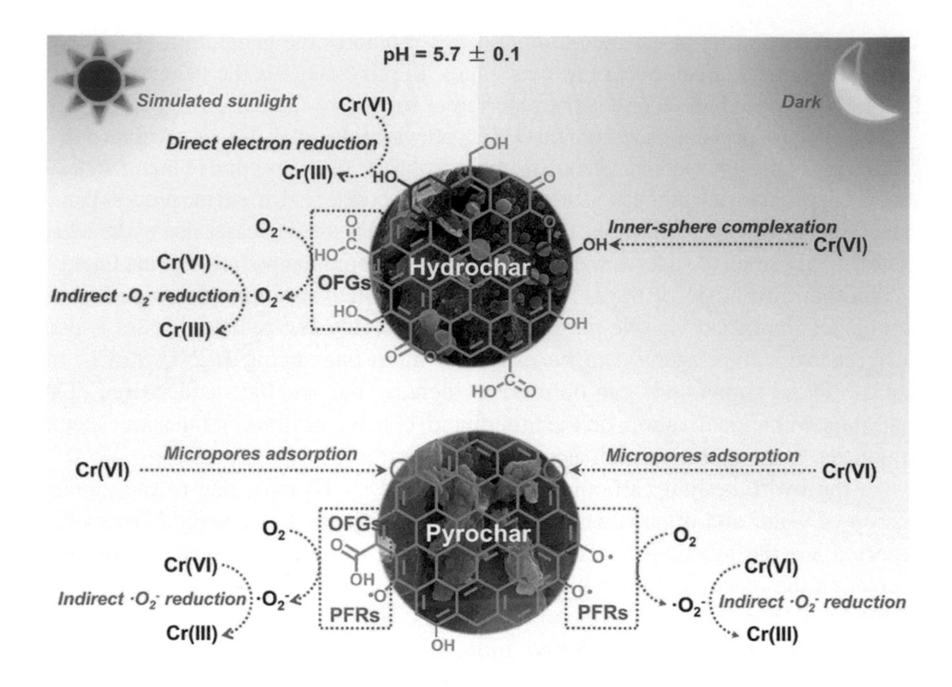

FIGURE 2.9 Schematic illustration for the structure-dependent removal processes of Cr(VI) from hydrochar and pyrochar in the dark and in sunlight (Chen et al. 2021).

content than pyrochar due to the surface-enriched oxygen functional groups (OFGs) generated from the hydrolysis of biomass components in water (Chen et al. 2021). The hydrochar can transfer electrons to dissolved oxygen in daylight to form $-O_2^-$ which subsequently reacts with H^+ to produce H_2O_2 (Chen et al. 2017). In the presence of light, H_2O_2 can decompose to generate -OH as shown in Figure 2.9.

In general, pyrochar synthesized at high temperatures shows high residual mass by thermogravimetric analysis (Hwang et al. 2021), unlike hydrochar, because the synthesis temperatures of the latter are not high enough to break down all biomass components, such as the crystalline region of cellulose or the remaining lignin complex (Hwang et al. 2021).

2.5 PHYSICAL–CHEMICAL CHARACTERISTICS OF BIOCHAR

Hydrochar and pyrochar can be characterized by different techniques, which provide great information about the materials obtained such as relevant information on specific surface area (SSA) and pore volume and functional groups. The main techniques to characterize chars are X-ray diffraction (XRD), X-ray fluorescence (XRF), scanning electron microscopy (SEM), Fourier transform infrared spectroscopy (FTIR), SSA, and Brunauer–Emmett–Teller (BET) isotherm. These techniques will be described in the subsequent sections.

2.5.1 X-Ray Diffraction (XRD)

Discovered in 1895 by Roentgen, X-rays, with a wavelength on the order of 0.5 to 2.5 Å, became useful in the study of crystal structure through the phenomenon of diffraction (Callister 2006). The use of the XRD allows one to know the type of crystalline structure of materials from their respective lattice parameters (Culpin 1989). Based on the diffraction patterns obtained, the chemical and morphological composition of the crystal can be established (Callister 2006).

The diffraction phenomenon is observed in structures whose crystalline network has a high regularity, unlike amorphous materials. When an X-ray beam is incident on a solid material, a fraction of this beam will be diffracted in all directions by electrons, which are associated with atoms or ions that are in the beam path (Culpin 1989). If the difference in the lengths of the incident and reflected paths is equal to an integer, a constructive interference of the scattered rays will occur enhancing the intensity of the wave. Thus, the phenomenon of X-ray diffraction can be better visualized in the scheme in Figure 2.10. The X-ray diffraction behavior is described by Bragg's law (Eq. 15).

$$n\lambda = 2d_{hkl}\sin(\theta) \quad \text{(Eq. 15)}$$

where λ is wavelength in nanometers; d_{hkl} corresponds to the interplanar distance, in nanometers; θ corresponds to the diffraction angle, given in degrees (°); and hkl corresponds to the Miller indices of the crystal.

If Bragg's law is not satisfied, the waves will interfere non-constructively, and a diffraction beam of very low intensity will be produced (Nagata et al. 2010). Bragg's law is a necessary but not sufficient condition for diffraction by real crystals. It specifies when diffraction will occur for unit cells that have atoms positioned at vertices. However, atoms at other positions act as additional center positions, which can produce out-of-phase dispersions at certain Bragg angles (Nagata et al. 2010).

Wang et al. (2019) characterized biochar produced from corn straw by XRD in which they observed peaks at $2\theta = 30$ and 55°, which were attributed to silica, as

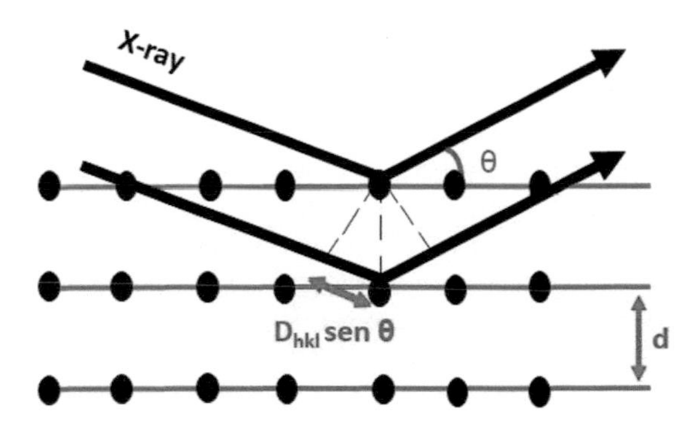

FIGURE 2.10 Schematic of the X-ray diffraction phenomenon.

well as broad-ranging peaks (2θ = 25 – 30°), indicating an amorphous character of the material (Wang et al. 2019). Similar results were found by Félix et al. 2017 who produced a biochar from sugarcane bagasse and performed its structural characterization (Félix et al. 2017). The main XRD results consisted of the identification of two peaks referring to cellulose centered at 2θ ~15° and 22° for the biomass, indicating the presence of crystalline cellulose and traces of amorphous silica in the range of 2θ = 25 to 45° for the biochar.

2.5.2 X-Ray Fluorescence (XRF)

XRF is based on the production and detection of characteristic X-rays, which are high-frequency electromagnetic radiations with a wavelength in the range 0.003 to 3 nm. These are produced by the photoelectric phenomenon, emitted by the constituent elements of the sample when irradiated. When electrons in the innermost layer of the atom (e.g., K and L) interact with photons of energy in the X-ray region, photo ejection of these electrons can occur, creating a vacancy. To promote stability, the electron vacancies are immediately filled by electrons from adjacent layers (Cave and Talens-Alesson 2005).

As a result, there is an excess of energy in the process, which is emitted in the form of X-rays characteristic of each atom present in the sample. The technique can be used for qualitative or quantitative purposes (Callister 2006), besides having advantages in determining chemical elements, generally, quickly, without destroying the matrix, with low operational cost and little sample (Callister 2006).

A BC produced from oak wood and used in the removal of Cr (VI) was characterized by X-ray fluorescence (Liu et al. 2020). Among other techniques, XRF was used for characterization. Confocal micro-X-ray fluorescence imaging (CMXRFI) results indicated that total Cr (tCr) was heterogeneously distributed in the image area, with a higher intensity near the particle surface. Such results contributed to the elucidation of the removal mechanism, which likely involve electrostatic attraction and diffusion within the particle, followed by reduction and ion exchange reactions.

2.5.3 Scanning Electron Microscopy (SEM)

The SEM technique allows to generate images of a solid surface with high resolution, be it porous or regular (Steter et al. 2016). The images obtained can be presented in three-dimensional form, obtaining information about the morphology of the material surface, homogeneity, and the presence of pores. The technique is based on the application of an electron beam on a surface in order to generate micro-processed images (Culpin 1989). This beam of electrons interacts with the sample, which must present appreciable electrical conductivity. For this, nonconductive samples are subjected to a surface coating with conductive material such as platinum (Pt) or gold (Au), and this process is called metallization (Culpin 1989).

Elemental analysis by energy dispersive X-ray spectroscopy (EDS) is commonly coupled with SEM (SEM-EDS), and the principle of the technique is similar to traditional XRF. The electrons fired at the sample are able to expel electrons from the innermost layers of its component atoms and thereby cause the fluorescence effect

(Nagata et al. 2010). As the electrons of a given atom have different energies and characteristics of that atom, it is possible to identify which chemical elements are present in the area of beam incidence.

SEM can be used to obtain morphological information from the biochar, for example, Wang et al. (2019) produced BC from willow waste and also the material after containing nanoparticles of zero-valent iron (nZVI). According to the authors, the biochar presents an amorphous morphology (Figure 2.11a), while the nZVI have a spherical shape (Figure 2.11B). According the authors, the morphology of the nZVI/BC composite is completely different from the morphology of nZVI and BC, and these differences are attributed to the formation of new functional groups on the surface of the composite (Wang et al. 2019).

Dong et al. (2011) used a biochar produced from beet syrup for the removal of Cr (VI) in aqueous systems, obtaining a maximum removal capacity of 123 mg g^{-1}. The surface characteristics of biochar, before and after Cr (VI) adsorption, were investigated by SEM-EDS, among others. Such analyses showed that most of the Cr bound to the biochar was Cr (III). These results indicated that the removal of Cr (VI) by biochar occurs by electrostatic attraction of Cr (VI) with the positive surface of biochar, followed by reduction of Cr (VI) to Cr (III) (Dong et al. 2011).

FIGURE 2.11 Scanning electron microscopy images of (a) BC, (b) iron nanoparticles (nZVI), and (c) composite (nZVI/BC) (Wang et al. 2019).

2.5.4 Fourier Transform Infrared Spectroscopy (FTIR)

Infrared radiation corresponds to a region of the electromagnetic spectrum that lies between the visible and microwave regions. The region between 14,000 and 4000 cm^{-1} is called near infrared (NIR). The region between 4000 and 400 cm^{-1} is called mid-infrared (Mid-IR). While the region between 700 and 200 cm^{-1} is called far infrared (FIR) (Hatanaka et al. 2010).

Infrared spectroscopy measures the binding energies between vibrational and rotational states when a molecule absorbs energy in the infrared region. The different functional groups in a molecule have different infrared absorption frequencies and intensities (Hatanaka et al. 2010). Mid-IR spectra consist of several absorption bands, which are related to the functional groups of a molecule. Consequently, the Mid-IR spectrum is characteristic and is widely used in structural identification and nowadays in quantitative analysis (Hatanaka et al. 2010).

The Fourier transform is a mathematical tool that can improve the resolution of the infrared spectra obtained (Frost et al. 2003). Thus, FTIR is considered to be one of the most important experimental techniques for the characterization of different compounds in terms of identification and/or determination of structural characteristics (Gomes et al. 2017). In addition to qualitative information, the technique allows the semi-quantitative determination of components in a sample or mixture.

FTIR analyses indicate the presence of a band in the region of 3420 cm^{-1}, attributed to OH stretching of hydroxyl groups, more pronounced for hydrochar than pyrochar. Hydrochar also shows a characteristic peak in the region of 1710 cm^{-1} attributed to the C=O stretching of carboxyl groups. Compared to pyrochar, hydrochar also exhibits a more pronounced band at 1160–1030 cm^{-1}, attributed to the stretching of C–O bond of the aliphatic. In addition, hydrochar presents a higher amount of carbon CH_3, presenting a more pronounced band in the region of 2923–2853 cm^{-1} compared to pyrochar (Chen et al. 2017).

Liu et al. (2019) characterized by FTIR a biochar obtained from corn straw modified with silicon, which was used in the adsorption of copper. This biochar showed a sharp band at 1610 cm^{-1} and a broad band at 3340–3450 cm^{-1} in the spectrum, which were attributed to the bending vibration and stretching vibration of hydroxyl groups. The FTIR spectra showed that the vibration of hydroxyl groups on the surface of biochar could be enhanced by Si modification. In addition, the characteristic bands at 467, 880, and 1030 cm^{-1} were associated with Si–O–Si.

Biochar can be used for changes in the sorption and bioavailability of herbicides in soil, as in the work developed by Gámiz et al. (2019) shows in which fresh and aged biochar was used to alter the soil. Both biochars exhibited similar absorption spectra, suggesting that, under the conditions studied, aging did not produce significant differences in surface functionality. Peaks appearing between 746 and 872 cm^{-1} were attributed to aromatic C–H single bond outside the plane strain; the broad band centered at 1188 cm^{-1} was indicative of single bond C–O elongation vibration, distinct from cellulose-derived products; and a band appearing at 1700 cm^{-1} is assigned to the single bond C=O extending from the carboxyl groups (Gámiz et al. 2019).

Gupta et al. (2015) also characterized by FTIR a biochar produced from tea waste that was used for the simultaneous adsorption of Cr(VI) and phenol. FTIR studies demonstrated functional groups such as aliphatic primary amines, aliphatic alcohols, and C=O, CO, and SiO_2 bonds. Furthermore, bands between 600 and 500 cm^{-1} were assigned to SiO_2 present on the biochar surface (Gupta and Balomajumder 2015). After adsorption of Cr (VI) and phenol, the strong band from 3500 to 3000 cm^{-1} becomes weak and broad, and it is also possible to observe changes in the vibrational frequencies around 1700 to 1200 cm^{-1}, which indicate the binding of Cr (VI) and phenol to the functional groups of the biochar.

2.5.5 SSA and BET Isotherm

Surface properties are fundamental to many applications, such as catalysis, for example. Among these properties, the specific surface area is the most important (Koh et al. 2009). In solid particles, this parameter has usually been obtained through nitrogen adsorption and desorption analysis, using the model developed by BET (Koh et al. 2009). The adsorption process usually occurs at high pressures and over wide temperature ranges (Brunauer et al. 1938).

Based on the Langmuir and BET mathematical models of adsorption equilibrium, it was possible to develop a new, more comprehensive isotherm model, called the multilayer model (Eq. 16). The equations obtained are able to predict average compatible values for the heat of adsorption and the volume of gas required to complete the monolayer in a solid (Brunauer et al. 1938).

$$v = \frac{v_m Cp}{(p - p_o)\left[1 + (c - 1)\frac{p}{p_o}\right]} \qquad \text{(Eq. 16)}$$

wherein v is the volume of gas adsorbed (m^3), v_m is the volume of gas adsorbed when the surface of the solid is completely covered by a monolayer, c is the BET constant, p is the pressure measured at the equilibrium state, and p_0 is the system pressure (Pa).

Copper/biochar as Cu_2O–CuO/biochar composite form was prepared via hydrothermal process by Khataee et al. (2019), who used it as a photocatalyst in the degradation and mineralization of the dye reactive orange 29 (RO29) (Khataee et al. 2019). According to the authors, the specific surface area and total pore volume of biochar were 78.63 m^2/g and 0.75 m^3/g, respectively. Biochar presented itself as an excellent support for the immobilization of Cu_2O–CuO, and, consequently, as expected by the authors, the incorporation of Cu_2O–CuO particles on the biochar surface improved the characteristics of Cu_2O–CuO in terms of specific surface area (32.91 m^2/g) and pore volume (0.37 cm^3/g).

In turn, Foo and Hameed (2012) prepared, characterized, and evaluated the adsorptive properties of a biochar produced from orange peel via microwave-induced K_2CO_3 activation (Foo and Hameed 2012). N_2 adsorption/desorption analysis was used to determine the SSA and total pore volume, which were identified to be 1104.45 m^2/g and 0.615 m^3/g, respectively.

A summary of the main results found for characterization of the biochars, as discussed before, is presented in Table 2.5.

TABLE 2.5
Different Techniques for Characterizing Chars.

XPS		FTIR		Raman		DRX	
		Wavenumber (cm^{-1})	Vibration type	Wavenumber (cm^{-1})	Vibration type	2θ degrees	Atribution
284.6	sp2-graphítico ou ligação C – C	3447–3434	Stretching of O – H	1270 – 1500	D-band represented disordered structures	20 to 25	Amorphous carbon
286.0	C–O	2930–2890	Stretching vibration C–H	1500–1600	G-band represented graphitized structure	20 and 40	SiO_2
286.6	C=O	2372	Stretching of – CH_2			21.6	110
288.9	O=C–O	1590–1705	Carbonyl/carboxyl C=O stretching vibration			24	200
289.5	Carbonatos	1621–1635	Vibration of =C				
		1388–1420					
		1383	COOH				
		1374–1377	Aliphatic -CH				
		1033–1103	Vibration of C–O				

References: Ortiz et al. (2020), Zhu et al. (2021), Pereira Lopes and Astruc (2021), Manna et al. (2020), Kalaiselvi et al. (2015), Guimarães et al. (2020), Ganie et al. (2021), Yin et al. (2020), Chen et al. (2021), Chu et al. (2020), Debalina et al. (2017), Meili et al. (2019), Sewu et al. (2020), Zhang et al. (2020), Diao et al. (2021), Veiga et al. (2021).

The evolution of biochar studies needs standardization of analyzes and methodologies, in addition to the establishment of the properties of greater interest for full characterization and comparison of the final product, because given the variation of properties that can be obtained from the production of biochar, the determination of the factors that influence its final characteristics is extremely important to obtain a suitable material in relation to its destination, which will only be possible if there is conciliation of the relationships between the raw material and the biochar, and the production conditions.

2.6 CONCLUDING REMARKS

Due to the large availability of biomass and the possibility of producing not only biochar, but also other materials, such as bio-oil and biogas, from these multiple biomass sources, studies focusing on the structural characteristics of biochar and its use for renewable energy production, agronomic applications, and environmental protection are expected to be explored more. In this chapter, we have summarized the aspects of the choice of feedstock, the effect of temperature, the relationships of biomass components with the structural characteristics of biochar, as well as its potential agronomic applications, with biochar being a material with a broad field of environmental application. Understanding the physicochemical characteristics of each biochar produced is essential to predict possible reaction mechanisms with certain herbicides, to be used in different environmental applications.

REFERENCES

Ahmad, M., A. U. Rajapaksha, J. E. Lim, M. Zhang, N. Bolan, D. Mohan, M. Vithanage, S. S. Lee, and Y. S. Ok. 2014. Biochar as a sorbent for contaminant management in soil and water: A Review. *Chemosphere* 99:19–23.

Aliyu, A., J. G. M. Lee, and A. P. Harvey. 2021. Microalgae for biofuels: A review of thermochemical conversion processes and associated opportunities and challenges. *Bioresource Technology Reports* 15:100694.

Alonso, D. M., S. G. Wettstein, and J. A. Dumesic. 2012. Bimetallic catalysts for upgrading of biomass to fuels and chemicals. *Chemical Society Reviews* 24: 8075–8098.

Anukam, A., S. Mamphweli, P. Reddy, E. Meyer, and O. Okoh. 2016. Pre-processing of sugarcane bagasse for gasification in a downdraft biomass gasifier system: A comprehensive review. *Renewable and Sustainable Energy Reviews* 66:775–801.

Arpia, A. A., W. H. Chen, S. S. Lam, P. Rousset, and M. D. G. Luna. 2021. Sustainable biofuel and bioenergy production from biomass waste residues using microwave-assisted heating: A comprehensive review. *Chemical Engineering Journal* 403:126233.

Brassard, P., S. Godbout, and V. Raghavan. 2016. Soil biochar amendment as a climate change mitigation tool: Key parameters and mechanisms involved. *Journal of Environmental Management* 181:484–497.

Brindhadevi, K., S. Anto, E. R. Rene, M. Sekar, T. Mathimani, N. Thuy Lan Chi, and A. Pugazhendhi. 2021. Effect of reaction temperature on the conversion of algal biomass to bio-oil and biochar through pyrolysis and hydrothermal liquefaction. *Fuel* 285:119106.

Brunauer, S., P. H. Emmett, and E. Teller. 1938. Adsorption of gases in multimolecular layers. *Journal of the American Chemical Society* 60:309–319.

Callister, W. 2006. Ciência e Engenharia de Materiais: Uma Introdução. *Rio de Janeiro: LTC* 589: 249.

Cao, X., K. S. Ro, J. A. Libra, C. I. Kammann, I. Lima, N. Berge, ang Li, et al. 2013. Effects of biomass types and carbonization conditions on the chemical characteristics of hydrochars. *Journal of Agricultural and Food Chemistry* 61:9401–9411.

Cave, K., and F. I. Talens-Alesson. 2005. Comparative effect of Mn(II) and Fe(III) as activators and inhibitors of the adsorption of other heavy metals on calcite. *Colloids and Surfaces A: Physicochemical and Engineering Aspects* 268:19–23.

Chandra, S., and J. Bhattacharya. 2019. Influence of temperature and duration of pyrolysis on the property heterogeneity of rice straw biochar and optimization of pyrolysis conditions for its application in soils. *Journal of Cleaner Production* 215:1123–1139.

Chen, N., Y. Huang, X. Hou, Z. Ai, and L. Zhang. 2017. Photochemistry of hydrochar: Reactive oxygen species generation and sulfadimidine degradation. *Environmental Science and Technology* 51:11278–11287.

Chen, Z., J. Wang, Y. Wang, B. Li, and M. Wang. 2021. Rapid formation of pyrogenic char (biochar) with high and low sorption capacity towards organic chemicals. *Environmental Pollution* 273:116472.

Chu, J. H., J. K. Kang, S. J. Park, and C. G. Lee. 2020. Application of magnetic biochar derived from food waste in heterogeneous sono-fenton-like process for removal of organic dyes from aqueous solution. *Journal of Water Process Engineering* 37:101455.

Culpin, B. 1989. The role of tetrabasic lead sulphate in the lead/acid positive plate. *Journal Power Sources* 25:305–311.

Debalina, B., R. B. Reddy, and R. Vinu. 2017. Production of carbon nanostructures in biochar, bio-oil and gases from bagasse via microwave assisted pyrolysis using Fe and Co as susceptors. *Journal of Analytical and Applied Pyrolysis* 124:310–318.

Diao, R., M. Sun, Y. Huang, and X. Zhu. 2021. Synergistic effect of washing pretreatment and co-pyrolysis on physicochemical property evolution of biochar derived from bio-oil distillation residue and walnut Shell. *Journal of Analytical and Applied Pyrolysis* 155:105034.

Dong, X., L. Q. Ma, and Y. Li. 2011. Characteristics and mechanisms of hexavalent chromium removal by biochar from sugar beet tailing. *Journal of Hazardous Materials* 190:909–915.

El Hanandeh, A., A. Albalasmeh, and M. Gharaibeh. 2021. Effect of pyrolysis temperature and biomass particle size on the heating value of biocoal and optimization using response surface methodology. *Biomass and Bioenergy* 151:106163.

Fakayode, O. A., E. A. A. Aboagarib, C. Zhou, and H. Ma. 2020. Co-pyrolysis of lignocellulosic and macroalgae biomasses for the production of biochar – a review. *Bioresource Technology* 297:122408.

Fang, J., L. Zhan, Y. S. Ok, and B. Gao. 2018. Minireview of potential applications of hydrochar derived from hydrothermal carbonization of biomass. *Journal of Industrial and Engineering Chemistry* 57:15–21.

Fazal, T., A. Razzaq, F. Javed, A. Hafeez, N. Rashid, U. S. Amjad, M. S. Ur Rehman, A. Faisal, and F. Rehman. 2020. Integrating adsorption and photocatalysis: A cost effective strategy for textile wastewater treatment using hybrid biochar-TiO_2 composite. *Journal of Hazardous Materials* 390:121623.

Félix, C. R. O., A. F. A. Júnior, C. C. Freitas, C. A. M. Pires, V. Teixeira, R. Frety, and S. T. Brandão. 2017. Pirólise Rápida de Biomassa de Eucalipto Na Presença de Catalisador Al-MCM-41. *Revista Materia* 22:0251.

Feng, D., Y. Zhao, Y. Zhang, Z. Zhang, H. Che, and S. Sun. 2017. Experimental comparison of biochar species on in-situ biomass Tar H_2O reforming over biochar. *International Journal of Hydrogen Energy* 42:24035–24046.

Foo, K. Y., and B. H. Hameed. 2012. Preparation, characterization and evaluation of adsorptive properties of orange peel based activated carbon via microwave induced K_2CO_3 activation. *Bioresource Technology* 104:679–686.

Frost, R. L., W. Martens, Z. Ding, J. T. Kloprogge, and T. E. Johnson. 2003. The role of water in synthesised hydrotalcites of formula $MgxZn_6$-$xCr_2(OH)16(CO_3) \cdot 4H_2O$ and $NixCo_6$-$xCr_2(OH)16(CO_3) \cdot 4H_2O$ – An infrared spectroscopic study. *Spectrochimica Acta Part A: Molecular and Biomolecular Spectroscopy* 59:291–302.

Gámiz, B., P. Velarde, K. A. Spokas, R. Celis, and L Cox. 2019. Changes in sorption and bioavailability of herbicides in soil amended with fresh and aged biochar. *Geoderma*, 337:341–349.

Ganie, Z. A., N. Khandelwal, E. Tiwari, N. Singh, and G. K. Darbha. 2021. Biochar-facilitated remediation of nanoplastic contaminated water: Effect of pyrolysis temperature induced surface modifications. *Journal of Hazardous Materials* 417:126096.

Gomes, A. S. O., N. Yaghini, A. Martinelli, and E. Ahlberg. 2017. A micro-Raman spectroscopic study of $Cr(OH)_3$ and Cr_2O_3 nanoparticles obtained by the hydrothermal method. *Journal of Raman Spectroscopy* 48:1256–1263.

Guimarães, T., A. F. De Oliveira, R. P. Lopes, and A. P. C. Teixeira. 2020. Biochars obtained from arabica coffee husks by a pyrolysis process: Characterization and application in Fe(II) removal in aqueous systems. *New Journal of Chemistry* 44:3310–3322.

Guizani, C., F. J. E. Sanz, and S. Salvador. 2014. Effects of CO_2 on biomass fast pyrolysis: Reaction rate, gas yields and char reactive properties. *Fuel* 116:310–320.

Gupta, A., and C. Balomajumder. 2015. Simultaneous adsorption of Cr(VI) and phenol onto tea waste biomass from binary mixture: Multicomponent adsorption, thermodynamic and kinetic study. *Journal of Environmental Chemical Engineering* 3:785–796.

Hatanaka, S., Y. Obora, and Y. Ishii. 2010. Iridium-catalyzed coupling reaction of primary alcohols with 2-alkynes leading to hydroacylation products. *Chemistry – A European Journal* 16:1883–1888.

Hung, C. M., C. P. Huang, S. L. Hsieh, M. L. Tsai, C. W. Chen, and C. D. Dong. 2020. Biochar derived from red algae for efficient remediation of 4-nonylphenol from marine sediments. *Chemosphere* 254:126916.

Hwang, H., J. H. Lee, M. A. Ahmed, and J. W. Choi. 2021. Evaluation of pyrochar and hydrochar derived activated carbons for biosorbent and supercapacitor materials. *Journal of Environmental Management* 298:113436.

Islam, A., M. J. Ahmed, W. A. Khanday, M. Asif, and B. H. Hameed. 2017. Mesoporous activated coconut shell-derived hydrochar prepared via hydrothermal carbonization-NaOH activation for methylene blue adsorption. *Journal of Environmental Management* 203:237–244.

Jais, F. M., C. Y. Chee, Z. Ismail, and S. Ibrahim. 2021. Experimental design via NaOH activation process and statistical analysis for activated sugarcane bagasse hydrochar for removal of dye and antibiotic. *Journal of Environmental Chemical Engineering* 9:104829.

Jellali, S., B. Khiari, M. Usman, H. Hamdi, Y. Charabi, and M. Jeguirim. 2021. Sludge-derived biochars: A review on the influence of synthesis conditions on pollutants removal efficiency from wastewaters. *Renewable and Sustainable Energy Reviews* 144:111068.

Jiang, Q., C. Zhang, P. Wu, P. Ding, Y. Zhang, M. H. Cui, and H. Liu. 2021. Algae biochar enhanced methanogenesis by enriching specific methanogens at low inoculation ratio during sludge anaerobic digestion. *Bioresource Technology* 338:125493.

Junior, E., and H. de Jesus. 2015. Caracterização físico-química da biomassa produzida em sistemas florestais de curta rotação para geração de energia. XIII-118.

Kalaiselvi, A., S. M. Roopan, G. Madhumitha, C. Ramalingam, and G. Elango. 2015. Synthesis and characterization of palladium nanoparticles using catharanthus roseus leaf extract and its application in the photo-catalytic degradation. *Spectrochimica Acta – Part A: Molecular and Biomolecular Spectroscopy* 135:116–119.

Kan, T., V. Strezov, and T. J. Evans. 2016. Lignocellulosic biomass pyrolysis: A review of product properties and effects of pyrolysis parameters. *Renewable and Sustainable Energy Reviews* 57:1126–1140.

Khan, T. A., A. S. Saud, S. S. Jamari, M. H. A. Rahim, J. W. Park, and H. J. Kim. 2019. Hydrothermal carbonization of lignocellulosic biomass for carbon rich material preparation: A review. *Biomass and Bioenergy* 130:105384.

Khataee, A., D. Kalderis, P. Gholami, A. Fazli, M. Moschogiannaki, V. Binas, M. Lykaki, and M. Konsolakis. 2019. Cu_2O-CuO@biochar composite: Synthesis, characterization and its efficient photocatalytic performance. *Applied Surface Science*. 498:143846.

Koh, K., A. G. Wong-Foy, and A. J. Matzger. 2009. A porous coordination copolymer with over 5000 m^2/g BET surface area. *Journal of the American chemical society* 131:4184–4185.

Küçük, M. M., and A. Demirbaş. 1997. Biomass conversion processes. *Energy Conversion and Management* 38:151–165.

Kumar, V., A. K. Chopra, and A. Kumar. 2017. A review on sewage sludge (Biosolids) a resource for sustainable agriculture. *Archives of Agriculture and Environmental Science* 2:340–347.

Leng, L., L. Yang, S. Leng, W. Zhang, Y. Zhou, H. Peng, H. Li, Y. Hu, S. Jiang, and H. Li. 2021. A review on nitrogen transformation in hydrochar during hydrothermal carbonization of biomass containing nitrogen. *Science of the Total Environment* 756:143679.

Leng, L., and H. Huang. 2018. An Overview of the Effect of Pyrolysis Process Parameters on Biochar Stability. *Bioresource Technology* 270:627–42.

Li, Q., H. Lin, S. Zhang, X. Yuan, M. Gholizadeh, Y. Wang, J. Xiang, S. Hu, and X. Hu. 2021. Co-hydrothermal carbonization of swine manure and cellulose: Influence of mutual interaction of intermediates on properties of the products. *Science of the Total Environment* 791:148134.

Li, Y., B. Xing, Y. Ding, X. Han, and S. Wang. 2020. A critical review of the production and advanced utilization of biochar via selective pyrolysis of lignocellulosic biomass. *Bioresource Technology* 312:123614.

Liu, J., W. Cheng, X. Yang, and Y. Bao. 2019. Modification of biochar with silicon by one-step sintering and understanding of adsorption mechanism on copper ions. *Science of the Total Environment* 704:135252.

Liu, P., C. J. Ptacek, D. W. Blowes, Y. Z. Finfrock, and Y. Liu. 2020. Characterization of chromium species and distribution during Cr(VI) removal by biochar using confocal micro-X-ray fluorescence redox mapping and X-ray absorption spectroscopy. *Environment international* 134:105216.

Liu, P., D. Rao, L. Zou, Y. Teng, and H. Yu. 2021. Capacity and potential mechanisms of Cd(II) adsorption from aqueous solution by blue algae-derived biochars. *Science of the Total Environment* 767:145447.

Lopes, R. P., and D. Astruc. 2021. Biochar as a support for nanocatalysts and other reagents: Recent advances and applications. *Coordination Chemistry Reviews* 426:213585.

Lopes, R. P., T. Guimarães, and D. Astruc. 2021. Magnetized biochar as a gold nanocatalyst support for p-nitrophenol reduction. *Journal of the Brazilian Chemical Society* 32:1680–1686.

Lu, L., X. Gao, A. Gel, G. M. Wiggins, M. Crowley, B. Pecha, M. Shahnam, W. A. Rogers, J. Parks, and P. N. Ciesielski. 2021. Investigating biomass composition and size effects on fast pyrolysis using global sensitivity analysis and CFD simulations. *Chemical Engineering Journal* 421:127789.

Madhubashani, A. M. P., D. A. Giannakoudakis, B. M. W. P. K. Amarasinghe, A. U. Rajapaksha, P. B. T. P. Kumara, K. S. Triantafyllidis, and M. Vithanage. 2021. Propensity and appraisal of biochar performance in removal of oil spills: A comprehensive review. *Environmental Pollution* 288:117676.

Marzbali, M. H., S. Kundu, P. Halder, S. Patel, I. G. Hakeem, J. Paz-Ferreiro, S. Madapusi, A. Surapaneni, and K. Shah. 2021. Wet organic waste treatment via hydrothermal processing: A critical review. *Chemosphere* 279:130557.

Meili, L., P. V. Lins, C. L. P. S. Zanta, J. I. Soletti, L. M. O. Ribeiro, C. B. Dornelas, T. L. Silva, and M. G. A. Vieira. 2019. MgAl-LDH/Biochar composites for methylene blue removal by adsorption. *Applied Clay Science* 168:11–20.

Melo, C. A., F. H. S. Junior, M. C. Bisinoti, A. B. Moreira, and O. P. Ferreira. 2017. Transforming sugarcane bagasse and vinasse wastes into hydrochar in the presence of phosphoric acid: An evaluation of nutrient contents and structural properties. *Waste and Biomass Valorization* 8:1139–1151.

Miliotti, E., L. Rosi, L. Bettucci, G. Lotti, A. M. Rizzo, and D. Chiaramonti. 2020. Characterization of chemically and physically activated carbons from lignocellulosic ethanol. *Energies* 13:1–17.

Moreira, M. T., I. Noya, and G. Feijoo. 2017. Bioresource technology the prospective use of biochar as adsorption matrix – A review from a lifecycle perspective. *Bioresource Technology* 246:135–141.

Nagata, N., P. G. Peralta-Zamora, L. T. Kubota, and M. I. M. S. Bueno. 2000. Extraction properties of modified silica gel for metal analysis by energy dispersive X-ray fluorescence. *Analytical Letters* 33:2005–2020.

Nunes, N., C. Ragonezi, C. S. S. Gouveia, and M. Â. A. P. de Carvalho. 2021. Review of sewage sludge as a soil amendment in relation to current international guidelines: A heavy metal perspective. *Sustainability* 13:1–20.

Othmani, A., J. John, H. Rajendran, A. Mansouri, M. Sillanpää, and P. V. L. Chellam. 2021. Biochar and activated carbon derivatives of lignocellulosic fibers towards adsorptive removal of pollutants from aqueous systems: Critical study and future insight. *Separation and Purification Technology* 274:119062.

Rasa, K., A. V. Aarnio, P. Rytkönen, J. Hyväluoma, J. Kaseva, H. Suhonen, and T. Jyske. 2021. Quantitative Analysis of Feedstock Structural Properties Can Help to Produce Willow Biochar with Homogenous Pore System. *Industrial Crops and Products* 166:113475.

Ribeiro, M. R., Y. M. Guimarães, I. F. Silva, C. A. Almeida, M. S. V. Silva, M. A. Nascimento, U. P. Silva, E. V. Varejão, R. N. Santos, A. P. C. Teixeira, and R. P. Lopes. 2021. Synthesis of value-added materials from the sewage sludge of cosmetics industry effluent treatment plant. *Journal of Environmental Chemical Engineering* 9:105367.

Rodriguez, L. O., E. Torres, D. Zalazar, H. Zhang, R. Rodriguez, and G. Mazza. 2020. Influence of pyrolysis temperature and bio-waste composition on biochar characteristics. *Renewable Energy* 155:837–847.

Rulkens, W. 2008. Sewage sludge as a biomass resource for the production of energy: Overview and assessment of the various options. *Energy and Fuels* 22:9–15.

Sajjadi, B., W. Y. Chen, and N. O. Egiebor. 2019. A comprehensive review on physical activation of biochar for energy and environmental applications. *Reviews in Chemical Engineering* 35:735–776.

Sajjadi, B., T. Zubatiuk, D. Leszczynska, J. Leszczynski, and W. Y. Chen. 2019. Chemical activation of biochar for energy and environmental applications: A comprehensive review. *Reviews in Chemical Engineering* 35:777–815.

Sarker, T. R., F. Pattnaik, S. Nanda, A. K. Dalai, V. Meda, and S. Naik. 2021. Hydrothermal pretreatment technologies for lignocellulosic biomass: A review of steam explosion and subcritical water Hydrolysis. *Chemosphere* 284:131372.

Saxe, J. P., J. H. Boman, M. Bondi, U. Norton, T. K. Righetti, A. H. Rony, and B. Sajjadi. 2019. Just or bust? Energy justice and the impacts of siting solar pyrolysis biochar production facilities. *Energy Research and Social Science* 58:101259.

Sewu, D. D., H. N. Tran, G. Ohemeng-Boahen, and S. H. Woo. 2020. Facile magnetic biochar production route with new goethite nanoparticle precursor. *Science of the Total Environment* 717:137091.

Shyam, S., J. Arun, K. Panchamoorthy, G. Ribhu, M. Ashish, and S. Ajay. 2022. Biomass as source for hydrochar and biochar production to recover phosphates from wastewater: A review on challenges, commercialization, and future perspectives. *Chemosphere* 286:131490.

Silva, C. M. S., A. de C. O. Carneiro, B. R. Vital, C. G. Figueiró, L. de F. Fialho, M. A. de Magalhães, A. G. Carvalho, and W. L. Cândido. 2018. Biomass torrefaction for energy purposes – definitions and an overview of challenges and opportunities in Brazil. *Renewable and Sustainable Energy Reviews* 82:2426–2432.

Solarte-Toro, J. C., J. A. González-Aguirre, J. A. P. Giraldo, and C. A. C. Alzate. 2021. Thermochemical processing of woody biomass: A review focused on energy-driven applications and catalytic upgrading. *Renewable and Sustainable Energy Reviews* 136:110376.

Steter, J. R., E. Brillas, and I. Sirés. 2016. On the selection of the anode material for the electrochemical removal of methylparaben from different aqueous media. *Electrochimica Acta* 222:1464–1474.

Sun, S.-F., H.-Y. Yang, J. Yang, and Z.-J. Shi. 2021. Structural characterization of poplar lignin based on the microwave-assisted hydrothermal pretreatment. *International Journal of Biological Macromolecules* 190:360–367.

Swetha, A., S. Shri Vigneshwar, K. P. Gopinath, R. Sivaramakrishnan, R. Shanmuganathan, and J. Arun. 2021. Review on hydrothermal liquefaction aqueous phase as a valuable resource for biofuels, bio-hydrogen and valuable bio-chemicals recovery. *Chemosphere* 283:131248.

Tekin, K., S. Karagöz, and S. Bektaş. 2014. A review of hydrothermal biomass processing. *Renewable and Sustainable Energy Reviews* 40:673–687.

Tripathi, M., J. N. Sahu, and P. Ganesan. 2016. Effect of process parameters on production of biochar from biomass waste through pyrolysis: A review. *Renewable and Sustainable Energy Reviews* 55:467–481.

Tumuluru, J. S., S. Sokhansanj, J. R. Hess, C. T. Wright, and R. D. Boardman. 2011. A review on biomass torrefaction process and product properties for energy applications. *Industrial Biotechnology* 7:384–401.

Veiga, P. A. da S., M. H. Cerqueira, M. G. Gonçalves, T. T. da S. Matos, G. Pantano, J. Schultz, J. B. de Andrade, and A. S. Mangrich. 2021. Upgrading from batch to continuous flow process for the pyrolysis of sugarcane bagasse: Structural characterization of the biochars produced. *Journal of Environmental Management* 285:112145.

Venkatachalam C. D., M. Sengottian, and S. R. Ravichandran. 2020. Hydrothermal conversion of biomass into fuel and fine chemicals. *Bioprocess Engineering for Bioremediation: Valorization and Management Techniques* 201–224.

Wang, S., M. Zhao, M. Zhou, Y. C. Li, J. Wang, and B. Gao. 2019. Biochar-supported nZVI (nZVI /BC) for contaminant removal from soil and water : A critical review. *Journal of Hazardous Materials* 373:820–834.

Wang, Y., Y. Yu, H. Huang, C. Yu, H. Fang, C. Zhou, X. Yin, W. Chen, and X. Guo. 2021. Efficient conversion of sewage sludge into hydrochar by microwave-assisted hydrothermal carbonization. *Science of the Total Environment* 803:149874.

Wilk, M., M. Śliz, and B. Lubieniecki. 2021. Hydrothermal co-carbonization of sewage sludge and fuel additives: Combustion performance of hydrochar. *Renewable Energy* 178:1046–1056.

Wu, P., Z. Wang, A. Bhatnagar, P. Jeyakumar, H. Wang, Y. Wang, and X. Li. 2021. Microorganisms-carbonaceous materials immobilized complexes: Synthesis, adaptability and environmental applications. *Journal of Hazardous Materials* 416:125915.

Ye, K., Y. Liu, S. Wu, and J. Zhuang. 2021. A review for lignin valorization: Challenges and perspectives in catalytic hydrogenolysis. *Industrial Crops and Products* 172:114008.

Yin, Z., S. Xu, S. Liu, S. Xu, J. Li, and Y. Zhang. 2020. A novel magnetic biochar prepared by K_2FeO_4-promoted oxidative pyrolysis of pomelo peel for adsorption of hexavalent chromium. *Bioresource Technology* 300:122680.

Yoder, J., S. Galinato, D. Granatstein, and M. Garcia-Pérez. 2011. Economic tradeoff between biochar and bio-oil production via pyrolysis. *Biomass and Bioenergy* 35:1851–1862.

Yogalakshmi, K. N., D. T. Poornima, P. Sivashanmugam, S. Kavitha, K. R. Yukesh, V. Sunita, S. A. Kumar, G. Kumar, and R. Banu J. 2022. Lignocellulosic biomass-based pyrolysis: A comprehensive review. *Chemosphere* 286:131824.

Yuan, X., P. D. Dissanayake, B. Gao, W. J. Liu, K. B. Lee, and Y. S. Ok. 2021. Review on upgrading organic waste to value-added carbon materials for energy and environmental applications. *Journal of Environmental Management* 296:113128.

Zhang, Q., D. Zhang, W. Lu, M. U. Khan, H. Xu, W. Yi, H. Lei, et al. 2020. Production of high-density polyethylene biocomposites from rice husk biochar: Effects of varying pyrolysis temperature. *Science of the Total Environment* 738:139910.

Zhang, S., K. Sheng, W. Yan, J. Liu, E. Shuang, M. Yang, and X. Zhang. 2021. Bamboo derived hydrochar microspheres fabricated by acid-assisted hydrothermal carbonization. *Chemosphere* 263:128093.

Zhang, Y., J. Wang, and Y. Feng. 2021. The effects of biochar addition on soil physicochemical properties: A review. *Catena* 202:105284.

Zheng, W., M. Guo, T. Chow, D. N. Bennett, and N. Rajagopalan. 2010. Sorption properties of greenwaste biochar for two triazine pesticides. *Journal of Hazardous Materials* 181:121–126.

Zhu, J. Y., U. P. Agarwal, P. N. Ciesielski, M. E. Himmel, R. Gao, Y. Deng, M. Morits, and M. Österberg. 2021. Towards sustainable production and utilization of plant-biomass-based nanomaterials: A review and analysis of recent developments. *Biotechnology for Biofuels* 14:1–31.

Zhu, P., O. Y. Abdelaziz, C. P. Hulteberg, and A. Riisager. 2020. New synthetic approaches to biofuels from lignocellulosic biomass. *Current Opinion in Green and Sustainable Chemistry* 21:16–21.

Zhu, X., Y. Li, and X. Wang. 2019. Machine learning prediction of biochar yield and carbon contents in biochar based on biomass characteristics and pyrolysis conditions. *Bioresource Technology* 288:121527.

3 Influence of Herbicide Environmental Behavior on Weed Management

Rafael De Prado[1], Candelario Palma-Bautista[1], José Guadalupe Vázquez-García[1], and Ricardo Alcántara-de la Cruz[2]*

[1] Department of Agricultural Chemistry and Edaphology, University of Cordoba, 14071 Córdoba, Spain.

[2] Centro de Ciências da Natureza, Universidade Federal de São Carlos – Campus Lagoa do Sino, 18290–000 Buri, Brazil.

*Corresponding author: ricardo.cruz@ufscar.br

CONTENTS

DOI: 10.1201/9781003202073-3

3.1 INTRODUCTION

Herbicides are important management tools in agriculture because they can be used instead of tillage for weed control, thereby conserving soil and moisture (Miller and Westra 2004). The interaction of herbicides with the soil and/or the environment in general is the guideline for understanding how herbicides work.

The soil is the fate of herbicides, whether they are applied directly to the soil (pre-emergence – PRE) or to the aerial part of plants (post-emergence – POST) (Mendes et al. 2021b). The soil is a complex and dynamic system where the redistribution and degradation of herbicides occur, which can be a short process, as for some simple and non-persistent molecules, or can last months or years for highly persistent herbicides (Silva et al. 2007). Upon contact with soil, herbicides are subject to several processes (retention, leaching, volatilization, runoff, photodegradation, microbiological degradation, chemical degradation, and uptake by plants) (Figure 3.1) (Monquero and Silva 2021). The order and speed of occurrence of these processes regulate the persistence (time that a molecule remains in the soil) of herbicides in the soil environment, directly or indirectly affecting the efficiency of weed control. Persistence also drives the behavior of herbicides in the environment by regulating the potential negative consequences on nontarget organisms, carryover to subsequent crops as well as public health (Gavrilescu 2005).

Retention (immobilization of a herbicide in the soil matrix) is regulated by specific dissipation mechanisms such as absorption, precipitation, and adsorption (Figure 3.1) (Blasioli et al. 2011). Understanding retention processes requires knowing the composition and dynamics of the soil system (Miller and Westra 2004), since they also determine the persistence of herbicides in the soil (Colquhoun 2006). Herbicide persistence is important for residual herbicides, generally PRE, to extend the period of weed control; however, some herbicides can persist for a long time in the soil but may not be bioavailable to be absorbed by plants, i.e., they are retained in soil particles, but are no longer active as herbicides. Because soil herbicide retention and persistence process varies depending on the physical–chemical heterogeneity of the soils and their interrelationships with the biological, atmospheric and aquatic systems (Colquhoun 2006; Curran 2016), differentiating these two processes is important for better planning of herbicide applications and crop rotation, which reduces the risk of carryover injuries and environmental pollution issues (Monquero and Silva 2021).

Biotic and abiotic factors interfere with herbicide activity in plants and dissipation process (Wang et al. 2015; Mendes et al. 2021b). The dissipation of herbicides into the environment via biological, chemical, or physical degradation is influenced by physicochemical properties of the soil (pH, clays, organic matter, etc.), biological

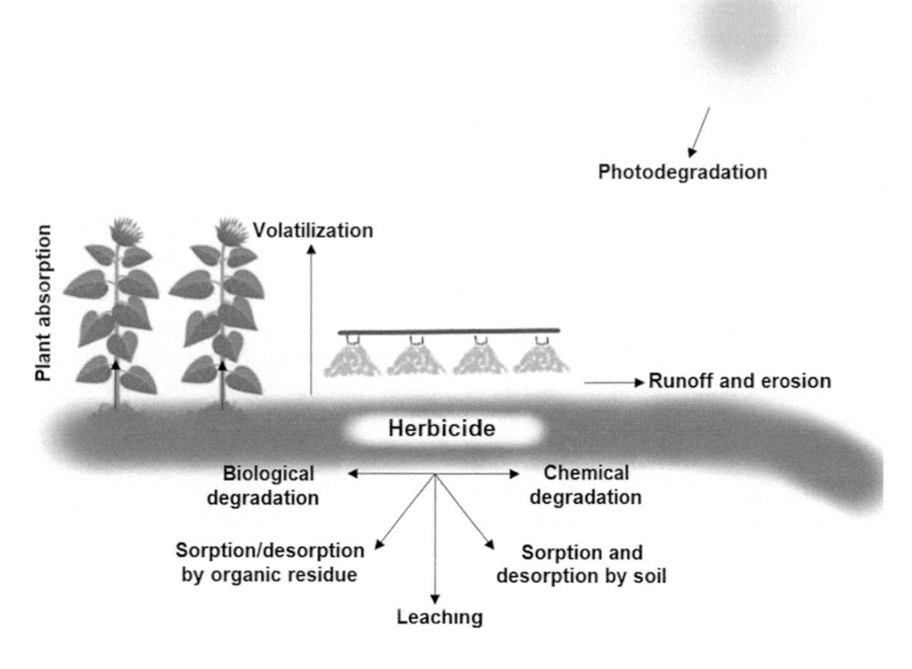

FIGURE 3.1 Processes and factors involved in the fate of herbicides in the environment.

properties (microorganism activities), and environmental conditions (rainfalls, sunlight, temperature, and moisture content) (Curran 2016). On the other hand, the rate and pathway of degradation also depend on the characteristics of the herbicide such as water solubility (S_w), vapor pressure (VP), octanol–water partition coefficient (K_{ow}), acid (pK_a) or base (pK_b) ionization equilibrium constant, Henry's law constant (K_H), and half-life time of degradation ($t_{1/2}$). Knowledge of the physicochemical properties of molecule herbicides can guide actions to improve weed control, preventing environmental factors from reducing the efficacy of products (Miller and Westra 2004; Blasioli et al. 2011).

The behavior (retention, degradation, and transport) of herbicides in the complex soil system depends on biotic and abiotic factors. To better understand such behavior, this chapter details the main characteristics of the soil, environmental conditions, and physicochemical characteristics inherent to herbicides, the main interactions between these factors, which can directly or indirectly affect weed control, as well as the magnitude of impacts on the environment and public health.

3.2 UNDERSTANDING THE SOIL SYSTEM

The structure of the soil is important for its proper functioning and can be defined as the spatial arrangement of the particles and solid components, their aggregation, and mechanical state (Romero-Ruiz et al. 2018). Favorable soil structure and stability are important to maintain soil fertility, increase agricultural productivity, improve porosity, and decrease erodibility (Bronick and Lal 2005; Romero-Ruiz et al. 2018).

The soil system is composed of solid, liquid, and gaseous phases. The main components of soil are water, air, mineral, and organic materials. The approximate proportions of solid, liquid, and gaseous components in a loamy textured soil are 45% mineral matter, 5% organic matter (OM), and 50% of the total volume of water and air (25% each) (Snakin et al. 2001; Wang et al. 2015). The solid phase of the soil is made up of mineral matter and OM (plant and animal residues). The organic fraction is in a continuous process of renewal, but it is reduced after the introduction of agricultural systems (Carpio et al. 2021). The liquid phase of the soil (fraction where the herbicide molecules are deposited) is made up of a mixture of water, mineral salts, and low molecular weight molecules, such as amino acids, peptides, sugars, and humic substances. Therefore, herbicide molecules may be subject to numerous reactions, depending on the properties of the soil solution, such as pH, ionic strength, and redox potential, influencing the fate of herbicides in the environment (Gavrilescu 2005; Wang et al. 2015).

According to FAO (2021) records, there are 28 types of soil in the world where each one has different subtypes. The main types of soil globally are Alfisols, Andisols, Aridisols, Entisols, Inceptisols, Oxisols, Spodosols, Ultisols, and Vertisols (Bronick and Lal 2005). Some soils have agronomically desirable properties such as good depth, stable structure, good porosity, and high permeability. However, some of those soils also have undesirable properties such as high toxicity by cations such as Al and Fe or Ca and Mg, low nutrient reserves, low cation exchange capacity (CEC), high anion sorption capacity, and high pH-PCZ value (pH at the point of zero charge) (Zhang and Norton 2002; Pachepsky and Rawls 2003). These undesirable properties are directly or indirectly related to the development and balance of electrical charges on the surface of the particles that make up their colloidal systems.

In addition to the variation between soils, the physical, chemical, and biological properties of the same soil can change with depth. These changes along the profile characterize alterations in the soil structure, its moisture retention capacity, and microbiological activity, among others (Berisso et al. 2012; Wang et al. 2015; Nawaz et al. 2020). The relative proportion of these components affects the fate and behavior of the herbicides in the soil.

3.3 PHYSICOCHEMICAL PROPERTIES OF HERBICIDES

When a herbicide is applied for weed control, it will enter a process by which it may lose effectiveness as a control method. This process is influenced by the environment in which it is to be found, from soil type and environmental conditions. The herbicide may vaporize and enter in the atmosphere or break down via microbial or chemical pathways into other less toxic compounds.

Most herbicides are synthetically obtained organic molecules, so that several herbicides have a similar chemical structure. Despite the similarity between compounds of the same chemical group, their retention in the soil and their selectivity for plants may vary. The main physicochemical properties that may influence the retention process of herbicides in the soil are: acid/base dissociation coefficients (pK_a or pK_b), octanol–water partition coefficient (K_{ow}), water solubility (S_w), vapor pressure (VP), coefficients of partition sorption–desorption (K_d), coefficient of air–water partition or constant of Henry's law (K_H), and degradation in the environment (half-life time, $t_{1/2}$). Knowledge of these physicochemical properties allows a more rational use of herbicides.

3.3.1 Acid/Base Dissociation Coefficients (pK_A or pK_B)

The electrolytic dissociation potential (pK) indicates the ability of a herbicide to form ions in solution in relation to the pH of the medium. Ionized forms of herbicides behave differently from non-ionized (neutral) forms (Mendes et al. 2021b). The higher the pK value of the herbicide, the weaker its acidic strength; therefore, the lower the possibility of the herbicide becoming anionic. Neutral molecules of acid herbicides (pK_a) such as dicamba, 2,4-D, glufosinate, glyphosate, mesotrione, oryzalin, some imidazolinones, among others, are able of donating a proton (negative charged ions); while the neutral forms of basic herbicides (pK_b) such as ametrine, atrazine, hexazinone, metribuzin, and simazine are capable of receiving protons (positive charged ions). The basic, acidic, or nonionizable (flumioxazin, linuron, diuron, metamitron, oxyfluorfen, thiobencarb, etc.) herbicide properties determine the ability of the herbicide to exist in the soil water or be retained onto soil particles. Thus, the acid or base strength of both the herbicide and the soil determines potential herbicide movement and plant uptake (Miller and Westra 2004).

3.3.2 Octanol–Water Partition Coefficient (K_{ow})

Herbicide lipophilicity is measured as the partition coefficient (K_{ow}) of a given molecule between non-aqueous (octanol, o) and aqueous (water, w) phases. The K_{ow} values are dimensionless, are expressed in logarithmic form (log K_{ow}), and are constant for each molecule at a given temperature. K_{ow} indicates the lipophilicity of herbicides with organic compounds – i.e., the affinity of a herbicide molecule for the polar phase (represented by water) and the non-polar phase (represented by n-octanol) of the soil solution (Pereira et al. 2016b; Amézqueta et al. 2019). A low K_{ow} value allows the mobility and transport of the herbicide due to its good solubility, and easy metabolization, and biodegradation – i.e., its persistence in the soil is short. On the contrary, a high K_{ow} value indicates possible absorption in OM, soil, and sediments – i.e., herbicide persistence in the soil is prolonged increasing environmental pollution.

3.3.3 Water Solubility (S_w)

From a practical point, this parameter determines to what extent a herbicide is in a solution phase (available to plants) or in the solid phase. S_w refers to how readily the herbicide dissolves in water and plays a major role in determining the fate of herbicides in water, soil, and air. The more hydrophilic groups (more polar) have greater affinity for water (>solubility) (Martins et al. 2013). S_w is measured in mg L^{-1} or parts per million (ppm) at controlled pH and temperatures from 20 to 25°C at 1 atmospheric pressure (Pereira et al. 2016b). S_w less than 50 mg L^{-1} (for atrazine and pendimethalin) is considered low; S_w in a range from 50 to 500 mg L^{-1} (ametryn and fomesafen) moderate; and S_w greater than 500 mg L^{-1} (acifluorfen and glyphosate) is considered high. Herbicides with high S_w values are not retained by the soil because they have greater number of hydrophilic groups (low lipophilicity – K_{ow}). In addition, they are more likely to experience runoff, seepage, and leaching. These herbicide transport processes will be discussed later. On the other hand, high soluble molecules have low adsorption coefficients and bioconcentration factors and also tend to biodegrade more rapidly in soil and water (Pereira et al. 2016b).

3.3.4 Vapor Pressure (VP)

Vapor pressure (VP) is the tendency of a herbicide to volatilize in the environment from its solid or liquid state – i.e., its tendency to be lost to the atmosphere in the form of gas (Mendes et al. 2021b). The factors that control volatilization from water are S_w, molecular weight, VP, and the nature of the air ± water interface through which a chemical must pass (Schmuckler et al. 2000). Herbicides with high PV values persist less time in the soil. Herbicide volatilization is greater as temperature increases and relative humidity decreases, resulting in transporting the molecule to sensitive areas or losing its effectiveness. The problem of volatilization is important for some herbicide groups such as synthetic auxins, dinitroanilines, and thiocarbamates, causing losses ranging from 10% to 90%, compared to a typical loss by leaching and runoff (Monquero and Silva 2021). PRE herbicides such as trifluralin, pendimethalin, butylate, norflurazon, among others, should be incorporated at this time of application to avoid loss by volatilization.

3.3.5 Henry's Law Constant (K_H)

The Henry's law constant (K_H) (also called the air–water partition coefficient) is the combination between S_w and VP and indicates the volatilization potential of the herbicide as a function of soil moisture. K_H is a parameter used to model the diffusive exchange of chemicals between aqueous phases and the atmosphere. Accurate knowledge of K_H is essential to predict the environmental behavior, transport, and fate of many types of herbicides (Delgado and Alderete 2003). In addition to considering S_w and VP, K_H also considers the molecular weight of herbicide molecules and indicates the relationship between their partial pressure in air and the concentration of that herbicide in water at a given volatility temperature of the same herbicide in solution (Rice et al. 1997). The S_w of herbicides in the gaseous state is less important than that of herbicides in the liquid and solid phase, and high K_H values indicate that the herbicides solutes are highly volatile. The application of a herbicide generates a high concentration on the soil surface, and if they are incorporated there may be losses due to volatilization, even of low-volatile herbicides. However, in humid soils, herbicides such as trifluralin, pendimethalin, and butylate can be lost by volatilization even when they are incorporated (Silva et al. 2007).

3.3.6 Sorption–Desorption Coefficient (K_D)

These coefficients estimate the partitioning tendency of the herbicide from the liquid phase to the solid phase of the soil, i.e., they indicate the mobility rate of the herbicide in the soil matrix. The herbicide sorption rate is usually expressed by a linear coefficient (K_d) which relates the amount of a herbicide found in solution to the amount sorbed by the solid phase after system reaches equilibrium (Gerstl 1990; Curran 2016). However, the soil organic carbon (OC) apparently has a better adsorption capacity for herbicides; therefore, the K_d has been normalized in relation to the OC content of the soil. The normalization of K_d makes it possible to estimate K_{oc}, which is the relationship between the partition coefficient of soil OC and the concentration of a herbicide absorbed by the soil and its concentration in the soil water (Reddy and Locke 1994). In other words, these coefficients indicate the affinity of the herbicide for the solid phase of the soil: $K_{d,}$ affinity for the mineral fraction, and K_{oc}, affinity for the OC fraction. These sorption coefficients do not depend on the type of soil, and, regardless of which one is estimated, the higher the K_d or K_{oc} value, the greater the sorption of the herbicide, i.e., it is less mobile in the soil. The relationship between K_d or K_{oc} is: $K_{oc} = K_d/OC \times 100$.

The K_{oc} values of mobile herbicides in the soil such as penoxsulam and sulfentrazone range from 15 to 75 mL g^{-1}, while the values of moderately mobile herbicides such as atrazine and mesotrione vary from 75 to 500 mL g^{-1}. Herbicides with higher K_{oc} values (> 500 mL g^{-1}) such as carfentrazone, glyphosate, paraquat, and pendimethalin adhere strongly to soil particles, making them less likely to move unless soil erosion occurs (Gavrilescu 2005). On the other hand, weak acid herbicides (imazetapyr, metsulfuron, nicosulfuron, and sulfometuron) are less likely to be sorbed into the soil, while weak-based herbicides (atrazine, hexazinone, simazine) and non-ionic ones (alachlor) tend to be more sorbed (Silva et al. 2007). Sorbed herbicides accumulate in soil particles and are not available for absorption, degradation, or uptake of plants from the soil environment, except when the soil particle moves by runoff (Colquhoun 2006).

3.3.7 Half-Life Time of Degradation ($t_{1/2}$)

The residual activity or the time required for the dissipation by 50% of herbicide applied is defined as half-life ($t_{1/2}$) (Curran 2016; Mendes et al. 2021b). The $t_{1/2}$ is a constant, which is independent of concentration, only for first order kinetic reactions, restricting its use to these conditions only. The $t_{1/2}$ of herbicides can be from a few days to years, depending on the type of soil, soil pH, climatic conditions, and cropping systems (Colquhoun 2006). In addition, increasing the herbicide application rate within the agricultural use limits increases the amount of herbicide available in the environment, which can change the $t_{1/2}$ (Silva et al. 2007); therefore, it is difficult to predict the $t_{1/2}$ with the climatic variation from one year to another.

Herbicides can be classified as non-persistent (alachlor, glyphosate, metribuzin), moderately persistent (ametryn, atrazine, chlorimuron), persistent (acifluorfen, imazethapyr, pendimethalin, trifluralin), and very persistent (indaziflam, paraquat, oxadiazon, sulfentrazone, tebuthiuron), depending on whether their $t_{1/2}$ is ≤30 days,

>30 and ≥100 days, ≤100 and ≥365 days, and >365 days, respectively (Gavrilescu 2005). Because $t_{1/2}$ is generally determined in the laboratory, they should not be considered for field recommendations, because, although this parameter is used to determine crop rotation restrictions, these recommendations often do not coincide (Colquhoun 2006).

3.4 SOIL PROPERTIES

Persistence is the most important interaction between herbicides and the soil (Furmidge and Osgerby 1967). This characteristic is sometimes desirable, especially for PRE herbicides, to continue to control weeds during the application season. However, persistence is not exclusive to PRE herbicides – i.e., some post-emergent (POST) or dual-activity herbicides (PRE and POST) also show persistence, and it is not desirable that this persistence extends to the next cycle as it may affect subsequent culture (Curran 2016). The potential for persistence in soil of herbicides is influenced by the physicochemical characteristics of the molecules themselves, the environment conditions, as well as the edaphic conditions; therefore, the interactions between these factors, together with the physicochemical characteristics of the molecules, determine the degradation and transport processes of herbicides (Curran 2016; Pereira et al. 2016b; Blasioli et al. 2011).

Among the main edaphic factors that influence the persistence of herbicides in the soil, texture, chemical composition, and microbial activity can be highlighted. Soil texture is determined by the proportion of sand, silt, and clay in the soil, as well as OM (Curran 2016). The CEC that occurs between soil colloids (clay) and/or OC with herbicides influence sorption (Silva et al. 2007). In certain cases, soil pH as well as soil mineralogy also affect the sorption of herbicides in the soil. Soil pH can also influence both the leaching rate and the chemical and microbial decomposition rate of herbicides. Microbial aspects of the soil environment include the types and abundance of soil microorganisms present in the soil (van Eerd et al. 2003; Blasioli et al. 2011).

3.4.1 Organic Matter and Microbial Activity

All materials containing carbon, alive or dead in the most different stages of decomposition, present on the surface or incorporated into the soil, conform to the OM. These materials interfere with all herbicide sorption processes that may occur in the soil, especially non-ionic or cationic PRE herbicides. The soil OM is a binding site for herbicide molecules and the adsorption of their metabolites to soil particles; therefore, OM plays an essential role in the fate of herbicides in the environment (Mehdizadeh et al. 2021), as well as in cation retention processes and complexation of toxic elements and micronutrients. Furthermore, OM is essential in the productive capacity of the soil, influencing the stability of the soil structure, infiltration, and aeration and microbial activity (Bucka et al. 2019). Therefore, the great variability and continuous renewal of OM affect the retention, transformation, and transport of herbicides in the soil (Silva et al. 2007).

Soil OM is divided into humified substances, composed of fulvic, humic, and humin acids and non-humidified substances composed by solid residues from plants, animals, and soil microorganisms (Bucka et al. 2019). The humic fraction of the soil OM is the most active, showing the highest correlation with herbicide sorption. However, the herbicide sorption rate depends on the type of herbicide and the proportion of the humified substances. For example, ionic and non-ionic herbicides correlate better with humin than with humic or fulvic acids (Stevenson 1972).

The organic source, climate, soil minerals, and microbial population affect the polarity, aromaticity, availability of hydrophobic sites, and functional groups of soil organic compounds (Silva et al. 2007). The greater or lesser number of hydrophobic sites of organic molecules can also influence the sorption of lipophilic herbicides (Martins et al. 2013). On the other hand, the source material of the soil, the predominant types of minerals in the clay fraction, the presence of saturating ions of the functional groups of OM, and the specificity of the ions can also affect the sorption of herbicides to OM (Pachepsky and Rawls 2003; Nawaz et al. 2020).

Herbicide sorption is more stable in OM than in mineral soil components, i.e., in soils with high OM contents, herbicide leaching is lower reducing the risk of groundwater contamination (Carpio et al. 2021). Furthermore, the amount of chemically protected OM, forming clay-organic complexes, is responsible for variations in the sorption coefficients (K_d/K_{oc}) of herbicides (Silva et al. 2007). In addition, CTC and OC content are highly correlated with the sorption of basic and non-ionic herbicides (Gerstl 1990), because in many soils, CTC is related to OM content. For some non-ionic herbicides such as alachlor, atrazine, dicamba, imazetaphyr, metsulfuron, nicosulfuron, and simazine, it is possible to calculate the sorption coefficients (K_d) based on the soil OC contents (Silva et al. 2007).

Microbial activity, a crucial factor in herbicide biodegradation, is influenced by the OM content of the soil. High OM soils generally have larger populations and greater biodiversity of microorganisms (Sadegh-Zadeh et al. 2017), improving herbicide biodegradation in soil. The degradation of metolachlor was influenced by the OM content of the soil (Wu et al. 2011); the addition of OM reduced the $t_{1/2}$ of metribuzin (Mehdizadeh et al. 2021); and the addition of pasty sugarcane residue (vinasse) accelerated the mineralization of ametryn (Prata et al. 2001). In most cases, the addition of OM to the soil accelerates the biodegradation of herbicides; however, the physical and chemical protection of herbicide molecules in OM can limit microbial activity, which can increase the persistence of herbicides in the environment (Silva et al. 2007).

The relationship of soil texture with the amount of OM present is important in the dosage of PRE herbicides. Clay soils generally have higher OM contents; however, the increase in herbicide dosage for these soils is not always correct, as OM under- or over-estimation may influence the amount of herbicide sorbed. Therefore, suitability of herbicide doses, besides determining OM accurately, must consider the presence of low-activity clays, specific sorption of molecules by some (Curran 2016; Silva et al. 2007).

3.4.2 Soil Texture

The proportion of sand, silt, and clay determines the texture and classification of soils. The proportion of these three aggregates defines the pore space, the transport of water, solutes, and gases in the soil – that is, the proportion of sand, silt, and clay regulates the phases (solid, liquid, and gaseous) of the soil (Moldrup et al. 2001; Snakin et al. 2001), influencing the activity and persistence of herbicides through soil–herbicide binding (sorption), leaching and vapor loss (volatilization), as well as the degradation of herbicide molecules (Cox et al. 1997; Curran 2016). Understanding these events in different types of soils is complex, but generally the degradation of herbicides in heavy soils, with a high clay content, organic matter or both, is faster than in light soils (Wlodarczyk and Siwek 2017). In addition, the potential to bind or retain the herbicide to soil particles is greater in medium and fine textured soils with high OM content (> 3%) (Curran 2016). This greater transport potential due to increased herbicide binding to soil particles reduces leaching and volatilization loss (Yates 2006; Larsbo et al. 2009); however, it can also decrease initial plant uptake and herbicide activity. On the other hand, coarse-to-medium-textured soils with OM less than 3% are less likely to retain herbicides and have transport problems (Curran 2016). However, under the right circumstances, herbicide transport can take place in any type of soil.

The persistence of herbicides may be greater in heavy soils as they are rapidly sorbed into clay mineral particles and OC, which may affect future susceptible crops due to desorption of these herbicide molecules (Curran 2016). Weak basic (atrazine, hexazinone, simazine) and nonionic (alachlor) herbicides have higher sorption rates than weak acid herbicides (imazethapyr, metsulfuron, nicosulfuron, and sulfometuron) (Silva et al. 2007; Monquero and Silva 2021).

Soil microbial activity can also be influenced by soil texture due to interactions between microorganisms and soil particles, such as the adhesion of microorganisms to soil particle surfaces, adsorption of microbial metabolites on soil aggregates and the effects of adsorption of organic substrates on their biodegradation, substrate accessibility, among others (Larsbo et al. 2009; Mendes et al. 2021b). The degradation of ethofumesate was faster in medium clayey soil than in sandy clayey soil (Mehdizadeh et al. 2021). Sulfur oxidation of 18 out of 25 herbicides was up to three times higher in silty clay soil than in sandy clay soil due to different types of microflora type of each soil (Lewis et al. 1978). The degradation of clomazone, metazachlor, and pendimethalin was not affected by soil texture (Wlodarczyk and Siwek 2017). Therefore, soil texture plays an essential role in the degradation and persistence of herbicides in the soil (Mehdizadeh et al. 2021).

3.4.3 Mineralogy

Soil texture is fundamental to determine the dosage of PRE herbicides. However, the sorption rate of acid herbicides in the soil can be influenced by clay diversity and the formation of clay–mineral compounds. In addition, high concentrations of OC can also affect herbicide adsorption on clays of the soil (Silva et al. 2007). For example, glyphosate sorption is more related to the soil mineral fraction than to OM (Piccolo et al. 1996).

Clay minerals such as montmorillonite and vermiculite have greater herbicide sorption capacity due to their expansion capacity and greater specific surface area (SSA) (Pachepsky and Rawls 2003). These properties give rise to strong attraction forces, increasing the sorption of herbicides. The formation of charges in these minerals occurs by isomorphic substitution in the tetrahedral and octahedral layers, allowing water, herbicides, and other molecules to penetrate between the basal planes and cause great expansion of the material (Moldrup et al. 2001; Blasioli et al. 2011). The formation of charges of non-expandable minerals such as kaolinite, typical of highly weathered regions, occurs at the edges of the mineral due to the dissociation of H^+ protons from the OH^- groups, retaining cations (Silva et al. 2007).

Iron and aluminum oxides, present in large concentrations in most weathered soils such as oxisols, also act in the sorption of several herbicides, especially those with the ability to dissociate protons (weak acid herbicides). Among these oxides, iron has a greater capacity to sorb herbicides than aluminum oxides (Silva et al. 2007).

3.4.4 Soil pH

pH, defined as the negative logarithm of the hydrogen ion concentration $[H^+]$ ($pH = -\log [H^+]$), is probably the most informative measure that can be taken to determine soil characteristics. Soil pH influences chemical solubility and availability of essential plant nutrients, pesticide performance, OM decomposition, as well as the microbial activity (Cho et al. 2016; Neina 2019). Acidic soils have greater sorption capacity for herbicides with acidic properties and high ionization capacity; therefore, soil pH affects bioavailability and persistence of herbicides, depending on the nature and structure of the molecules and its metabolites (Mehdizadeh et al. 2021).

The influence of soil pH on the herbicide retention and degradation process is related to the electrolyte dissociation capacity (pK_a) of the compounds. Soil pH affected the persistence and dissipation mesotrione in the soil (Quan et al. 2015). The sorbed amount of 2,4-D and imidazolinones decreases as the pH of acidic soils increases until reaching neutrality (7.0), as the pH influences the binding of herbicide molecules to the soil's humic and mineral fractions (Loux et al. 1993; Silva et al. 2007), i.e., more herbicide remains available in the soil solution. The reduction of the soil's sorb capacity with increasing pH increases the potential leaching of ionic herbicides into the soil profile, in addition to increasing the residual effect of herbicides with long persistence in future crops. Chlorsulfuron leaching is higher in soils with high pH and low MO content; therefore, operations such as liming to increase the pH of acidic soils can favor the leaching of highly soluble and ionic herbicides (Silva et al. 2007).

Soil pH can also influence the population of soil microorganisms, affecting the biodegradation speed of herbicides (Cho et al. 2016). Chemical and microbial degradation, two ways in which herbicides degrade in soil, tend to be slower in higher pH soils (Curran 2016). Chlorsulfuron degradation is faster in acidic soils (Silva et al. 2007). The optimal degradation of metribuzin by *Bacillus* sp. occurred at a soil pH of 7.0 (Zhang et al. 2014), whereas that of 2,4-D by *Cupriavidus gilardii* occurred from a slightly acidic soil pH of 5.0 to basic soil pH of 8.0 (Wu et al. 2017).

3.5 ENVIRONMENTAL CONDITIONS

Humidity, temperature, and sunlight also play an essential role in the fate and degradation of herbicides (Curran 2016). Herbicides degrade faster as temperature and humidity increase due to intensification of biological and chemical processes in the soil such as hydrolysis, oxidation, hydroxylation, and volatilization (Mehdizadeh et al. 2021). The degradation of metsulfuron-methyl and thiencarbazone-methyl in cold seasons was less than in warm seasons (Cessna et al. 2017). However, only high temperatures may increase herbicide persistence and affect future crops after drought (Curran 2016). On the other hand, excess moisture (heavy rainfall) can reduce soil microbial activity and favor both leaching and runoff of soluble herbicides (Hall et al. 1989; Lewan et al. 2009).

Degradation catalyzed by sunlight (photolysis) occurs for many herbicides, especially in liquid solution (water) or on plant leaf surfaces (Katagi 2004). In most of the more persistent herbicides, once contact with the soil is made, losses due to photolysis are small (Curran 2016). The exception is dinitroanilines that can be photodegraded if they remain on the soil surface even in the absence of water. Therefore, dinitroanilines such as trifluralin and pendimethalin must be incorporated into the soil at the time of application to reduce losses by photodegradation and volatilization (Curran 2016).

3.6 INTERACTIONS BETWEEN ENVIRONMENT AND HERBICIDES

Soils play different roles in the provision of essential ecosystem services for agriculture. The fertility is responsible for the cycles of nutrients, filters, and reservoirs that purify and store water, a structural function that is the growth substrate of plants, the regulation of the climate through the capture of carbon and greenhouse gases, and the conservation of biodiversity since soils are a reservoir of biological diversity (Dominati et al. 2010).

There are several methods used for weed control in agricultural fields, all of which have the potential to influence soil ecosystem services. Herbicides play an important role in weed management, and their use has increased since their introduction to the market due to their effectiveness and reduced cost compared to other weed control methods. Herbicides currently occupy the largest share (53%) of the pesticide market globally (Figure 3.2) (FAOSTAT 2021). Likewise, the introduction of genetically modified herbicide resistant crops has also contributed to the use of this chemical weed control tool (Green 2014).

Most of the sprayed herbicides deposit and integrate into the different phases of the soil. Once herbicides are in the soil matrix, where they can be retained by colloids or remain in the soil solution (Monquero and Silva 2021), they are subject to one of three potential fates (retention, transformation, or transport) (Colquhoun 2006). Herbicide retention consists of its sorption in the soil, decreasing its bioavailability in the solution (Gerstl 1990; Reddy and Locke 1994). The retention of herbicides in the soil controls and is controlled by chemical and biological transformations, which determine their persistence and residual activity (Kawakami et al. 2007). Both the herbicide retained in the soil, as well as that available in the liquid and gaseous phases, is exposed to the dissipation processes, transport, and transformation (Mehdizadeh et al. 2021). Herbicides can be removed from the soil by volatilization, runoff, runin, leaching of the area into soil water, and absorption by plant roots or foliage and degraded into non-toxic compounds by microorganisms, sunlight, or chemical reaction (Blasioli et al. 2011; Curran 2016).

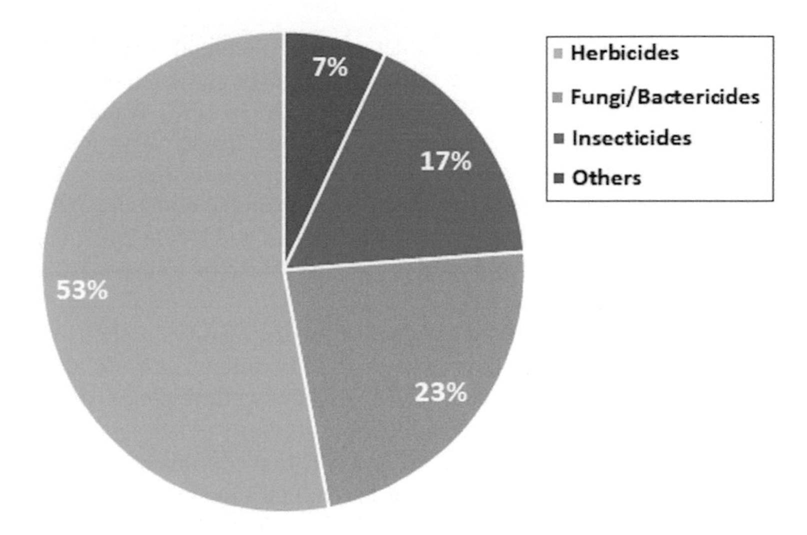

FIGURE 3.2 Global pesticide market share in 2019. *Source*: FAOSTAT (2021).

The order, intensity, and occurrence or not of any event involved in the retention, transformation, and transport of herbicides are variable, depending on the physicochemical characteristics of each molecule; type of soil; soil pH; climatic conditions before, during, and after application; presence and richness of microorganisms in the soil; type of cultivation, among other factors (Monquero and Silva 2021). In addition, agricultural management is another important factor in the dynamics and destination of herbicides in the environment, since any agricultural task (fallow, sowing, fertilization, addition of OM, irrigation, pesticide application, etc.), directly or indirectly in the soil, can affect its physical (density and texture), chemical (pesticides, fertilizers, water, MO), and biological (diversity and activity of microorganisms) properties (Moore 1994; Ponge et al. 2013), mainly the surface layer of the soil (0–30 cm). Understanding the behavior and fate of herbicides in the soil through the multiple and complex interactions of these variables, which directly or indirectly affect the efficiency of a herbicide in controlling weeds, allows reducing the risk of damage to susceptible crops and the environment.

3.6.1 Retention Process

3.6.1.1 Sorption/Desorption

Sorption is a general process that includes adsorption, absorption, precipitation, and hydrophobic partition processes and refers to the herbicide's ability to bind to soil colloids. Several types of intermolecular interactions may be involved in this phenomenon such as Van der Waals forces, hydrophobic bonds and water structure, hydrogen bonds, charge transfer, ion exchange, induced interactions, ion–pole, dipole–dipole, and magnetic interactions, among others (Kah and Brown 2006; Wang et al. 2020).

The type of soil is very important in determining potential sorption. The physicochemical properties of the soil are influenced by colloidal components (mineral and organic), which have variable SSAs (Shein and Devin 2007; Wang et al. 2020). In addition, CEC also is used to measure the sorption capacity of various colloidal fractions. High content of MO and/or clay minerals (montmorillonite and vermiculite) increase soil sorption of herbicides due to their chemical reactivity and the large number of binding sites, largest SSAs, and CEC (Shein and Devin 2007; Gavrilescu 2014); therefore, fewer herbicide molecules are available to be transported, absorbed by plants and/or degraded.

The release of herbicidal molecules adsorbed on the SSAs of colloids to the soil solution is known as desorption (Gavrilescu 2014). The amount of herbicide sorbed is always higher than that desorbed (Curran 2016). Knowing the balance between sorption and desorption is essential to determine the availability of herbicides in the solution alone and can be calculated using the soil partition coefficients K_d, K_{oc}, and K_f (absorption coefficient through isotherm of Freundlich's model) (Reddy and Locke 1994). In addition, other characteristics of herbicides that influence the sorption and desorption are the size and configuration of the molecule: pK_a, pK_b, S_w, and K_{ow}.

The sorption of some herbicides is not necessarily correlated with the OC content, and, in these cases, other properties of the soil, such as CEC and soil pH, may be better than OC (Bin et al. 2011). In addition, soil moisture also affects the sorption of herbicides, as water competes with herbicides for binding sites (Sebastian et al. 2016). In this sense, humid soils sorb less molecules than dry soils, and highly soluble herbicides do not adsorb well to the soil (Coulqhuon 2006). Therefore, leaching and volatilization also are influenced by the sorption/desorption balance in the soil/water system (Gavrilescu 2014). Therefore, associating the information on the sorption/desorption coefficients with the soil attributes (OM, clay content, moisture and pH) may contribute to predict the distribution of herbicides in the environment.

The sorption and desorption of indaziflam, a long persistence herbicide, were lower in soils with higher OM content (Alonso et al. 2011) – i.e., the mobility of indaziflam decreased as the OM content increased. For a soil with 16.8% OM, 10 to 100 times higher doses of indaziflam were needed to reduce the growth by 50% of *Kochia scoparia* compared to a soil with 0.4% OM (Sebastian et al. 2016). In relation to pH, in soils with low pH, generally the sorption is higher because there are fewer positive particles to compete for the negatively charged binding sites (McCauley et al. 2009). Herbicides such as imidazolinones have this behavior because most of their molecules are in dissociated form (Kraemer et al. 2009) – i.e., sorption/desorption of these herbicides increases or decreases depending on the pH of the soil.

3.6.2 Transformation Process

3.6.2.1 Degradation

The ideal herbicide is one that persists actively in the environment long enough to control the target weeds and then rapidly degrades so as not to represent some type of carryover to susceptible crops that come in rotation/succession or environmental risk (Mendes et al. 2021b). Carryover is herbicide persistence longer than desired; however, some herbicides can persist in the soil for a long time but are not active as herbicides – i.e., they do not pose a risk of carryover (Colquhoun 2006). The possibility of finding active herbicidal molecules depends on their solubility, mobility in the soil, sorption and desorption capacity, as well as climatic conditions, soil type, and cultivation practices. Most herbicides indicate crop rotation safety guidelines; however, restrictions often include some crops and/or specific environmental conditions (Colquhoun 2006). Therefore, the degradation of herbicides is important because it reduces the risk of carryover and environment issues and is produced by chemical changes in the molecule due to physical changes (photodecomposition) and chemical (oxidation–reduction, hydrolysis, formation of insoluble salts in water, etc.) or biological processes (microbiological degradation) (Noshadi and Homaee 2018; Singh and Singh 2014).

3.6.2.1.1 Photodegradation

Photodegradation by sunlight, especially ultraviolet light (UV), is one of the most important pathways for herbicides after their release into the environment (Katagi 2004). This occurs due to the sorption of UV light, which is the most destructive, by the herbicide causing the excitation of its electrons, breaking certain bonds in the molecules. The extent of sunlight photolysis is highly dependent on UV sorption profiles of the herbicide, the surrounding medium, and the emission spectrum of sunlight, since the energy to break chemical bonds in herbicide molecules usually ranges from 250 to 400 nm (Katagi 2004). Photolysis can be a direct process is which a substance is transformed by absorbing light energy or an indirect process in which other substances absorb energy, transform, and then alter the primary substance (Torrents et al. 1997; Pereira et al. 2016b). This characteristic is specific to each herbicide and can be beneficial, reducing the excessive persistence of residues in the soil. However, soil-acting herbicides that rapidly photodegraded in the top 3-mm layer must be incorporated into the soil at the time of application, because photodegradation can reduce the effectiveness of weed control (Monquero and Silva 2021). For example, trifluralin, napropamide, and paraquat are herbicides that can photodegrade rapidly, while phenylureas can photodegrade only when subjected to long periods of light.

3.6.2.1.2 Biological Degradation

Biodegradation is a transformation of a compound by the action of living organisms (Sharma et al. 2020). Soil microorganisms and plants accounts are of particular importance for herbicide degradation (van Eerd et al. 2003; Mendes et al. 2021b). The microbiological degradation of herbicides into less toxic or non-toxic products is carried out by certain bacteria, fungi, and algae and can occur due to an adaptive action or accidentally (Singh and Singh 2014). In degradation by adaptive action, the presence of a certain herbicide stimulates the microbiota to produce enzymes capable of degrading the herbicide molecules. In accidental degradation, herbicides are not the main source of food for the microbiota, and microorganisms do not receive a particular advantage in this (Ortiz-Hernández et al. 2013). In this case, there are no changes in the microbiota population, and the microorganisms degrade the herbicides if they get in their way. Microorganisms can degrade herbicides by catabolism (use the herbicide as a food for their development), co-metabolism (use another compound as food source, but degrade the herbicide due to its activity), or both (Lewis et al. 1978; Mendes et al. 2021b).

Soils with high levels of microbiological activity have reduced herbicide persistence (Sadegh-Zadeh et al. 2017). However, the activity of microorganisms is influenced by environmental factors, such as the OM content, soil pH, fertility, temperature, and soil moisture, the last two factors being the most important (Leweis et al. 1978; Mendes et al. 2021b). The microbial degradation of atrazine can be carried out by a wide variety of microorganisms (Singh and Singh 2014). However, some microorganisms are specific, and their population in the soil is related to the amount of herbicide available for consumption; therefore, the repeated use of a herbicide can increase the population of microorganisms but reduce weed control (Ramakrishnan et al. 2011). Microorganisms are more active in soils rich in OM. Soil pH influences depending on the species, but extreme pHs reduce the population and diversity of microorganisms. Soil moisture that varies from 50% to 100% of field capacity favors microbial activity, but in flooded soils, activity is reduced by a lack of oxygen. Warm soils also favor microbial activity, but at temperatures ≤5 °C, it becomes negligible (Colquhoun 2006).

The use of herbicides is capable of influencing the soil microbiota (Monquero and Silva 2021). Microbial biomass generally decreases in the first years after the application of a new herbicide; but the microbial flora (some strains) adapt to the new condition (Santos et al. 2005). Glyphosate favored bacterial density but reduced the fungal population in a red clay soil with pH 4.3 and 1.3% MO, but the diversity of both microorganisms returned to the initial stage within 30 days of application (Castro Jr et al. 2006). Bentazon doses ranging from 10 to 100 ppm temporarily reduced populations of bacteria but increased populations of fungi and actinomycetes (Marsh et al. 1978).

On the other hand, weeds and cultivated plants also reduce herbicide residues in the soil (van Eerd et al. 2003). When resistant plants take up the herbicide, it can be stored or metabolized, but most of the molecules are commonly conjugated to glycosides and peptides (Dimaano and Iwakami 2021). In addition, some plants have an enhanced ability to remove and/or immobilize herbicides from the soil. The use of these plants in combination with certain agronomic practices to reduce the carryover effect in susceptible cultures is known as phytoremediation (Santos et al. 2018). *Urochloa brizantha* reduced the concentration of picloram of the soil layer surface. The phytoremediation activity of *U. brizantha* was higher as the soil pH was more acidic (Braga et al. 2016). Evaluating tree species, *Calophyllum brasiliense* showed the greatest potential for phytoremediation of soils contaminated with ametryn and hexazinone (Hussain et al. 2009). Four species of green manure (*Crotalaria spectabilis, Canavalia ensiformis, Stizolobium térimum,* and *Lupinus albus*) phytoremediated (absorbed) up to 23% of quinclorac and 14% of tebuthiuron at 21 days (Mendes et al. 2021a). However, conjugation may not be a definitive route of degradation, as conjugated herbicides can hydrolytically break down being released to the environment again. The release of herbicides from plant residues increases their concentration in the soil and can represent carryover risks (Melo et al. 2016).

3.6.2.1.3 Chemical Degradation

The non-biological degradation occurs through various reactions, mainly hydrolysis, hydroxylation, mutilation and oxidation, reduction, and hydrolysis. High temperatures and good moisture facilitate these chemical reactions. In addition, extreme pH values can lead to increased hydrolysis of some herbicides (Mendes et al. 2021b).

Hydrolysis: In soil solution, hydrolysis is one of the most common abiotic transformation and significant degradation processes. This reaction (also considered as a physicochemical property) introduces a hydroxyl group to the initial molecule. The hydrolysis occurs chemically or enzymatically as hydrolysis of ester, hydrolysis of epoxides, hydrolysis of amine, and hydrolysis of halogens (Sarmah and Sabadie 2002). Amides are converted to acids and amines ($R–CO–NH–R_1 \rightarrow R–COOH + R_1–NH_2$); carbamates in amines and alcohols ($R–NH–CO–O–R_1 \rightarrow R–NH_2 + R_1–OH$); esters in acids and alcohols ($R–CO–O–R_1 \rightarrow R–COOH + R_1–OH$); and nitriles in carboxylic acids ($R–CNH \rightarrow R–CO–NH_2 \rightarrow R–COOH$). The resulting product is more easily degraded (biodegradation and photolysis), metabolized, and less toxic than the original compound (Blasioli et al. 2011). In addition, the hydroxyl group makes the compound more soluble in water reducing its bioconcentration potential. Hydrolysis is an important route of degradation of sulfonylureas and chloroacetamides by acid-catalyzed cleavage and base-catalyzed contraction/rearrangement (Sarmah and Sabadie 2002; Carlson et al. 2006), but not for imidazolines (Yavari et al. 2019).

Hydroxylation and Methylation: First reaction involves the introduction of hydroxyl groups into aromatic or aliphatic structures (more susceptible to attack), and the second reaction involves the addition of the methyl group to an alcohol or phenol with the formation of a methylene ether. Both reactions are generally mediated by hydroxylases of cytochrome P450 (Cyt-P450) (Nie et al. 2014; Dimaano et al. 2021). The main reactions (alkyl-hydroxylation, aryl-hydroxylation, *N*-demethylation, *O*-demethylation) may occur in the primary and secondary metabolism of many organisms and are dependent on NADPH and/or O_2 (Dimaano et al. 2021). Hydroxylation, *N* demethylation, and mineralization by microorganisms, mainly fungi, are important degradation mechanisms of isoproturon (Rønhede et al. 2005). Cyt-P450 mediated the alkyl-hydroxylation and *N*-demethylation of chlorotoluron in tobacco plants (Yamada et al. 2000); *O*-demethylation is common in sulfonylureas; and Cyt-P450 *O*-demethylated bensulfuron-methyl in *Echinochloa phyllopogon* (Iwakami et al. 2014). Aryl-hydroxylation is a typical reaction of plant metabolism, and crops such as wheat, barley, and corn oxidize the four positions of the phenyl ring of flumetsulam (Frear et al. 1993).

Oxidation: Many organic herbicides are susceptible to oxidation, where free radicals such as RO – or RO_2 produced by some reaction (e.g. photolytic) extract hydrogen-forming ROH or ROOH, leaving the compound in the form of a radical that can undergo further reaction (Oturan et al. 2011). Oxidation is an important process of herbicide degradation, as in many cases there are no hydroxylation and methylation without oxidation. OH radicals are a key oxidizer of most organic species, and herbicide oxidation can be biotic or abiotic (Murschell and Farmer 2019). In plants and microorganisms, oxidation is a primary metabolism process usually mediated by Cyt-P450 (van Eerd et al. 2003; Dimaano and Iwakami 2021). A oxidized molecule undergoes alterations in its structure and properties, mainly in S_w and chemical and biological activities (Nawaz et al. 2020).

3.6.3 Transport Process

Overall, there are five processes that can move herbicides, which are the diffusion, volatilization, leaching, erosion and runoff, assimilation by microorganisms, and absorption by plants (Blasioli et al. 2011; Mehdizadeh et al. 2021). Diffusion can be verified in the gaseous and liquid phases or in the air of the inter-solid phase. The pesticide is transferred through the soil from one zone where it is more concentrated to another where it is less (Pérez-Lucas et al. 2018). Assimilation by microorganisms and absorption by plants are also considered as biological degradation routes, as previously described. In this section, volatilization, leaching, and run-off will be described, since these transport processes are responsible for the greatest losses of herbicides ranging from 10% to 90% (average ~40–80%), 0% to 4% (~1%), and 0% to 10% (~5%), respectively.

3.6.3.1 Leaching

Leaching process is the upward or downward movement of the herbicide in the soil and is the main form of transport of non-volatile and water-soluble herbicides in the soil (Cox et al. 1997; Gehrke et al. 2021). To be leached, the herbicide must be in soil solution, free or adsorbed on small particles, such as clays, low molecular weight fulvic and humic acids, amino acids, peptides and sugars, among others (Stevenson 1972; Pérez-Lucas et al. 2018). The intensity of leaching depends on the physico-chemical properties of the herbicide (S_w, K_{oc}, $t_{1/2}$, formulation, and additives) and on soil (texture) and climatic characteristics (soil pH and water content, soil persistence and remobilization potential, topography or slope of the area, intensity of rainfall or irrigation after application and soil management) (Hall et al. 1989; Lewan et al. 2009; Berisso et al. 2012; Sadegh-Zadeh et al. 2017). The higher the herbicide sorption by soil colloids, the lower the leaching. In sandy soils, leaching will be even higher than in silty or clay soils (Rossi et al. 2005). For most herbicides, especially non-ionic, the OM content of the soil is the constituent most correlated with their sorption and movement (Larsbo et al. 2009; Carpio et al. 2021). On the other hand, frequent and heavy rainfall can lead to more herbicide losses through leaching (Hall et al. 1989; Gish et al. 2011).

The movement of herbicides in the soil has a major influence on their performance in the field. Little leaching is desirable, as it can make the herbicide more efficient. For example, it is desirable that PRE herbicides leach down to 5-cm depth, as most weed seeds with germination potential are usually found in this range. Leaching can account for the selectivity of a herbicide, but, when excessive, the herbicide can be washed into the water table, causing undesirable contamination.

Leaching potential of a herbicide (Groundwater Ubiquit Score – GUS), also known as groundwater vulnerability index, establishes the relationship between $t_{1/2}$ and K_{oc} ($GUS = \log t_{1/2} (4 - \log K_{oc})$) (Gustafson 1989). Herbicides with GUS less than 1.8 are considered low or non-leachable (glyphosate, paraquat), those with GUS between 1.8 and 2.8 are considered moderately leachable (diuron, S-metolachlor), and those with GUS greater than 2.8 are leachable (atrazine, mesotrione). However, the leaching of some herbicides also depends on soil factors such as soil pH, OM content, and moisture. Diuron and tebuthiuron leached up to 50-cm deep as a function of OM content

(Matallo et al. 2005). The second herbicide was found in water samples collected at a depth of 53 m (Monquero and Silva 2021). Imidazolinones reached depths of 25 cm in irrigated rice in a floodplain soil (Refatti et al. 2017). Clomazone leaching was higher in soil with lower pH and lower OM content (Pereira et al. 2016a). In the case of fomesafen, the most important factor in reducing the potential for leaching is OM, followed by the texture and pH of the soil (Silva et al. 2014).

3.6.3.2 Volatilization

Volatilization is a process by which herbicide molecules change from a liquid state to a vapor form, which can be lost to the atmosphere (Penner and Michael 2014). Volatilization is another important source of herbicide loss, causing losses of up to 90%. The factors that most influence volatilization are high temperature and humidity in combination with the VP, S_w, K_H, and structure and molecular weight of the herbicides (Majewski et al. 1989; Gish et al. 2011). For some herbicides, this process can be so intense that, after application, they must be immediately incorporated into the soil (Yates 2006), as discussed in Sections 3.3.4 and 3.3.5.

Once in the atmosphere, the herbicide in vapor form can be transported long distances and be re-deposited to the surface by wind (dry deposition) and/or water (rain, dew, snow, and fog), contaminating nontarget environments and organisms (Gavrilescu 2005). The introduction of crops resistant to synthetic auxins has allowed these herbicides to be used on a larger scale; however, their use has caused great losses in adjacent areas with sensitive crops in the United States due to drift due to volatilization of dicamba and 2,4-D (Egan et al. 2014). In the state of Minnesota, the United States, dicamba applications are prohibited when the temperature exceeds 29 °C (Monquero and Silva 2021). In Brazil, the introduction of crops resistant to synthetic auxins has caused great concern (Oliveira et al. 2021), since volatilization drift problems have been reported in vineyards (Alcántara-de la Cruz et al. 2020). On the other hand, the level of tolerance of crops to volatilized herbicides is variable. For example, the lowest tolerance to clomazone in the vapor phase was showed by sorghum, followed by maize and rice (monocotyledons), while melon and cucumber (dicotyledons) were the most tolerant (Schreiber et al. 2013).

3.6.3.3 Runoff and Runin

The movement of herbicides on the soil surface and subsurface from treated to untreated areas after heavy rainfall with the downpour is called run-off and ruin-in, respectively. This process affects with great-intensity PRE herbicides, which are applied directly to the soil, since they are generally exposed to the elements, before or shortly after sowing (Gish et al. 2011). High S_w and low sorption potential favor run-off; however, heavy rains can also carry soil particles (erosion) into water bodies, i.e., herbicides with high sorption potential can also be transported by runoff over long distances, compromising water quality (Monquero and Silva 2021). Peak concentrations of herbicides in surface water are recorded soon after heavy rain events, which decrease as a result of their susceptibility to hydrolysis. Therefore, the risk of runoff may be reduced by localized application (Pappalardo et al. 2016).

In dryland areas of the United States, of 130 surface water samples analyzed, 72% was detected to be imazetaphyr, 95% acetochlor, 89% alachlor, and 99% atrazine

(Battaglin et al. 2000). In Brazil, 50% of water samples collected from rice-growing areas contained clomazone, quinclorac, and/or propanil (Marchesan et al. 2010). Acetochlor, ametryn, atrazine, clomazone, diuron, glyphosate hexazinone, isoxaflutole, pendimethalin, simazine, sulfentrazone, tebuthiuron, and trifluralin were detected in samples in surface water and sediments of the Corumbataí River, SP, Brazil, with triazines being the herbicides with the greatest persistence (Armas et al. 2007). Atrazine runoff generally is higher in soils with low OM content (Correia et al. 2007).

3.7 CONCLUDING REMARKS

The behavior of herbicides in the environment, mainly in the soil, depends on the sum of several processes involved, which are responsible for the final destination of these compounds. Herbicide transport, retention, and transformation processes are often studied separately; however, many of them occur simultaneously at different intensities, depending on the characteristics of the application environment and the physicochemical properties of the products. Therefore, knowledge of the properties of the soil, the climatic factors involved, and the mechanisms of interaction of herbicides with the environment is essential for greater effectiveness of weed control with herbicides in a technically and economically viable way and/or for a better final fate of these products (meaning less impact) on the environment.

REFERENCES

Alcántara-de la Cruz R, Oliveira GM, Carvalho LB, Silva MFGF. 2020. Herbicide resistance in Brazil: Status, impacts, and future challenges. In *Pests, weeds and diseases in agricultural crop and animal husbandry production*, eds. Kontogiannatos D, Kourti A, Mendes KF, 1–25 (1st ed.). London: IntechOpen.

Alonso DG, Koskinen WC, Oliveira RS, Constantin J, Mislankar S. 2011. Sorption-desorption of indaziflam in selected agricultural soils. *J Agric Food Chem*. 59: 13096–13101.

Amézqueta S, Subirats X, Fuguet E, Roses M, Rafols C. 2019. Octanol-water partition constant. In *Liquid-Phase Extraction*, ed. Poole CF, 183–208 (1st ed.). Amsterdam: Elsevier.

Armas EDD, Monteiro RTR, Antunes PM. 2007. Spatial-temporal diagnostic of herbicide occurrence in surface waters and sediments of Corumbataí River and main affluents. *Quím Nova*. 30: 1119–1127.

Battaglin WA, Furlong ET, Burkhardt MR, Peter CJ. 2000. Occurrence of sulfonylurea, sulfonamide, imidazolinone, and other herbicides in rivers, reservoirs and ground water in the Midwestern United States, 1998. *Sci Total Environ*. 248: 123–133.

Berisso FE, Schjønning P, Keller T, et al. 2012. Persistent effects of subsoil compaction on pore size distribution and gas transport in a loamy soil. *Soil Tillage Res*. 122: 42–51.

Bin G, Cao X, Dong Y, Luo Y, Ma LQ. 2011. Colloid deposition and release in soils and their association with heavy metals. *Crit Rev Environ Sci Technol*. 41: 336–372.

Blasioli S, Braschi I, Gessa CE. 2011. The fate of herbicides in soil. In *Herbicides and environment*, ed. Kortekamp A, 175–194 (1st ed.). London: IntechOpen.

Braga RR, Santos JB, Zanuncio JC, et al. 2016. Effect of growing *Brachiria brizantha* on phytoremediation of picloram under different pH environments. *Ecol Eng*. 94: 102–106.

Bronick CJ, Lal R. 2005. Soil structure and management: A review. *Geoderma*. 124: 3–22.

Bucka FB, Kölbl A, Uteau D, Peth S, Kögel-Knabner I. 2019. Organic matter input determines structure development and aggregate formation in artificial soils. *Geoderma*. 354: 113881.

Carpio MJ, Sánchez-Martín MJ, Rodríguez-Cruz MS, Marín-Benito JM. 2021. Effect of organic residues on pesticide behavior in soils: A review of laboratory research. *Environments*. 8: 32.

Carlson DL, Than KD, Roberts AL. 2006. Acid-and base-catalyzed hydrolysis of chloroacetamide herbicides. *J Agric Food Chem*. 54: 4740–4750.

Castro Jr JV, Selbach PA, Záchaayub MA. 2006. Avaliação do efeito do herbicida glifosato na microbiota do solo. *Pestic: R Ecotoxicol Meio Ambiente*. 16: 21–31.

Cessna AJ, Knight D, Ngombe D, Wolf TM. 2017. Effect of temperature on the dissipation of seven herbicides in a biobed matrix. *Can J Soil Sci*. 97: 717–731.

Cho SJ, Kim MH, Lee YO. 2016. Effect of pH on soil bacterial diversity. *J Ecol Environ*. 40: 10.

Colquhoun, J. (2006). Herbicide persistence and carryover. Available from: http://corn.agronomy.wisc.edu/Management/pdfs/A3819.pdf (accessed October 26, 2021).

Correia FV, Macrae A, Guilherme LRG, Langenbach T. 2007. Atrazine sorption and fate in a Ultisol from humid tropical Brazil. *Chemosphere*. 67: 847–854.

Cox L, Celis R, Hermosin MC, Becker A, Cornejo J. 1997. Porosity and herbicide leaching in soils amended with olive-mill wastewater. *Agric Ecosyst Environ*. 65: 151–161.

Curran WS. 2016. Persistence of herbicides in soil. *Crops Soils*. 49: 16–21.

Delgado EJ, Alderete JB. 2003. Prediction of Henry's law constants of triazine derived herbicides from quantum chemical continuum solvation models. *J Chem Inform Comput Sci*. 43: 1226–1230.

Dimaano NG, Iwakami S. 2021. Cytochrome P450-mediated herbicide metabolism in plants: Current understanding and prospects. *Pest Manag Sci*. 77: 22–32.

Dominati E, Patterson M, Mackay A. 2010. A framework for classifying and quantifying the natural capital and ecosystem services of soils. *Ecological Ec*. 69(9): 1858–1868.

Egan JF, Barlow KM, Mortensen DA. 2014. A meta-analysis on the effects of 2, 4-D and dicamba drift on soybean and cotton. *Weed Sci*. 62: 193–206.

FAO. 2021. Soil map of the world. Available from: www.fao.org/soils-portal/data-hub/soil-maps-and-databases/faounesco-soil-map-of-the-world/en/ (accessed October 26, 2021).

FAOSTAT. 2021. Pesticide use. Available from: www.fao.org/faostat/en/#data/RP (accessed October 26, 2021).

Frear DS, Swanson HR, Tanaka FS. 1993. Metabolism of flumetsulam (DE-498) in wheat, corn, and barley. *Pest Biochem Physiol*. 45: 178–192.

Furmidge CGL, Osgerby JM. 1967. Persistence of herbicides in soil. *J Sci Food Agric*. 18: 269–273.

Gavrilescu M. 2005. Fate of pesticides in the environment and its bioremediation. *Eng Life Sci*. 5: 497–526.

Gavrilescu M. 2014. Colloid-mediated transport and the fate of contaminants in soils. In *The role of colloidal systems in environmental protection*, ed. M. Fanun, 397–451 (1st ed.). Amsterdam: Elsevier.

Gehrke VR, Fipke MV, Avila LAD, Camargo ER. 2021. Understanding the opportunities to mitigate carryover of imidazolinone herbicides in lowland rice. *Agriculture*. 11: 299.

Gerstl Z. 1990. Estimation of organic chemical sorption by soils. *J Contam Hydrol*. 6: 357–375.

Gish TJ, Prueger JH, Daughtry CS, et al. 2011. Comparison of field-scale herbicide runoff and volatilization losses: An eight-year field investigation. *J Environ Qual*. 40: 1432–1442.

Green JM. 2014. Current state of herbicides in herbicide-resistant crops. *Pest Manag Sci*. 70: 1351–1357.

Gustafson DI. 1989. Groundwater ubiquity score: A simple method for assessing pesticide leachability. *Environ Toxicol Chem*. 8: 339–357.

Hall JK, Murray MR, Hartwig NL. 1989. Herbicide leaching and distribution in tilled and untilled soil. *J Environ Qual*. 18: 439–445.

Hussain S, Siddique T, Arshad M, Saleem M. 2009. Bioremediation and phytoremediation of pesticides: Recent advances. *Crit Rev Environ Sci Technol*. 39: 843–907.

Iwakami S, Endo M, Saika H, Okuno J, et al. 2014. Cytochrome P450 CYP81A12 and CYP81A21 are associated with resistance to two acetolactate synthase inhibitors in *Echinochloa phyllopogon*. *Plant Physiol*. 165: 618–629.

Kah M, Brown CD. 2006. Adsorption of ionisable pesticides in soils. *Rev Environ Contam Toxicol*. 188: 149–217.

Katagi T. 2004. Photodegradation of pesticides on plant and soil surfaces. *Rev Environ Contam Toxicol*. 182: 1–78.

Kawakami T, Eun H, Ishizaka M, et al. 2007. Adsorption and desorption characteristics of several herbicides on sediment. *J Environ Sci Health B*. 42: 1–8.

Kraemer AF, Marchesan E, Avila LA, Machado SLO, Grohs M. 2009. Environmental fate of imidazolinone herbicides: A review. *Planta Daninha*. 27: 629–639.

Larsbo M, Stenström J, Etana A, Börjesson E, Jarvis NJ. 2009. Herbicide sorption, degradation, and leaching in three Swedish soils under long-term conventional and reduced tillage. *Soil Tillage Res*. 105: 200–208.

Lewan E, Kreuger J, Jarvis N. 2009. Implications of precipitation patterns and antecedent soil water content for leaching of pesticides from arable land. *Agric Water Manag*. 96: 1633–1640.

Lewis JA, Papavizas GC, Hora TS. 1978. Effect of some herbicides on microbial activity in soil. *Soil Biol Biochem*. 10: 137–141.

Loux MM, Reese KD. 1993. Effect of soil type and pH on persistence and carryover of imidazolinone herbicides. *Weed Technol*. 7: 452–458.

Majewski MS, Glotfelty DE, Seiber JN. 1989. A comparison of the aerodynamic and the theoretical-profile-shape methods for measuring pesticide evaporation from soil. *Atmos Environ*. 23: 929–938.

Marchesan E, Sartori GMS, Avila LAD, et al. 2010. Residues of pesticides in the water of the Depression Central rivers in the State of Rio Grande do Sul, Brazil. *Ciên Rural*. 40: 1053–1059.

Marsh JAP, Winfield GI, Da Vies HA, Grossbard E. 1978. Simultaneous assessment of various responses of the soil microflora to bentazone. *Weed Res*. 18: 293–300.

Martins CR, Lopes WA, De Andrade JB. 2013. Organic compound solubility. *Quím Nova*. 36: 1248–1255.

Matallo MB, Spadotto CA, Luchini LC, Gomes MA. 2005. Sorption, degradation, and leaching of tebuthiuron and diuron in soil columns. *J Environ Sci Health B*. 40: 39–43.

McCauley A, Jones C, Jacobsen J. 2009. Soil pH and organic matter. *Nutr Manag Module*. 8: 1–12.

Mehdizadeh M, Mushtaq W, Siddiqui SA, et al. 2021. Herbicide residues in agroecosystems: Fate, detection, and effect on non-target plants. *Rev Agric Sci*. 9: 157–167.

Melo CAD, Dias RDC, Mendes KF, Assis ACLP, Reis MR. 2016. Herbicides carryover in systems cultivated with vegetable crops. *R Bras Herbic*. 15: 67–78.

Mendes KF, Maset BA, Mielke KC, et al. 2021a. Phytoremediation of quinclorac and tebuthiuron-polluted soil by green manure plants. *Int J Phytoremediation*. 23: 474–481.

Mendes KF, Mielke KC, Barcellos LHJ, Alcántara-de la Cruz R, Nogueira de Sousa R. 2021b. Anaerobic and aerobic degradation studies of herbicides and radiorespirometry of microbial activity in soil. In *Radioisotopes in weed research*, ed. Mendes KF, 95–126 (1st ed.). Boca Raton: CRC Press Taylor & Francis Group.

Miller P, Westra P. 2004. Herbicide behavior in soils. Available from: https://erams.com/static/wqtool/PDFs/bmps_colorado/00562.pdf (accessed October 26, 2021).

Moldrup P, Olesen T, Komatsu T, Schjønning P, Rolston DE. 2001. Tortuosity, diffusivity, and permeability in the soil liquid and gaseous phases. *Soil Sci Soc Am J*. 65: 613–623.

Monquero PA, Silva PV. 2021. Comportamento de herbicidas no ambiente. In *Matologia: Estudos sobre Plantas Daninhas*, eds. Barroso AAM, Murata AT, 253–294 (1st ed.). Jaboticabal: Fábrica da Palavra.

Moore JC. 1994. Impact of agricultural practices on soil food web structure: Theory and application. *Agric Ecosyst Environ*. 51: 239–247.

Murschell T, Farmer DK. 2019. Atmospheric OH oxidation chemistry of trifluralin and acetochlor. *Environ Sci Process Impacts*. 21: 650–658.

Nawaz H, Waheed R, Nawaz M, Shahwar D. 2020. Physical and chemical modifications in starch structure and reactivity. In *Chemical properties of starch*, ed. Emeje M, 1–21 (1st ed.). London: IntechOpen.

Neina D. 2019. The role of soil pH in plant nutrition and soil remediation. *Appl Environ Soil Sci*. 2019: 5794869.

Nie Y, Chi CQ, Fang H, et al. 2014. Diverse alkane hydroxylase genes in microorganisms and environments. *Sci Rep*. 4: 4968.

Noshadi E, Homaee M. 2018. Herbicides degradation kinetics in soil under different herbigation systems at field scale. *Soil Tillage Res*. 184: 37–44.

Oliveira MC, Lencina A, Ulguim AR, Werle R. 2021. Assessment of crop and weed management strategies prior to introduction of auxin-resistant crops in Brazil. *Weed Technol*. 35: 155–165.

Ortiz-Hernández ML, Sánchez-Salinas E, Dantán-González E, Castrejón-Godínez ML. 2013. Pesticide biodegradation: Mechanisms, genetics and strategies to enhance the process. In *Biodegradation – life of science*, ed. Chamy R, 253–287 (1st ed.). London: IntechOpen.

Oturan MA, Oturan N, Edelahi MC, Podvorica FI, El Kacemi K. 2011. Oxidative degradation of herbicide diuron in aqueous medium by Fenton's reaction based advanced oxidation processes. *Chem Eng J*. 171: 127–135.

Pachepsky YA, Rawls WJ. 2003. Soil structure and pedotransfer functions. *Eur J Soil Sci*. 54: 443–452.

Pappalardo SE, Otto S, Gasparini V, Zanin G, Borin M. 2016. Mitigation of herbicide runoff as an ecosystem service from a constructed surface flow wetland. *Hydrobiologia*. 774: 193–202.

Penner D, Michael J. 2014. Bioassay for evaluating herbicide volatility from soil and plants. In *Pesticide formulation and delivery systems*, ed. Sesa C, 36–43 (vol. 33). West Conshohocken: ASTM International.

Pereira GAM, Barcellos L, Gonçalves VA, et al. 2016a. Sorption of clomazone in Brazilian soils with different physical and chemical attributes. *Planta Daninha*. 34: 357–364.

Pereira VJ, Cunha JPAR, Morais TP, Oliveira JPR, Morais JB. 2016b. Physical-chemical properties of pesticides: Concepts, applications, and interactions with the environment. *Bioscience J*. 32: 627–641.

Pérez-Lucas G, Vela N, El Aatik A, Navarro S. 2018. Environmental risk of groundwater pollution by pesticide leaching through the soil profile. In *Pesticides – Use and misuse and their impact in the environment*, ed. Larramendy ML, 1–27(1st ed.). London: IntechOpen.

Piccolo A, Celano G, Conte P. 1996. Adsorption of glyphosate by humic substances. *J Agric Food Chem*. 44: 2442–2446.

Ponge JF, Pérès G, Guernion M, et al. 2013. The impact of agricultural practices on soil biota: A regional study. *Soil Biol Biochem*. 67: 271–284.

Prata F, Lavorenti A, Regitano JB, Tornizielo VL. 2001. Degradation and sorption of ametryne in two soils with vinasse application. *Pesq Agropec Bras*. 36: 975–981.

Quan G, Yin C, Chen T, Yan J. 2015. Degradation of herbicide mesotrione in three soils with differing physicochemical properties from China. *J Environ Qual*. 44: 1631–1637.

Ramakrishnan B, Megharaj M, Venkateswarlu K, Sethunathan N, Naidu R. 2011. Mixtures of environmental pollutants: Effects on microorganisms and their activities in soils. *Rev Environ Contam Toxicol*. 211: 63–120.

Reddy KN, Locke MA. 1994. Prediction of soil sorption (K_{oc}) of herbicides using semiempirical molecular properties. *Weed Sci*. 42: 453–461.

Refatti JP, Avila LAD, Noldin JA, Pacheco I, Pestana RR. 2017. Leaching and residual activity of imidazolinone herbicides in lowland soils. *Ciên Rural*. 47: e20160705.

Rice CP, Chernyak S, McConnell LL. 1997. Henry's law constants for pesticides measured as a function of temperature and salinity. *J Agric Food Chem*. 45: 2291–2298.

Romero-Ruiz A, Linde N, Keller T, Or D. 2018. A review of geophysical methods for soil structure characterization. *Rev Geophysics*. 56: 672–697.

Rønhede S, Jensen B, Rosendahl S, et al. 2005. Hydroxylation of the herbicide isoproturon by fungi isolated from agricultural soil. *Appl Environ Microbiol*. 71: 7927–7932.

Rossi CVS, Alves PLCA, Marques Jr J. 2005. Mobility of sulfentrazone in red latosol and chernosol. *Planta Daninha*. 23: 701–710.

Sadegh-Zadeh F, Abd Wahid S, Jalili B. 2017. Sorption, degradation and leaching of pesticides in soils amended with organic matter: A review. *Adv Environ Technol*. 3: 119–132.

Santos JB, Jakelaitis A, Silva AA, et al. 2005. Microbial activity in soil after herbicide application under no-tillage and conventional planting systems. *Planta Daninha*. 23: 683–691.

Santos NMC, Costa VAM, Araújo FV, et al. 2018. Phytoremediation of Brazilian tree species in soils contaminated by herbicides. *Environm Sci Poll Res*. 25: 27561–27568.

Sarmah AK, Sabadie J. 2002. Hydrolysis of sulfonylurea herbicides in soils and aqueous solutions: A review. *J Agric Food Chem*. 50: 6253–6265.

Schmuckler ME, Barefoot AC, Kleier DA, Cobranchi DP. 2000. Vapor pressures of sulfonylurea herbicides. *Pest Manag Sci*. 56: 521–532.

Schreiber F, Avila LAD, Scherner A, Moura DDS, Helgueira DB. 2013. Plants sensitive to clomazone in vapor phase. *Ciên Rural*. 43: 1817–1823.

Sebastian DJ, Nissen SJ, Westra P, Shaner DL, Butters G. 2016. Influence of soil properties and soil moisture on the efficacy of indaziflam and flumioxazin on *Kochia scoparia* L. *Pest Manag Sci*. 73: 444–451.

Sharma A, Kumar V, Kohli SK, et al. 2020. Pesticide metabolism in plants, insects, soil microbes and fishes. In *Pesticides in crop production: Physiological and biochemical action*, eds. Srivastava KP, Singh VP, Singh A. et al., 35–53 (1st ed.). Hoboken: John Wiley & Sons, Inc.

Shein EV, Devin BA. 2007. Current problems in the study of colloidal transport in soil. *Eurasian Soil Sci*. 40: 399–408.

Silva, A.A. Viviam R, Oliveira Jr, RS. 2007. Biologia de plantas daninhas. In *Tópicos em manejo de plantas daninhas*, eds. Silva AA, Silva JF, 155–209 (1st ed.). Viçosa: Editora UFV.

Silva GR, D'Antonino L, Faustino LA, et al. 2014. Mobility of fomesafen in Brazilian soils. *Planta Daninha*. 32: 639–645.

Singh B, Singh K. 2014. Microbial degradation of herbicides. *Crit Rev Microbiol*. 42: 245–261.

Snakin VV, Prisyazhnaya AA, Kovács-Láng E. 2001. *Soil liquid phase composition*, 316p (1st ed.). Amsterdam: Elsevier.

Stevenson FJ. 1972. Role and function of humus in soil with emphasis on adsorption of herbicides and chelation of micronutrients. *BioScience*. 22: 643–650.

Torrents A, Anderson BG, Bilboulian S, Johnson WE, Hapeman CJ. 1997. Atrazine photolysis: Mechanistic investigations of direct and nitrate-mediated hydroxy radical processes and the influence of dissolved organic carbon from the Chesapeake Bay. *Environ Sci Technol*. 31: 1476–1482.

van Eerd LL, Hoagland RE, Zablotowicz RM, Hall JC. 2003. Pesticide metabolism in plants and microorganisms. *Weed Sci*. 51: 472–495.

Wang C, Wang R, Huo Z, Xie E, Dahlke HE. 2020. Colloid transport through soil and other porous media under transient flow conditions – A review. *Wiley Interdiscip Rev Water*. 7: 1–33.

Wang E, Cruse RM, Zhao Y, Chen X. 2015. Quantifying soil physical condition based on soil solid, liquid and gaseous phases. *Soil Tillage Res*. 146: 4–9.

Wlodarczyk M, Siwek H. 2017. The influence of humidity and soil texture on the degradation process of selected herbicides immobilized in alginate matrix in soil under laboratory conditions. *Pol J Soil Sci*. 50: 121–130.

Wu XM, Li M, Long YH, et al. 2011. Effects of adsorption on degradation and bioavailability of metolachlor in soil. *J Soil Sci Plant Nutr*. 11: 83–97.

Wu XM, Wang W, Liu J, et al. 2017. Rapid biodegradation of the herbicide 2, 4-dichlorophenoxyacetic acid by *Cupriavidus gilardii* T-1. *J Agric Food Chem*. 65: 3711–3720.

Yamada T, Kambara Y, Imaishi H, Ohkawa H. 2000. Molecular cloning of novel cytochrome P450 species induced by chemical treatments in cultured tobacco cells. *Pest Biochem Physiol*. 68:11–25.

Yates SR. 2006. Measuring herbicide volatilization from bare soil. *Environ Sci Technol*. 40: 3223–3228.

Yavari S, Sapari NB, Malakahmad A, Yavari S. 2019. Degradation of imazapic and imazapyr herbicides in the presence of optimized oil palm empty fruit bunch and rice husk biochars in soil. *J Hazardous Mat*. 366: 636–642.

Zhang H, Zhang Y, Hou Z, et al. 2014. Biodegradation of triazine herbicide metribuzin by the strain *Bacillus* sp. N1. *J Environ Sci Health B*. 49: 79–86.

Zhang X, Norton L. 2002. Effect of exchangeable Mg on saturated hydraulic conductivity, disaggregation and clay dispersion of disturbed soils. *J Hydrol*. 260: 194–205.

4 Interaction Mechanisms Between Biochar and Herbicides

Rodrigo Nogueira de Sousa[1], Matheus Bortolanza Soares[1], Felipe Hipólito dos Santos[1], Camille Nunes Leite[1], and Kassio Ferreira Mendes[2]*

[1] Department of Soil Science, Luiz de Queiroz College of Agriculture – University of São Paulo, Piracicaba, Brazil

[2] Department of Agronomy, Federal University of Viçosa, Viçosa, MG, Brazil

* Corresponding author: rodrigosousa@usp.br

CONTENTS

DOI: 10.1201/9781003202073-4

4.1 INTRODUCTION

Biochar is the solid product of thermochemical decomposition of biomass pyrolysis of vegetative biomass, bones, manure solids, or other plant-derived organic residues at moderate temperatures under oxygen-limiting conditions (Guo et al. 2016). Biochar is used as an additive in agricultural soils due to many interesting characteristics including high carbon content, high pH, high stability, high porosity, and high surface area (Brassard et al. 2019). The physical–chemical properties of biochars depend heavily on both production operational parameters and biomass feedstock. Thus, each biochar should be produced for their use in specific applications (Guo et al. 2016; Brassard et al. 2019). Therefore, biochar has often been applied in agriculture worldwide, with an agronomic focus on improving soil quality and with environmental focus on the mitigation of greenhouse gas emission and remediation tools for organic and inorganic pollutants (Xiao and Chen 2016).

On the other hand, the physical–chemical properties of herbicides also strongly interfere in the interaction of these products with biochars. It is known that herbicides can be applied in PRE- and POST-emergence of weeds and crops, and when they are applied in PRE directly on the soil surface, such as very long-chain fatty acid (VLCFA) inhibitors, microtubules, photosystem II (PSII), among other modes of action, they can interact with biochars and reduce control efficiency and be an effective remediation tool. However, it is important to point out that these interactions are governed according to the ionization (acids or bases, measured by the acid/base ionization constant – pK_a/pK_b) or non-ionization (neutral) of herbicides, in addition to other properties, such as water solubility (S_w), sorption coefficient (K_d) and octanol–water coefficient (K_{ow}). However, the remediation of herbicides in the soil is also common to the products applied in the POST, because residues of these herbicides reach the soil and can cause contamination, especially to crops in rotation/succession, a process known as carryover.

The exploration of biochar sorption mechanisms for organic pollutants, such as herbicides, and the understanding of the relationship between biochar structure and herbicide property in sorption will help in the analysis of bioavailability and fate of herbicides, in predicting the biochar sorption behavior in the soil, and in guiding the selection of specific biochar for specific product soil remediation (Xiao and Chen 2016; Mendes et al. 2019). As such, the interactions between biochar and herbicides have attracted great attention from the scientific community (Xiao and Chen 2016). The sorption and desorption of herbicides in the soil influence their biodegradation (Scow 1993). However, the biodegradation kinetics of sorbed herbicides in biochar considers the potential impacts of sorbate–sorbent interactions on microbial populations. That is, herbicides that are very sorbed in biochar may be immobilized but increase soil persistence; on the other hand, the addition of biochar as a soil amendment can increase water retention and be a nutrient source for microorganisms that are the main degraders of herbicides and consequently reduce their persistence in the soil.

Thus, this chapter is fundamental for understanding the mechanisms of interactions between biochar and herbicides, which will be essential for elucidating the behavior of herbicides in biochar-amended soil.

4.2 PHYSICOCHEMICAL PROPERTIES OF BIOCHARS

Biochar is a sort of black carbon composed of a wide range of carbonized organic materials exhibiting different physical and chemical properties (Pignatello 2013). The definition of biochar is not well established, and the material is defined according to its use, e.g., organic materials carbonized under limited oxygen conditions intended specifically for application to soil or water to obtain agricultural or environmental gains (Sohi et al. 2010; Mukherjee and Lal 2013).

The properties of biochar are mainly related to the raw material and production conditions (Joseph et al. 2021), and, through adjustments in these parameters, it is possible to obtain materials with desired characteristics for each type of soil or purpose (Abiven et al. 2014; Ippolito et al. 2020). Some thermochemical technologies of carbonization have been developed in the last decades, such as hydrothermal carbonization, microwave pyrolysis, as well as the improvement of well-established techniques such as slow, fast, and flash pyrolysis, although the last two are more used to producing bio-oil (Meyer et al. 2011; Ahmad et al. 2014).

Thermochemical reactions involving condensation, decarboxylation, demethylation, and removal of volatile components determine the relative production of biochar, bio-oil, and volatilized gases from biomass (Sohi 2012). However, regardless of the feedstock and pyrolysis technology, some production factors are critical for the properties and homogeneity of the biochars obtained, among which the following stand out: *(i)* Highest Treatment Temperature (HTT) reached during the process; *(ii)* residence time at HTT; *(iii)* ramp rate; *(iv)* moisture; *(v)* pressure; and *(vi)* particle size (Manyà 2012; Spokas et al. 2012). Furthermore, a series of reactions known as "aging" starts as soon as the material is applied to the soil (Wang, O'Connor et al. 2020). Aging includes dissolution–precipitation, acid–base, oxidation–reduction, and sorption–desorption reactions that can cause significant changes in the properties of biochars such as reactive surface development (Joseph et al., 2021) affecting their interaction with pesticides (Joseph et al. 2010; Li et al. 2019).

The properties that stimulate interest in the industry and research concerning biochars are its porous structure (honey-like comb) composed of polycondensed carbon rings with abundant functional groups. These can be important from an environmental point of view for *(i)* sequestration carbon in stable compounds with a long turnover period; *(ii)* improving bulk soil chemical and physical properties such as water and nutrients' retention; *(iii)* retention of potentially toxic elements (PTEs) in contaminated water; and *(iv)* sorption of PTEs and pesticides in soils, such as herbicides, insecticides, and fungicides (Kookana et al. 2011; Lopez-Capel et al. 2016). These properties and their relationship to the factors mentioned are discussed in subsequent sections.

4.2.1 Elemental Composition

Biochar production can be carried out using agricultural residues, wood, bones, and animal waste, each containing different concentrations of inorganic matter, holocellulose, and lignin variables (Sohi 2012; Li et al. 2019). In general, woody biomass contains more carbon (C) in the dry matter than in crop residues and manures (Lopes-Capel et al. 2016). Biochars produced with manure as feedstock have the highest N and P content and, therefore, the lowest C:N ratio, while wood-feedstock-produced biochars show the highest C:N ratio (Ippolito et al. 2020; Joseph et al. 2021). The remainder of the biomass is composed of hydrogen (H), oxygen (O), and nitrogen (N), and, to a lesser extent, sulfur (S), phosphorus (P), potassium (K), iron (Fe), zinc (Zn), magnesium (Mg), manganese (Mn), copper (Cu), and molybdenum (Mo) among others (Lopes-Capel et al. 2016; Ippolito et al. 2020).

During heating, H, O, and N are more easily removed from the biomass so that the remaining product is enriched in C (50–95%). The increase in the pyrolysis temperature causes an increase in the C content of the biochar, while the H and O content is reduced (Ahmad et al. 2014). The ratio between C:H, C:O, and C:N is different among biochars and has a significant impact on the interaction of the material with the soil (Manyà 2012). The International Biochar Initiative (IBI) and the European Biochar Certificate (EBC) recognize carbonized materials with a molar ratio H:C-org ≤ 0.7 and an O:C-org ratio ≤ 0.4 as biochar (Schimmelpfennig and Glaser 2012). The elemental ratio is considered a reliable measure of the extent of pyrolysis and the level of oxidative change to which feedstocks were exposed, in addition to establishing quality criteria for identification and comparison between materials (Sohi et al. 2010; Spokas et al. 2012). The H:C and O:C ratios decrease with increasing pyrolysis temperature, and it is assumed that the lower the ratio, the greater the degree of aromaticity and stability of the charcoal (Kookana et al. 2011).

In higher HTT (>550°C), there is a reduction in the biochar yield which on the other hand is characterized by a higher proportion of C condensed into highly stable aromatic rings (Joseph et al. 2010). The C content of biochars is negatively correlated with yield. In lower HTT (<350 °C), the biochar productivity is higher, but as the extent of the carbonization process is smaller, the resulting biochar lacks the stability associated with C-rich and highly aromatic wood-derived biochars (Table 4.1). The C:N ratio is one of the most important chemical parameters to predict N mineralization or immobilization in soils (Igalavithana et al. 2016). The C:N ratio of biochar presents values of 8:1 for materials produced with municipal sewage sludge to values of the order of 1500:1 for biochar produced with wood waste (Table 4.1). Hamer et al. (2004) reported that biochars obtained from maize straw (*Zea mays* L.) and rye (*Secale cereale* L.) were mineralized faster than biochar produced from wood residues, a result associated with H:C, O:C, and C:N. However, the validity of the C:N ratio for biochar may be limited due to the high stability of C bonds found in some materials (Mukome and Parikh 2015).

TABLE 4.1
Selected Properties of Biochars Produced from Different Feedstocks.

Feedstock	Carbonization Technology	Heating Rate (°C min^{-1})	HTT (°C)	C (%)	O (%)	H (%)	N (%)	O:C	C:N	Properties pH	SA (m^2 g^{-1})	% Ash	Reference
Poultry litter	Hydrothermal	-	250	51.33	8.38	4.61	5.05	0.16	10.16	7.38	7.10	30.21	Ghanim et al. (2017)
	Slow pyrolysis	2.5	350	51.07	15.63	3.79	4.45	0.30	11.48	8.7	3.9	30.7	Cantrell et al. (2012)
		8.3	700	45.91	10.53	1.98	2.07	0.22	22.18	10.3	50.9	46.2	
Cotton seed hull	Slow pyrolysis	-	200	51.90	40.50	6.00	0.60	0.78	86.50	-	-	3.1	Uchimiya et al. (2011)
		-	350	7.00	15.70	4.53	1.90	0.20	40.53	-	4.7	5.7	
		-	500	87.50	7.60	2.82	1.50	0.09	58.33	-	0.0	7.9	
		-	650	91.00	5.90	1.26	1.60	0.06	56.88	-	34.0	8.3	
		-	800	90.00	7.00	0.60	1.90	0.08	47.37	-	322.0	9.2	
Sewage sludge	Hydrothermal	-	195	107.9	-	-	2.24	-	48.17	5.0	21.0	36.9	Melo et al. (2019)
	Slow pyrolysis	7.0	300	30.72	11.16	3.11	4.11	0.36	7.47	6.8	4.5	56.6	Ahamad et al. (2014)
		7.0	400	26.62	10.67	1.93	4.07	0.40	6.54	6.6	14.1	67.1	
		7.0	500	20.19	9.81	1.08	2.84	0.49	7.11	7.3	26.2	71.9	
		7.0	600	24.76	8.41	0.83	2.78	0.34	8.91	8.3	35.8	74.6	
		7.0	700	22.04	7.09	0.57	1.73	0.32	12.74	8.1	54.8	76.6	
Sugarcane bagasse	Hydrothermal	-	200	69.15	25.74	5.11	0.54	0.37	128.06	4.0	10.7	-	Fang et al. (2015)
		-	250	75.08	19.28	5.64	0.76	0.26	98.79	5.3	3.9	-	
		-	300	79.31	15.35	5.34	0.88	0.19	90.64	5.8	4.9	-	
Sugarcane straw	Slow pyrolysis	5.0	350	34.0	-	-	0.6	-	56.6	8.6*	17.0	5.0	Soares et al. (2022)
		5.0	550	52.7	-	-	0.7	-	75.28	9.3*	129.0	10.3	
		5.0	750	58.3	-	-	0.6	-	97.16	9.8*	223.0	11.6	
Pig bones	Fast pyrolysis	40	400	29	-	3	4	-	7.25	7.7	3	-	Dela Piccolla et al. (2021)
		40	550	17	-	1	3	-	5.67	8.3	87	-	
		40	800	18	-	1	2	-	9.00	9.3	89	-	
Pine needles	Slow pyrolysis	7.0	300	84.19	7.57	4.37	3.88	0.09	21.70	6.4	4.1	7.2	Ahmad et al. (2014)
		7.0	500	90.10	3.74	2.06	4.10	0.04	21.98	8.1	13.1	11.8	
		7.0	700	93.67	2.07	0.62	3.64	0.02	25.73	10.6	390.5	18.7	
Pine wood chips	Gasification	-	750	65.29	8.99	0.63	-	0.14	-	11.1	426.0	33	Hansen et al. (2015)
Mixed sawdust	Fast pyrolysis	-	500	69.00	14.60	2.7	0.3	0.21	230	-	1.6	-	Spokas et al. (2009)

* pH of Dissolved Organic Carbon (DOC) extracted. – not measured.

Although elements such as N, K, and P are plant nutrients and chemically improve soil fertility, high concentrations of these elements in biomass make it difficult to convert feedstock into biochar (Mašek et al. 2016). Raw material components with low boiling points, such as N, S, and Cl, or with low melting points as Na and K can pose challenges for the thermal conversion process due to their tendency to cause corrosion (Mašek et al. 2016). These factors are particularly important in the conversion of animal waste and bones, components with higher inorganic matter content (Vassilev et al. 2013; Iriarte-Velasco et al. 2016).

Some pre-treatments of the feedstock were developed to reduce the content of contaminants or with the intention of making the thermal conversion more efficient, with emphasis on (i) moisture reduction; (ii) particle size reduction; (iii) pelletizing; (iv) washing biomass to remove chemicals; and (v) mixture with additives such as fertilizers, clays, or other biomass feedstocks (Zwetsloot et al. 2015; Mašek et al. 2016). Pretreatment of biomass with water, for example, is an effective method to remove high solubility chemicals (Cl, K, and Na) and reduce ash content in biochar (Manyà 2012). However, such measure has as a consequence the reduction in biochar productivity, given that alkali and alkaline earth metals catalyze biomass decomposition reactions.

In terms of contaminants, industrial waste or sewage sludge must be carefully considered as raw material for biochar production due to the constant presence of PTEs and/or organic contaminants (Steiner 2016). Chemically, biochar contaminants are organic or inorganic; and in terms of origin, the contaminants can be pyrolysis by-products formed during the production of biochar or components of the raw material used to produce biochar (Godlewska et al. 2021). Contamination by PTEs can be hampered by avoiding raw materials that contain elements such as As, Hg, and Cd (Steiner 2016), as most PTEs are not volatile at common carbonization temperatures of less than 700 °C and remain in the biochar as ashes. In contrast, organic contaminants such as polycyclic aromatic hydrocarbons (PAHs) and dioxins can be formed during carbonization through the fragmentation and recombination of organic compounds (Godlewska et al. 2021). These polyaromatic compounds are commonly present in the natural environment, and their toxicity ranges from being non-toxic to extremely toxic (Steiner 2016; Godlewska et al. 2021).

4.2.2 Specific Surface Area and Porosity

Biochars produced from the same biomass but by different carbonization processes have a very variable specific surface area (SSA) and porosity values. The SSA and porosity of biochars increase proportionally to the increase in the pyrolysis temperature up to a limit between 700 and 900°C when the structure collapses and the SSA falls again (Kookana et al. 2011). Pyrolysis at low temperatures also called roasting (200–320°C) usually results in products with lower SSA, while biochars resulting from gasification (>700°C) tend to have higher SSA values (Peterson et al. 2013). Destruction of aliphatic alkyl and ester groups and also exposure of the aromatic core of lignins through higher pyrolysis temperatures may account for an increase in surface area (Ahmad et al. 2014).

Biochars produced at an HTT of 450°C had a common SSA of less than 10 m^2 g^{-1} whereas materials produced at temperatures of 600–750°C had an SSA of approximately 400 $m^2 g^{-1}$ (Brown et al. 2006; Kookana 2011). Pyrolysis at high temperatures (>550°C) produces materials with high SSA (>400 $m^2 g^{-1}$), aromaticity, and, consequently, resistance to mineralization (Singh and Cowie 2010). The surface area studied by N_2-BET was relatively low (3.1–21.6 $m^2 g^{-1}$) for biochars produced in a fluidized bed reactor with switch grass and corn husk, and the reason given was the short residence time of solid particles (Manyà 2012).

Biochar produced from plant biomass often maintains inherent plant anatomy or internal structure (Peterson et al. 2013). In general, higher lignin content is responsible for a macroporous structure, whereas higher cellulose content produces a higher SSA microporous structure (Ioannidou and Zabaniotou 2007; Steiner 2016). The pore structure of particular biochar is strongly influenced by HTT. The HTT and the time the biomass is kept in the HTT are important as they determine the extent of pyrolysis or the degree to which volatile components in the feedstock will be thermally eliminated (Peterson et al. 2013). As volatile components are eliminated, micropores of diameter 0.2–2 nm are formed in their place, and the SSA of the material increases (Peterson et al. 2013). Studies have shown that regardless of the differences in ramp rate, the maximum SSA is obtained at a final temperature between 650 and 750 °C and that above this temperature the SSA falls again due to the collapse of micropores that are a key factor for high SSA (Brown et al. 2006; Kookana et al. 2011).

In general, biochars of high surface area, microporosity, and aromaticity exhibit exceptional sorption ability toward herbicides (Yang and Sheng 2003; Spokas et al. 2009; Kookana et al. 2011). The addition of wheat char (1% w/w) contributed 80–86% to the sorption of diuron and bromoxynil by char-amended soil and contributed 70% of ametryn by char-amended soil (Sheng et al. 2005). The addition of mixed sawdust biochar (5% w/w) to Minnesota soil increased the K_f value (Freundlich constant) from 2.25 $\mu g^{1-1/n} mL^{1/n} g^{-1}$ to 3.12 $\mu g^{1-1/n} mL^{1/n} g^{-1}$ for sorption of atrazine and resulted in decreasing dissipation rates of this herbicide (Spokas et al. 2009).

4.2.3 Acidity and Liming Potential

In general, the hydrogenionic potential (pH) of biochar is increased with increasing pyrolysis temperature due to the separation of alkaline elements and decomposition of acidic functional groups (Gaskin et al. 2009; Singh et al. 2017). The average pH value determined in biochar produced from various raw materials was 8.9, with a range from 5.9 to 12.3, with the lowest value for saw dust pyrolyzed at 450°C and the highest for buffalo weed at 700°C (Ahmad et al. 2014; Singh et al. 2017). The pH is a key factor for the sorption of ionizable herbicides due to their tendency to protonate or not in certain pH ranges. Some herbicides such as glufosinate and glyphosate have weak acid behavior ($pK_a < 3.0$), while others are weak bases such as atrazine and hexazinone ($pK_b < 3$), so the pH of the soil solution directly impacts herbicide speciation and its interaction with sorbents (Mendes, de Sousa et al. 2021).

Biochar holds a considerable fraction of inorganic material in the form of ash. Alkali and alkaline earth metals such as Ca, Mg, and K accumulate in the ash as oxides and/or carbonates with varying degrees of crystallinity (Singh et al. 2017). Minerals such as quartz (SiO_2) and calcite ($CaCO_3$) are among the most commonly found minerals in biochar ash and can be used as alkalizing correctives in acidic soils (Ahmad et al. 2014; Alam and Alessi 2019).

Soil pH elevation after biochar application in acidic soils is well reported in the literature (Ahmad et al. 2014; Singh et al. 2017; Alam and Alessi 2019). The ability of biochar to raise soil pH depends on biochar chemical properties, application rate, and soil buffering power (Camps-Arbestain et al. 2015). The liming potential obtained through the titration of biochar in the acidic solution can be used to compare the ability of different materials to correct the pH of soils (Singh et al. 2017). The result is expressed as $CaCO_3$ equivalent and separated into the following classes: Class 0 (<1% $CaCO_3$-eq), Class 1 (1–10% $CaCO_3$-eq), Class 2 (10–20% $CaCO_3$-eq), and Class 3 (> 20% $CaCO_3$-eq) (Camps-Arbestain et al. 2015). Studies have shown that biochar derived from a range of plant and animal biomass increases soil pH by up to 73% with an average increase of 28% depending on the rate of application (Mukherjee and Lal 2013). A previous study showed that the sorption of atrazine and acetochlor in amended soil is significantly enhanced when biochar is used as a soil amendment (Spokas et al. 2009).

4.2.4 Surface Functional Groups

Surface functional groups are defined as "chemically reactive molecular units at the interface region between a solid and the soil solution, which react with ions and molecules at that interface" (Sposito 2008). Surface functional groups can be organic or inorganic molecular units. The main organic functional groups are carboxyl, carbonyl, and phenol, while silanol, aluminol, and hydroxyl are the main inorganic groups (Sparks 2019).

Biochar contains a wide range of organic surface functional groups that play an important role in the sorption of herbicides in soil (Ahmad et al. 2014). Biochars derived from pyrolyzed biomass at temperatures between 300 and 400°C are only partially carbonized, have a high O content and H:C ratio, and a greater number of surface functional groups than biochar produced at a temperature of 500 to 700°C, which has lower O content, low H:C ratio, high SSA, and fewer surface functional groups (Chan and Xu 2009). However, the physisorption in meso- and micropores seems to be more relevant in soils since most herbicides applied are non-ionic ones.

The gathering of spectroscopic results such as nuclear magnetic resonance (NMR), Fourier transform infrared (FTIR), X-ray photoelectron spectroscopy (XPS), and X-ray absorption near edge structure (XANES) showed that with the progressive aging of biochar, artificial or natural, structural changes occur that lead to the formation of surface functional groups such as carboxylic and phenolic groups (Mia et al. 2017). Consequently, the O:C ratio, net negative charge, and CEC increase with the progress of the aging of biochar in the soil (Mia et al. 2017; Joseph et al. 2021).

4.2.5 Ion Exchange

Ion exchange capacity is defined as the maximum number of moles of ionic charge adsorbed and that can be desorbed, per unit mass of the sorbent under certain conditions of temperature, pressure, and composition of the surrounding solution (Sposito

2008). Biochar may have anion exchange capacity (AEC), but as most surface functional groups are weak acids dissociated in the relevant pH range in agricultural soils between 4.5 and 7.5, the cation exchange capacity (CEC) resulting from net negative charges is higher (Mukherjee and Lal 2013).

Most charges expressed by biochars are variable in nature (Silber et al. 2010; Mukherjee et al. 2011). Negative charges develop as a result of the dissociation of protons from oxygenated functional groups on biochar surfaces. Since dissociation increases with increasing pH, the negative charge of biochar increases with increasing pH and, therefore, CEC increases (Silber et al. 2010; Mukherjee et al. 2011).

The application of biochar can increase the CEC of soils in the medium to long term, and its use presents important perspectives for soils with low CEC such as sandy soils, and for highly weathered soils from tropical regions such as Oxisols and Ultisols (Gaskin et al. 2009; Kookana et al. 2011). Biochar from woody materials tends to have low CEC values, while non-woody plant materials such as sugarcane straw or leaves tend to have higher CEC values (Kookana et al. 2011). Studies have shown that the CEC of biochar produced from different biomasses presented a range of 50 to 200 $mmol_c\ kg^{-1}$ (Joseph et al. 2021) being responsible for an increase in the corrected soil CEC of up to two times (Mukherjee and Lal 2013).

Biochar oxidation during aging reactions is responsible for creating negative surface charges which increase CEC over time (Glaser et al. 2002; Steiner 2016). The point of zero charge for fresh biochar samples was significantly higher than for after-aging biochar samples, such that freshly applied biochar contributed to soil AEC, and aged biochar exhibited higher CEC at pH values ≥ 3 (Cheng et al. 2008).

4.2.6 Dissolved Organic Carbon

Biochar aging impacts the physicochemical properties of biochar resulting in the release of a variety of biochar-derived organic materials (BDOMs), ranging from biochar surface oxidation products to polycarboxylic acids (Mia et al. 2017). BDOMs can be separated into four fractions: (i) superficially oxidized carbon, which does not dissolve in acid, alkaline solution, or water; (ii) humic acids (HAs), which dissolve in alkaline solutions but not in acid; (iii) fulvic acids (FAs), which dissolve in both acid and alkaline solutions; and (iv) dissolved organic carbon (DOC) that is soluble in water (Mia et al. 2017). DOC is the most reactive fraction of dissolved organic matter (DOM), consisting of a mixture of heterogeneous organic materials often classified as amorphous and gel-like or as a "hard and glassy" condensed matrix (Bolan et al. 2011). These two domains have very different sorption efficiencies, and, therefore, the quantity and relative proportion of these materials in the SOC influence the sorption capacity of pesticides. The extent of pesticide sorption in soils increases with the degree of aromaticity of the SOC (Kookana et al. 2011).

The carbonization technology greatly influences the DOM content and consequently the biochar-derived SOC (Mašek et al. 2016; Melo et al. 2019). For example, hydrothermal carbonization (HTC), which consists of treating organic materials in an aqueous environment at temperatures between 150 and 350°C under autogenous pressure, releases more DOM after application to the soil than "conventional" pyrolysis biochar due to its more unstable/labile character of C, represented by aliphatic compounds of protein-like substances (Wu et al. 2020).

4.3 PHYSICOCHEMICAL PROPERTIES OF HERBICIDES

Herbicides are chemical compounds purposely synthesized to maintain optimum crop production in sufficient quantity and quality due to their ability to provide selective weed control without damaging the main crops (Tomczyk and Sokołowska 2019; Yvari et al. 2020). They also may be applied to crops to improve harvesting as desiccants. The potential effects of herbicides are strongly influenced by their toxic mode of action and their method of application (EPA 2021).

There are two main herbicide application forms, PRE- and POST-emergence: the first is applied directly to the soil in order to avoid the emergence of the weed seed bank, and the second is applied directly over the plant. When the herbicide is applied in PRE, the molecule must be bioavailable in the soil solution, sorbed, or the form of bound residue, so that it can be absorbed by weeds and has an efficient control (Chi et al. 2017). Moreover, PRE herbicides are absorbed by seeds, roots, hypocotyls, cotyledons, coleoptiles, or leaves before emergence.

These herbicides are soil-applied, known as residual herbicides, and their soil behavior can impact both weed control efficiency and environmental fate (Oliveira Jr et al. 2000). When in contact with soil, the herbicide can be sorbed by a wide range of physicochemical mechanisms in which its molecules may become unavailable. The formation of this fraction is an important mechanism for the dissipation of herbicides, and it can remain in the soil for a long time (Gevao et al. 2000). However, in specific cases, part of this fraction linked to the soil may become available to weeds, a process called desorption or remobilization in case of bound residue (Amondham et al. 2006).

There is a general concern about the indiscriminate use of herbicides in agriculture. Thus, the study about the comprehension of the behavior and interaction of these organic molecules in the environment is essential to the greater sustainability of the planet. The environment-threatening potential of herbicides is determined by climatic factors, the nature of the media, and the specific characteristics of each herbicide, especially its water solubility and degradation half-life time (Ruijuan et al. 2009; Stenrod 2015). The behavior of herbicides in the soil is driven by their physical–chemical properties combined with the chemical, physical, and biological attributes of the soil and its constituents, besides the climatic conditions (Mendes, Mielke et al. 2021).

The main herbicide properties that influence their soil retention process are (Mendes, de Sousa et al. 2021) are given in the subsequent sections.

4.3.1 N-OCTANOL–WATER PARTITION COEFFICIENT (K_{ow})

The K_{ow} is defined as the ratio of the concentration of an herbicide in the saturated n-octanol phase in water and its concentration in the saturated aqueous phase in n-octanol at equilibrium at the temperature of 25 °C (Figure 4.1) (Howard and Muir 2010). It is typically expressed in logarithmic form as log K_{OW}. This physicochemical

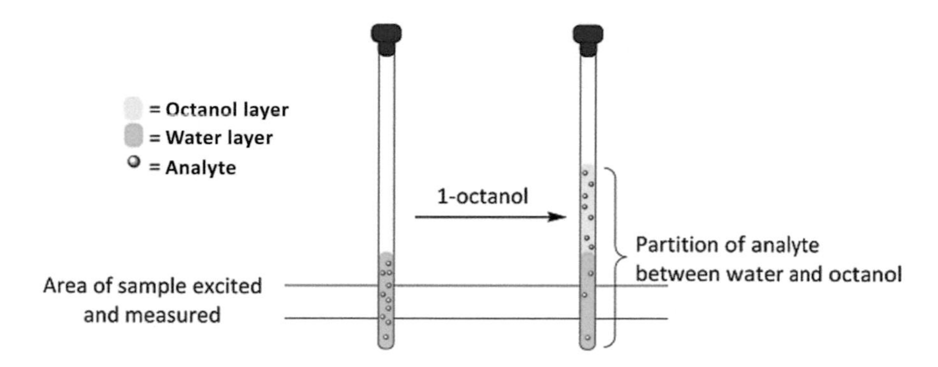

FIGURE 4.1 Schematic diagram of the analyte partition between the water and 1-octanol layers upon equilibration (Cumming and Rucker 2017).

property applied in evaluative mode is for the prediction of distribution among environmental compartments through equations for estimating bioaccumulation in animals and plants and in predicting the toxic effects of a substance (Mackay 1991; Finizio et al. 1997). Log K_{OW} represents a measure of the tendency of a chemical shifting from the aqueous phase into lipids in the following ways: 1) positive values of log K_{ow} indicate the hydrophobic property of compounds; 2) larger values indicate a greater hydrophobic property of compounds; and 3) if the value of log K_{OW} is greater than 3, the chemical product is considered to be very hydrophobic (Cumming and Rucker 2017; Zhu et al. 2022). The log K_{ow} values of the pendimethalin and glyphosate are 5.4 and –3.02, respectively (Table 4.2), and these indicate that pendimethalin could exhibit a strong tendency to sorb onto mulch and to bioaccumulate in the environment while glyphosate is highly hydrophilic and does not bioaccumulate.

4.3.2 Coefficients of Sorption (K_D)

The K_d refers to the ratio between the concentration of sorbed herbicide (Cs) in the soil and its concentration in the equilibrium solution (Ce) (Schwarzenbach et al. 1993). This conception, according to Weber (2004), consists of measuring the quantity of herbicide sorbed from a specified concentration by a specific mass of soil. Thus, the higher the K_d of a herbicide, the greater is the soil sorption capacity. It is important to highlight that K_d values are generally determined at herbicide concentrations that would occur when the compounds are applied at recommended rates followed by sufficient precipitation to bring the soil to field capacity (Weber et al. 2000). As a consequence, the mobility of the herbicide along the soil profile is inversely correlated with the K_d. For example, the values of K_d of pendimethalin, atrazine, and metribuzin are 228 (non-mobile), 3.2 (moderately mobile), and 0.874 L kg^{-1} (mobile) (Table 4.2). These K_d values are used in mathematical models (e.g., Linear Freundlich and Langmuir) to preview herbicides' mobility in soils. Usually, the sorption coefficients are normalized by the organic carbon content in the soil (K_{oc}).

TABLE 4.2
Structural Formulas and Physicochemical Properties of Herbicides.

Herbicide	Structural formula	Mode of action (chemical group)	S_w at 20°C (mg L^{-1})	Log K_{ow}	pK_a or pK_b at 25°C	VP at 25°C (mPa)	K_d or K_f (L kg^{-1})	DT50 (days)
2,4-dichlorophenol (2,4-D)		Synthetic auxin (alkylchlorophenoxy)	24300 (high)	–0.82 (low)	3.40 (strong acid)	0.009 (low)	0.70 (mobile)	4.4 (non-persistent)
Alachlor		Inhibition of VLCFA/ inhibition of cell division (Chloroacetamide)	240 (moderate)	3.09 (high)	0.62 (strong acid)	2.9 (low)	16.5 (slightly mobile)	14 (non-persistent)
Ametryn		Inhibits photosynthesis /photosystem II (triazine)	200 (moderate)	2.63 (low)	10.07 very weak acid)	0.365 (low)	76.81 (non-mobile)	37 (Moderately persistent)
Aminocyclopyrachlor		Synthetic auxin (pyrimidine carboxylic acid)	3130 (high)	–2.48 (low)	4.6 (weak acid)	6.92×10^{-3} (low)	0.39 (mobile)	31 (Moderately persistent)
Atrazine		Inhibits photosynthesis/ photosystem II (triazine)	35 (low)	2.7 (moderate)	1.7 (very weak base)	0.039 (low)	3.2 Moderately mobile)	75 (Moderately persistent)

Herbicide	Structural formula	Mode of action (chemical group)	S_w at 20°C (mg L^{-1})	Log K_{ow}	pK_a or pK_b at 25°C	VP at 25°C (mPa)	K_d or K_f (L kg^{-1})	DT50 (days)
Carfentrazone		Cell membrane disruption/PPO inhibitor (triazolone)	-	3.7 (high)	neutral	-	-	8.2 (non-persistent)
Chlorimuron-ethyl		Inhibits plant amino acid synthesis (Sulfonylurea)	1200 (high)	0.11 (low)	4.2 (weak acid)	4.9×10^{-07} (low)	-	40 (Moderately persistent)
Clethodim		acetyl CoA carboxylase inhibitor- ACCase (Cyclohexanedione)	5450 (high)	4.14 (high)	4.47 (weak acid)	2.08×10^{-03} (low)	0.39 (mobile)	0.55 (non-persistent)
Clomazone		disruption of chlorophyll and carotenes synthesis (Isoxazolidinone)	1212 (high)	2.58 (low)	neutral	27 (highly)	2.08 (moderately mobile)	22.6 (non-persistent)
Dicamba		Synthetic auxin (Benzoic acid)	250000 (high)	−1.80 (low)	1.87 (strong acid)	1.67 (low)	10.16 (very mobile)	9.62 (non-persistent)

(Continued)

TABLE 4.2 (Continued)

Herbicide	Structural formula	Mode of action (chemical group)	S_w at 20°C (mg L^{-1})	Log K_{ow}	pK_a or pK_b at 25°C	VP at 25°C (mPa)	K_d or K_f (L kg^{-1})	DT50 (days)
Diclosulam		Inhibits plant amino acid synthesis/ AHAS (Sulfonanilide)	6.32 (low)	0.85 (low)	4 (weak acid)	6.67×10^{-10} (low)	-	49 (Moderately persistent)
Diquat		Photosystem I (electron transport) inhibitor	718000 (high)	–4.6 (low)	ions +	1.00×10^{-03} (low)	23099 (non-mobile)	2345 (Very persistent)
Diuron		Photosynthesis inhibitor	35.6 (low)	2.87 (moderate)	neutral	1.5×10^{-03} (low)	12.8 (slightly mobile)	146.6 (persistent)
Flumioxazin		Inhibition of protoporphyrinogen oxidase/PPO (N-phenylphtalamides)	0.786 (low)	2.55 (low)	neutral	0.32 (low)	-	21.9 (non-persistent)
Glufosinate-ammonium		Glutamine synthetase inhibitor (Phosphinic acid)	500000 (high)	–4.01 (low)	9.15 (weak acid)	3.10×10^{-02} (low)	2.3 (slightly mobile)	7.4 (non-persistent)

Glyphosate	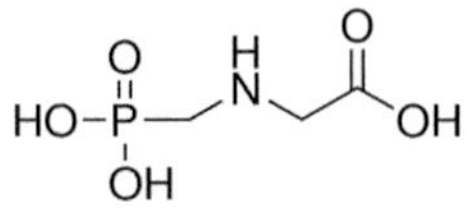	Inhibition of EPSP synthase (Phosphonoglycine)	10500 (high)	−3.02 (low)	2.34 (strong acid)	0.0131 (low)	209.4 (slightly mobile)	15 (non-persistent)
Hexazinone	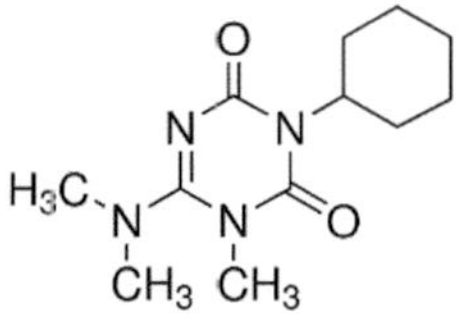	Inhibits photosynthesis/ photosystem II (Triazinone)	33000 (high)	1.17 (low)	2.2 (weak base)	0.03 (low)	-	105 Persistent)
Imazethapyr	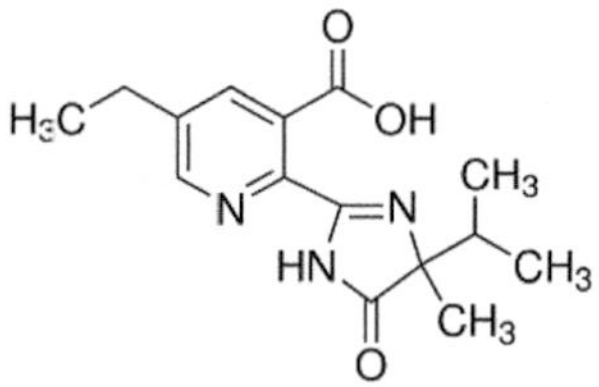	Inhibits plant amino acid synthesis/ AHAS (Imidazolinone)	1400 (high)	1.49 (low)	3.9 (weak acid)	1.33×10^{-02} (low)	-	90 (Moderately persistent)
Indaziflam	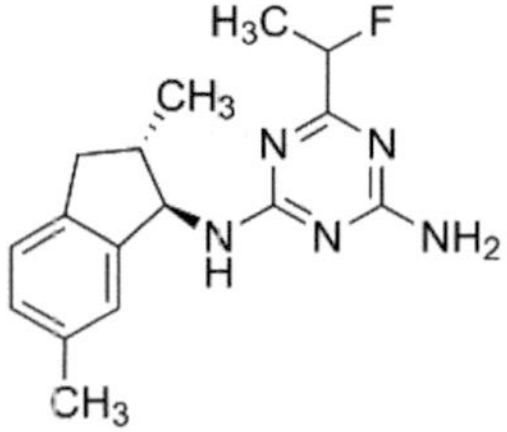	Inhibits cellulose biosynthesis/	2.3 (low)	2.8 (moderate)	3.5 (weak acid)	2.5×10^{-05} (low)	-	150 (persistent)

(Continued)

TABLE 4.2 (Continued)

Herbicide	Structural formula	Mode of action (chemical group)	S_w at 20°C (mg L^{-1})	Log K_{ow}	pK_a or pK_b at 25°C	VP at 25°C (mPa)	K_d or K_f (L kg^{-1})	DT50 (days)
Metribuzin		Inhibits photosynthesis/ photosystem II (Triazinone)	10700 (high)	1.75 (low)	12.8 (weak acid)	0.121 (low)	0.874 (mobile)	7 (non-persistent)
Nicosulfuron		Inhibits plant amino acid synthesis/AHAS (Sulfonylurea	7500 (high)	0.61 (low)	7.58 (weak acid)	8.00×10^{-07} (low)	0.29 (mobile)	26 (non-persistent)
Pendimethalin		Inhibition of mitosis and cell division (Dinitroaniline)	0.33 (low)	5.4 (high)	2.8 (strong acid)	3.34 (low)	228 (non-mobile)	182.3 (persistent)
Picloram		Synthetic auxin (Pyridine compound	560 (high)	–1.92 (low)	2.3 (strong acid)	8.0×10^{-05} (low)	0.14 (very mobile)	82.8 (Moderately persistent)
Sulfentrazone		Cell membrane disruption/PPO inhibitor (Aryl triazolinone)	80 (high)	0.991 (low)	6.56 (weak acid)	1.30×10^{-04} (low)	-	541 (very persistent)

(-) not available

Source: Adapted from PPDB (2022).

4.3.3 Degradation Half-Life Time in Soil (DT_{50})

In general, DT_{50} can be understood as the time required for the concentration of the product to decline to half of the initial value (Schwarzenbach et al. 1993; EPA 2021). With the same meaning but under controlled conditions in the laboratory, the DT_{50} that is equal to ln(2)/k (the rate constant per day, k) indicates the time required to reduce the concentration by 50% from any concentration point in the incubation time (EPA 2021). It is important to express the rate of decline of a first-order degradation. The degradation process occurs mainly by microorganisms; however, chemical degradation and photodegradation also influence the process. Also, it is possible to measure the dissipation time (DT_{50}) under field conditions, evaluating not only degradation but also off-target transport. The DT_{50} of sulfentrazone is 541 days, which is considered very persistent; the DT_{50} of diclosulam is 49 days being considered moderately persistent, and the DT_{50} of glyphosate is 15 days (non-persistent) (Table 4.2).

4.3.4 Acid/Base Dissociation Coefficients (pK_A or pK_B)

We can measure a solution of herbicide and tell how acidic or basic it is and can measure its strength by different scales in which pK_a or pK_b is the most usual even though the pH scale is more familiar. The "p" before K_a or K_b, like pH, i.e. pK_a and pK_b, implies that you are dealing with a -log of the value following the "p" (Helmenstine 2020). K_a, pK_a, K_b, and pK_b values indicate whether the herbicide solution will accept protons or donate at a particular pH value. They describe the degree of ionization of an acid or base and are real predictors of acid or base strength since adding water to herbicide will not change the equilibrium constant. K_a and pK_a are related to acids, while K_b and pK_b are related to bases. And then, when the pH value of the soil is greater than pK_b, the herbicides contain more of their molecules in molecular form (neutral/non-ionizable). On the other hand, acid herbicides present more of their molecules in molecular form when the pH value of the soil is lower than pK_a (Mendes, de Sousa et al. 2021). However, if the pH of the soil is equal to the pK_a or pK_b of herbicide, this means that 50% of the product molecules will be ionizable (anions and cations for acid and base herbicides, respectively) and 50% will be in molecule form. Weak acids and bases do not completely dissociate into their ions in water, and strong acids and bases dissociate completely into their ions in water, yielding one or more protons (hydrogen cations and hydroxyl anions, respectively) per molecule (Helmenstine 2020). In addition, there are non-ionizable (neutral) herbicides, as diuron and clomazone (Table 4.2). For example, the basic herbicides are atrazine and hexazinone, and acid herbicides are metribuzin and indaziflam. Under field conditions, knowing the ionization of the herbicide is fundamental to agronomic efficacy in weed control, since the practice of liming to raise the pH for agricultural cultivation directly interferes with the bioavailability of the herbicide in the soil solution.

4.3.5 Vapor Pressure (VP)

The VP is the pressure exerted by steam in equilibrium with a liquid at a certain temperature. The VP is among the most important physical properties as it plays an important role in governing the gas–phase concentration of herbicides and their tendency for long-range transport (Goel et al. 2007). The VP is measured in Pascal (Pa), commonly at 25°C, and it is highly dependent on temperature; hence, spatial latitude, longitude, and seasonal conditions greatly influence compound phase distribution and transport. Herbicide active ingredients, formulation type, ambient temperature, and humidity can influence volatility (Ouse et al. 2018). The higher the VP, the greater is the potential for herbicide volatilization. Synthetic auxin herbicides and carotenoid inhibitors (i.e., clomazone) are commonly volatilized in the field since they have high VP (Table 4.2). Techniques such as nanoparticle encapsulation, formulation types, and soil incorporation of herbicides contribute to minimizing volatilization losses. It is important to mention that the volatilization of herbicides also can be measured by Henry's law constant (K_H).

4.3.6 Water Solubility (S_w)

The S_w is a measurement of herbicide concentration that can dissolve in water at a specific temperature. The S_w is commonly measured in mg/L. According to Ney (1998), S_w is considered low when the value is less than 10 mg/L, moderate when the value ranges from 10 to 1,000 mg/L, and high when the value is higher than 1,000 mg L^{-1} (Figure 4.2). The S_w is one of the most important properties affecting bioavailability to weed control and the environmental fate of herbicides, mainly by transport through leaching. The higher S_w of the herbicide the higher its bioavailability in the soil solution, i.e. the lower its sorption into the soil colloids. On the other hand, in this

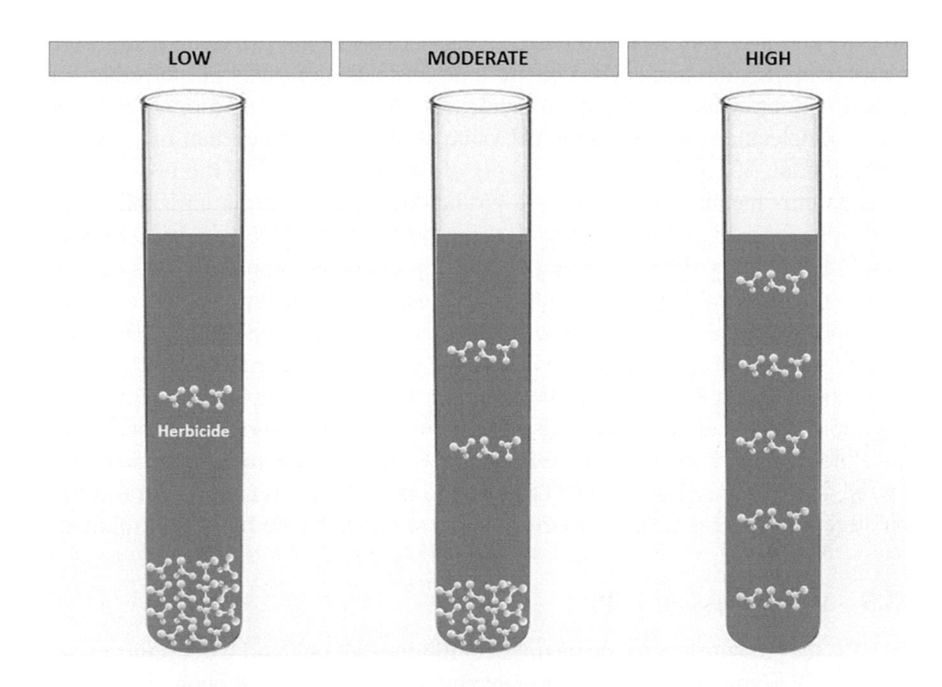

FIGURE 4.2 Water solubility (S_w) of herbicides schematized into low, moderate, and high.

condition, the herbicide can more easily be leachate and contaminate the groundwater. It is important to mention that leaching is necessary, only in the topsoil, where there is absorption of herbicide by the weed seed bank. Hexazinone and imazethapyr are herbicides applied directly to the soil and have high S_w; however, the pendimethalin is highly sorbed to the soil due to its low S_w (Table 4.2).

4.4 INTERACTION MECHANISMS

Due to the wide structural chemical diversity of herbicides and biochar, as well as the numerous processes for obtaining them (biochar), the possibilities of interaction between these molecules are varied, resulting in differentiated interaction mechanisms that have not yet been fully elucidated. The nature of these interactions can be exclusively physical, chemical, or both and result in the phenomenon of sorption. The predominance of one mechanism over another is the result of the combination of the properties and structural characteristics of both involved (porosity, SSA, hydrophobicity, ion exchange capacity, functional groups, aromaticity, etc.) and the conditions provided by the medium (concentration, pH, and temperature) (Kookana 2010; Yang et al. 2018). In order to predict the most likely mechanisms for a given sorption system, it is of fundamental importance to associate the properties and concepts of intermolecular interactions, thermodynamic aspects, structure, and composition of those involved (Hao et al. 2013; Sun et al. 2010; Xiao and Pignatello 2015b). As an example of the plurality of possible mechanisms, Figure 4.3 (Zhang, Sun, Ren et al. 2018) presents the main forms of interaction identified between herbicide and biochar.

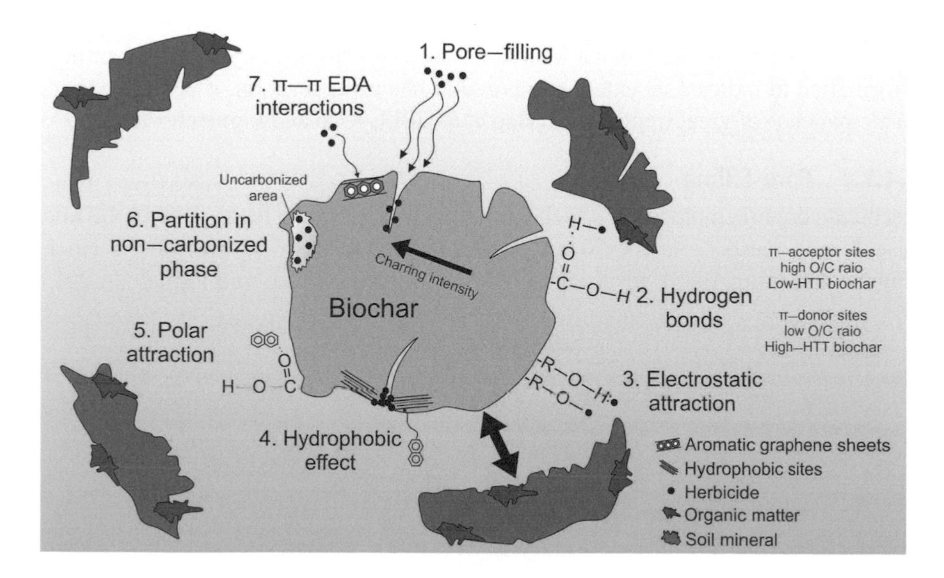

FIGURE 4.3 Interaction mechanisms between herbicides and biochar. HTT = Highest Heat Temperature.

In general consensus, carbonated sorbents can exhibit at least two distinct sorption domains: an 'organic', 'soft', 'rubbery domain', 'soft polymeric' or 'gel' species that are associated with linear, and other 'dark', 'hard', 'glassy' (glassy domain) 'polymeric' and/or 'microporous' sorption isotherms that exhibit nonlinear sorption isotherms (Ahn et al. 2005; Pignatello and Xing 1996). In the first case, the carbonated material is considered flexible enough to accommodate a solute similar to a liquid organic solvent through the partition mechanism (mass transfer occurring between two immiscible phases). In the second case, the carbonated matrix is notably more rigid, and sorption presumably occurs on the surface and in cavities and micropores of molecular size, as there is no homogeneity in this process, and the isotherms exhibit nonlinear behavior (Ahn et al. 2005; Pignatello and Xing 1996). In this chapter, greater focus will be given to the herbicide–biochar interaction mechanisms. However, a summary of the main thermodynamic sorption models (sorption isotherms) obtained in recent research can be found in Xiao and Chen (2016).

4.4.1 Physisorption

Physisorption, or physical sorption, refers to the process of surface interaction of an exclusively physical nature in which there is a predominance of entrapment due to the pore size ratio existing in the sorbent material and the sorbate molecules and weak intermolecular stabilization forces (Coulomb and van der Waals) and intermediate (Hydrogen bonds). As it is a process dependent on the number of pores and surface charges existing in the sorbent material, physisorption has a limited extent, being capable of reversal in particular cases, as it depends largely on electrostatic contact interactions (weak interactions). Physisorption is characterized by being a fast and poorly selective process – that is, very distinct molecules can interact with the sorbent simultaneously. It is noteworthy that although the available data on the participation of physical sorption in the herbicide–biochar interaction mechanisms do not yet provide sufficient information to indicate its extent compared to chemical sorption, it is known that its participation is of great importance (Hao et al. 2013; Xiao and Pignatello 2015b).

4.4.1.1 Pore-Filling

Herbicide organic molecules, which vary greatly in size, can be trapped in the micro (x < 2 nm), meso (2 < x < 50 nm), and macropores (x > 50 nm) of the biochar (Figure 4.4) (Sing 1985), resulting in physical sorption (Xiao and Pignatello 2015b;

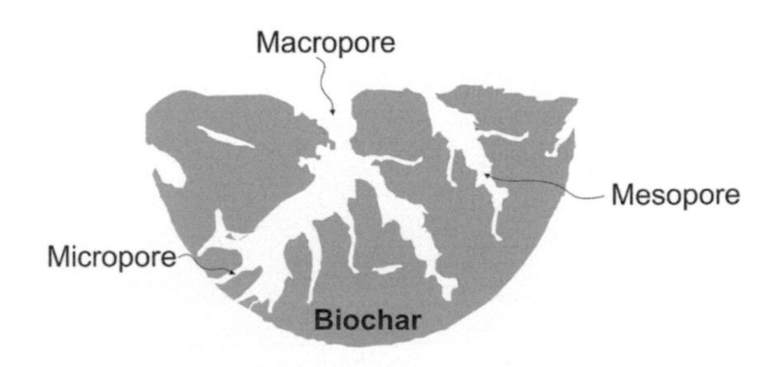

FIGURE 4.4 Porosity distribution in biochar (Beniak et al. 2001).

Yang et al. 2016; Yang et al. 2018). Obtaining a surface conducive to the interaction and accommodation of molecules have been extensively reported as a result of biochar production parameters and biomass type (Mohan et al. 2014; Tripathi et al. 2016; Xiao and Chen 2016). Among these parameters, the pyrolysis condition is the most focused.

The presence and distribution of pores can act as the first mechanism for capturing molecules through their filling. The significant increases in SSA, porosity, and the number of pores with increasing temperature and pyrolysis time (Chun et al. 2004; Yu et al. 2006; Sun et al. 2012; Mohan et al. 2014) reflect the increase in pyrolysis sorption capacity of the material. The composition in terms of carbon content and aromaticity also tends to increase (Hao et al. 2013; Smernik et al. 2006; Spokas 2010). It is also possible that pores participate in and favor the partition mechanism, as they are actively involved in increased diffusion (Ahn et al. 2005; Pignatello and Xing 1996; Xiao and Pignatello 2015b) and may influence the sorption of other molecules by competition and blocking action (Chen et al. 2012; Xiao and Pignatello 2015a, 2015b).

The pore-filling mechanism is strongly affected by its morphology. This property was verified in the study by Yang et al. (2018) by transmission electron microscopy (TEM) in which biochar pores derived from different materials (wood, rice husk, bamboo chips, cellulose, lignin, and chitin) were pyrolyzed at 700°C, and they presented the shape of cleft and played a significant role in the sorption of aromatic compounds (polycyclic aromatic hydrocarbon, phenols, anilines, nitrobenzenes) and that, after sorption, the deformation of the pores occurred making them narrower for other molecules. This characteristic corroborates another aspect generally pointed out, the sieving effect, in which the filling of pores can govern adsorbate sorption as a consequence of molecular size and steric effect, with biochar of smaller pores being more strongly influenced (Xiao and Pignatello 2015b; Yang et al. 2018). Sorbate concentration can also favor the occurrence of the pore-filling mechanism. At low concentrations of herbicide and other organic molecules in solution, biochar with low volatile compounds and small pores can dominate through this mechanism (Shen et al. 2020).

In short, understanding the structural aspects of both biochar and herbicides has led to a better understanding of the interaction behavior and imprisonment extension. As an example, in the study by Wei et al. (2017), the sorption of the herbicide metolachlor in rice-husk-derived biochar was predominated by the pore-filling mechanism and hydrogen bonds, whose filling mechanism was dominant in biochar produced at higher temperatures (500 and 750°C), a condition that provided a larger surface area with smaller pores, susceptible to herbicide trapping. Studies on this front are needed but still are scarce.

4.4.1.2 Coulombic Force

The trapping of the herbicide in the biochar pores can be considered a consequence of the flow or eventuality of contact between both molecules. In the existence of molecules containing charge, the sorption phenomenon opens up different possibilities and intensities of interaction, among them being the Coulomb force. Coulomb force refers to the effect of attraction (opposite charges) or repulsion (equal charges) caused by the electrical charge of particles, ions, or small molecules at a short distance.

The intensity of this interaction is dependent on particle charge and size. Larger loads and smaller radii result in a greater magnitude of interaction. In ionic systems, cations end up standing out from anions in terms of interaction magnitude even when they have the same charge since they are naturally deficient in electrons, and this happens because cations provide a greater distortion of the electronic cloud of neighboring particles (Chang 2010). Protonation (a term that will be better discussed in chemisorption) of atrazine and simazine molecules in the face of a reduction in pH is an example of the occurrence of the Coulomb attraction effect (Zheng et al. 2010). At low pH values, these herbicides express the net positive charge and the biochar negative charge over the widest pH range (Yuan and Xu 2011; Yang et al. 2020), favoring up to 1.5 times the sorption capacity (Figure 4.5) (Shows and Olesik 2000; Zheng et al. 2010; Xiao and Pignatello 2015b).

4.4.1.3 van der Waals Interactions

Coulomb forces explain the effect of attraction and repulsion between charges, but they do not explain the attraction behavior between large and unsaturated molecules. This happens because the molecules do not have a completely homogeneous charge. Each atom has a certain electronic configuration, and together they generate a vector of resulting value characteristic of the arrangement and balance of charges in the molecule – this vector is called net dipole moment. When the resulting dipole value is null, each vector in the different bonds vanishes, and the molecule is called nonpolar (uncharged). However, when the resulting dipole is not null, the molecule presents more electronegative regions (greater electron attraction), and, therefore, poles are generated, calling it the polar molecule. Paraquat, being an ion, and atrazine exemplifies this property (Figure 4.6).

The possibility of interaction of electronic clouds of nonpolar and polar molecules with each other characterizes the famous intermolecular forces of van der Waals:

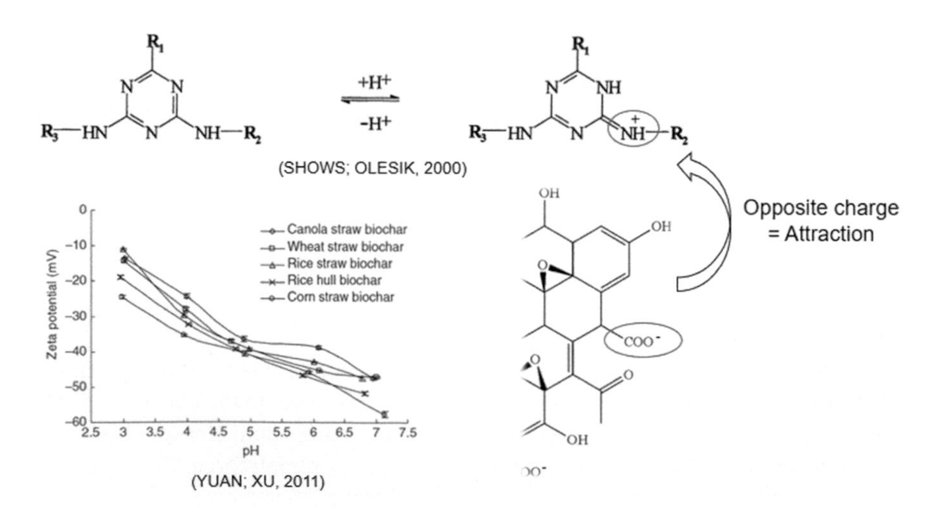

FIGURE 4.5 Coulomb attraction between atrazine and biochar in pH values below 7.

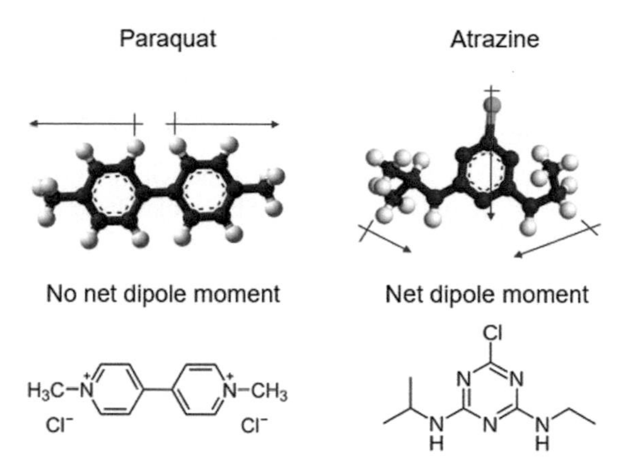

FIGURE 4.6 Electric dipole moment of atrazine and paraquat molecule.

FIGURE 4.7 van der Waals interactions between a biochar surface fragment and the paraquat and atrazine molecule.

dipole–dipole, induced dipole–dipole, and dispersion forces. In the dipole–dipole force, two polar molecules interact through the opposite charges of their resulting dipoles. In induced dipole–dipole force, the resulting dipole of a polar molecule induces a polarization in a nearby nonpolar molecule, forming an induced dipole. Scattering forces (or London forces) are interactions between nonpolar molecules that arise from the fluctuation in the distribution of electrons in atoms created from temporary induced dipoles (instant dipole), which arise from momentary spatial arrangements.

Ions can also induce the formation of dipoles, ion-dipoles. However, this induction is not considered a van der Waals force (Chang 2010). Figure 4.7 exemplifies the van der Waals interactions between a biochar surface fragment and the molecules of paraquat and atrazine.

In general, these types of interactions are frequently reported in the occurrence of physical sorption, justifying part of the sorption of neutral organic molecules to organic matter (Pignatello and Xing 1996), as has been reported between glyphosate and rice-husk-derived biochar activated by vapor (Herath et al. 2016) and of diverse organic molecules in carbon nanotubes (Yang and Xing 2010).

4.4.1.4 Hydrogen Bond

The presence of functional groups such as carboxylic acids (-COOH), alcohol/phenol (-OH), ammonium (-NH_3^+), and thiol traces (-SH), which have the elements oxygen and nitrogen bonded to hydrogen, enables biochar and herbicides to have a special type of intermolecular interaction, the hydrogen bond (Sander and Pignatello 2005). This type of bond is the one with the greatest strength and stability among the intermolecular interactions. This is a special type of dipole–dipole interaction that occurs between hydrogen atoms and more electronegative atoms like O, N, and F in polar bonds like N–H, O–H, or F–H. Hydrogen bonds justify deviations from the ideality of physical states and boiling points for various organic compounds.

Sun et al. (2010), in their study with atrazine molecules and their sorption in soil organic matter fractions, reported the occurrence of sorption to the large hydrophobic structure of organic matter as a result of the occurrence and stabilization via hydrogen bonding. In the same herbicide molecule (atrazine) sorbed in biochar derived from leonardite, a reduction in sorption was observed, as a consequence of the reduction of OH bonds and an increase in CH bonds with the increase in temperature process of the sorbent, which significantly reduced the possibility of hydrogen bonding (Chokejaroenrat et al. 2020; Sakulthaew et al. 2021). This type of interaction has also been reported for nicosulfuron in biochars derived from macadamia flakes and peanut husks (Trigo et al. 2014; Wang, Liu et al. 2020).

The determination of the existence or occurrence of hydrogen bonds between biochar and herbicide can be strongly influenced by competition with hydration water. Water molecules can form hydrogen bonds with the polar surface of the biochar and therefore can compete for sorption sites with contaminating molecules (Shen et al. 2020). It is noteworthy that this effect is difficult to characterize, requiring special attention to contaminant studies, and it is recommended to work with polar contaminants to ensure that this effect is studied or isolated when trying to optimize the sorption of polar herbicides such as glyphosate (Shen et al. 2020).

4.4.2 Chemisorption

Chemisorption, or chemical sorption, refers to the sorption process in which the forces involved are valence forces of the same type, which operate in the formation of compounds. This phenomenon is characterized by its specificity, change in electronic state that can be detected by different methods (e.g., infrared, microwave, electrical conductivity, and magnetic susceptibility), and the tendency of the sorbent to dissociate and reactions that can result in a change in the chemical nature of the sorption. The energy involved in the process is of the same magnitude as a chemical reaction, requiring a minimum energy (activation energy) for sorption (bonding) to occur. Because there is a chemical bond between the sorbent and the molecule, desorption becomes much more difficult, being referred to in many cases as an irreversible desorption process (Braida et al. 2003). The mechanisms for occurrence are varied and still not completely deciphered. However, some mechanisms are predominant in the literature.

4.4.2.1 Protonation of the Herbicide

Similar to organic matter, biochar has a variable surface charge with pH, with the negative charge predominating over a wide pH range, as can be seen in Figure 4.8 by the negative potentials of fine and coarse-textured biochar derived from (a) *Quercus lobata*, (b) sugarcane bagasse, and (c) peanut husks, whose negative potentials predominate at pH greater than 2, covering most soil types (Ahmad et al. 2014; Yang et al. 2020).

Formulations containing herbicide molecules can be found with a positive (cationic), negative (anionic), and/or neutral charge, as is already described in the item on physicochemical properties of herbicides. The intensification or increase in the occurrence of herbicide sorption to the biochar structure can occur with protonation.

Protonation is the addition of H^+ protons to a molecule or ion giving rise to a conjugate acid that is poorer in electrons than the parent molecule. This protonation effect can create different electrically positive sites on the herbicide molecules, as occurs for glyphosate (strong acid) (Figure 4.9), increasing the electrostatic attraction

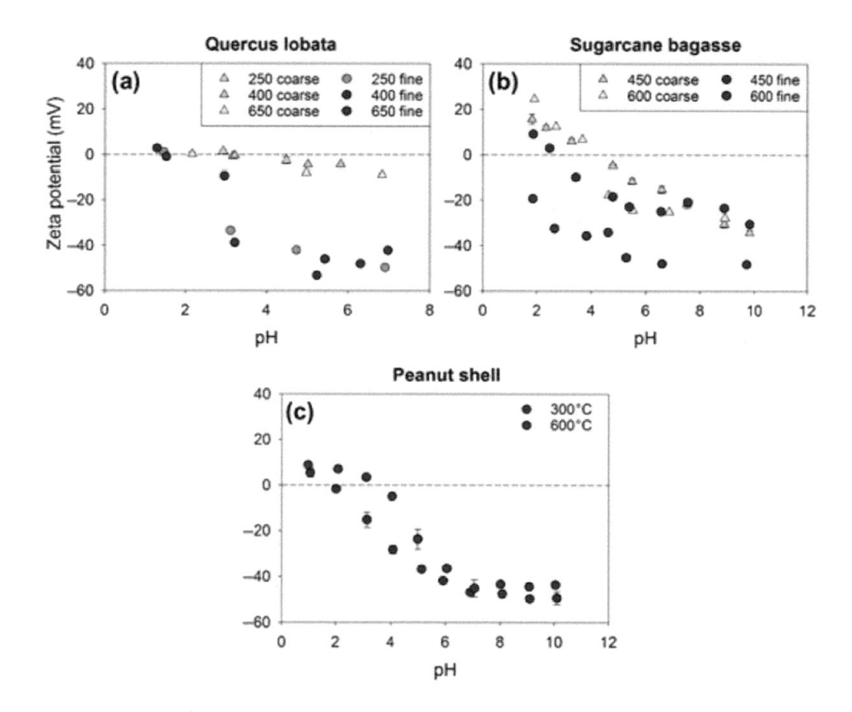

FIGURE 4.8 Zeta potential of biochar as a function of the solution pH (Yang et al. 2020).

FIGURE 4.9 Successive protonations of a fully deprotonated glyphosate molecule (Liu et al. 2016).

with the biochar. Protonation easily occurs in different environments (soil, water, etc.) as a result of changes in the pH of the medium, acid–base reaction, and microbial decomposition action (Villaverde et al. 2018). Naturally acidic soils, as they occur in tropical regions, have the strong protonation effect of herbicide molecules of the chemical group of triazines, increasing their sorption to soil organic matter and reducing the availability of the herbicide in the soil solution for absorption by the plant (Fuscaldo et al. 1999). The effect of this addition of protons can result in quite different behavior, such as increased solubility, greater potential for sorption to organic matter, variation in the herbicide degradation time, and reactivity of its decomposition by-products (Colombini et al. 1998; Song et al. 2018). The result of the Coulomb attraction between the opposite charges of the herbicide–biochar interaction often results in the formation of a covalent bond, establishing chemical sorption.

4.4.2.2 Electron Donor–Acceptor (EDA) Mechanisms

The interaction of the protonated herbicide with the biochar can result in chemical bonding, but, in other cases, the sorption of the herbicide on carbon-rich materials can occur through a complexation effect. A molecular complex is a defined stoichiometry association between two or more molecules, largely with closed-shell electronic structures (aromatic ring), and can easily be seen as being composed of an electron donor molecule and an electron acceptor molecule. Electron donor molecules can be either unsaturated or aromatic compounds (π-donor) or compounds carrying a surrogate containing a pair of free electrons (n-donor). The π-donor's ability to form a complex is enhanced by substituents that increase the electron density of the system by positive ring induction/polarization (Figure 4.10a) or resonance (Figure 4.10b). Electron acceptor molecules are generally classified as inorganic or π-acid acceptors and unsaturated or aromatic systems carrying an electron-withdrawing substitute (Figure 4.10c and d) (Hutzinger et al. 1971). Complexation has a great influence on

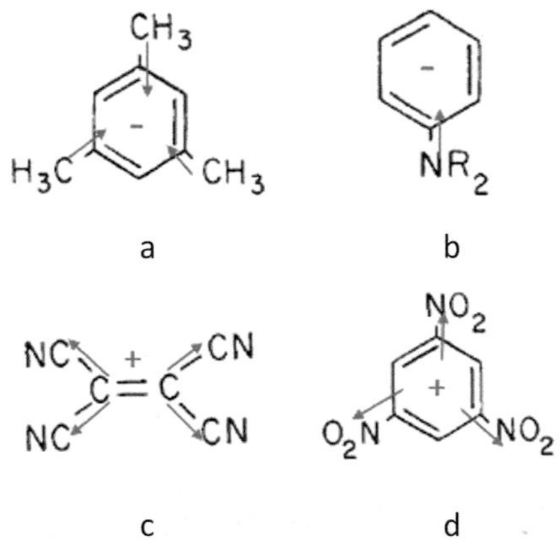

FIGURE 4.10 Examples of donors (a and b) (1,3,5-methyl benzene and dialkylarylamine) and electron acceptors (c and d) (tetracyanoethylene and 1,3,5-trinitrobenzene) (Hutzinger et al. 1971).

chemisorption as the lower activation energy involved in complexation induces further σ bonds (Hutzinger et al. 1971).

The mechanisms for the occurrence of EDA complex formation comprise four main categories: π–π, π–cation, π–anion, and polar–π interaction. The formation of these complexes has been reported in the interaction mechanisms between organic molecules, herbicides, and carbon-rich derivatives (Sander and Pignatello 2005; Zhu and Pignatello 2005; Lattao et al. 2014; Xiao and Pignatello 2015a; Herath et al. 2016; Yang et al. 2018; Wang, Liu et al. 2020).

4.4.2.3 π Interaction

The σ bonds are those that occur in atomic orbitals and occupy the same cartesian axis (Figure 4.11), while π bonds occur when atomic orbitals on different axes interpenetrate in parallel. Single sticky bonds are the result of σ bonds, and double (1 σ and 1 π) and triple (1 σ and 2 π) bonds add to the existing σ bond for the establishment of one or two new bonds (π). In general, π bonds are naturally weaker than σ bonds, and the compounds formed by this type of bond also end up being less stable (March 1992). However, the π bonds take considerable participation in the interaction mechanisms between organic molecules in the sorption mechanisms.

The conformation of the orbitals involved in the π bond gives the molecule a region susceptible to interaction. If the EDA mechanism is considered, unsaturated and aromatic molecules may tend to donate or receive electrons depending on the substituents that compose them. In the presence of slightly electronegative substituents, the orbitals involved in π bonds present polarization toward the center of the aromatic ring, resulting in a great capacity for interaction with other molecules, which can act as donors. Whereas molecules, with highly electronegative substituents at their end, tend to deform the electron cloud of the π bond, making it deficient in electrons and making the ring capable of acting as an electron acceptor (Figure 4.10). If the π bonds interact, the occurrence of a donor–acceptor interaction is very likely, resulting in the formation of an organic complex.

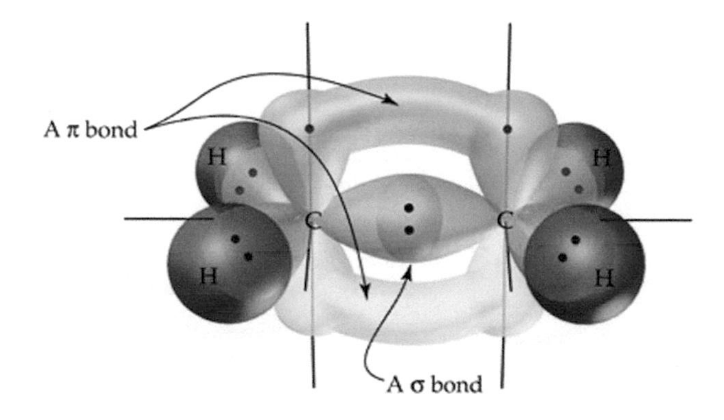

FIGURE 4.11 Sigma and pi bonds.

Source: https://socratic.org/questions/are-pi-bonds-easier-to-break.

4.4.2.3.1 π–π Interaction

Biochars are highly disordered, condensed, and highly disordered polycyclic aromatic layers (Harris 1999; Sander and Pignatello 2005). These layers vary in size, degree of functionality, and the extent to which pentagon and hexagon rings are distributed across the hexagonal carbon structure (Harris 1999; Sander and Pignatello 2005). Herbicides lack the structural complexity of biochars but are largely made up of aromatic rings or double and triple bonds, often associated with highly electronegative elements such as N, O, F, and Cl. The presence of π bonds, therefore, is a feature common to both.

When two π orbitals of a system of molecules interact, an interaction commonly symbolized by π–π or π^+–π is established. This very frequent interaction in molecules that have an aromatic ring can occur in three different ways: parallel displaced (displaced or offset-staked), edge-to-face or T-shaped, and overlapped (sandwich or stacked) (Figure 4.12a, b, and c) (Hohenstein and Sherrill 2009). The predominance of a given conformation to a system of molecules is strongly influenced by the energy resulting from complexation, as seen in Figure 4.12 between benzene rings (Banerjee et al. 2019). This type of interaction justifies the conformation of diamond, graphite, and DNA (West 2018).

The occurrence of π–π interaction has explained cases in which the sorption of organic molecules is greater than expected and whose possibilities of interaction by other mechanisms (van der Waals and hydrogen bonds) do not explain the observed behavior. This characteristic was pointed out by Sander and Pignatello (2005) in their study with the molecules of benzene, toluene, and nitrobenzene sorbed in biochar. While toluene and benzene molecules had practically the same sorption capacities, when a system calibration was performed by choosing an inert solvent, nitrobenzene indicated a certain interaction specificity, presenting a higher sorption capacity than the other two molecules (Sander and Pignatello 2005). The possibility of this specificity having been caused by hydrogen bonding was discarded with the study of the increase in pH (from 6.5 to 11), and the results of hydrogen nuclear magnetic resonance (H-NMR) showed complexation with polyaromatic units that make up the biochar (graphene: naphthalene, phenanthrene, and pyrin), showing strong evidence of complexation via π–π donor–acceptor interactions (Sander and Pignatello 2005).

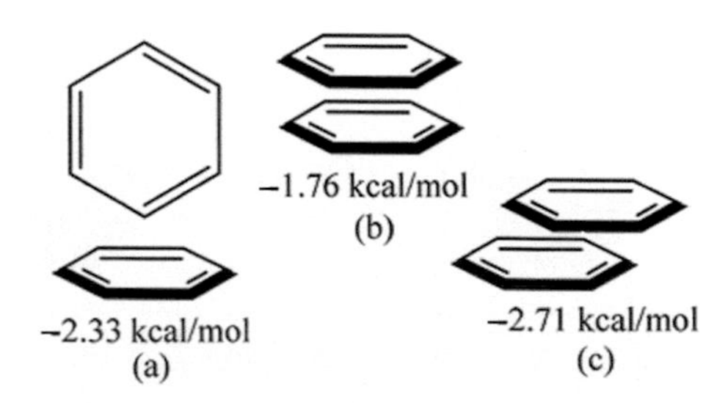

FIGURE 4.12 π–π interaction forms of the aromatic ring: (a) edge-to-phase (T-shaped), (b) sandwich (staked), and (c) displaced (offset-staked) (Banerjee et al. 2019).

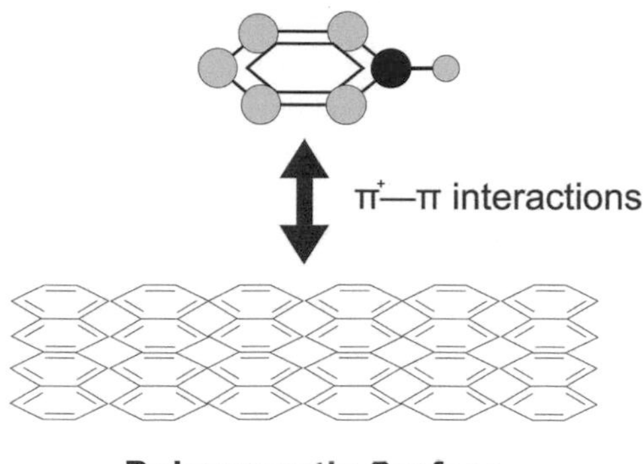

FIGURE 4.13 π–π interaction between herbicides containing amine groups and carbonated materials (Xiao and Pignatello 2015a).

Continuing the studies of this type of interaction, Xiao and Pignatello (2015a) found great potential for the interaction of protonated molecules containing an amine functional group (like many herbicides) to act as π-acceptors to form an EDA complex with polyaromatic surfaces of materials carbonates such as biochar (Figure 4.13). This same type of stabilization via π–π interaction was pointed out in the sorption of the atrazine molecule in corn-cob-derived biochar, whose higher pyrolysis temperatures (>400°C) indicated an increase in aromaticity, based on the FTIR results, enabling atrazine to act as a π-donor to biochar's aromatic C rings as a π-acceptor (Hao et al. 2013).

4.4.2.3.2 Cation–π Interaction

Another possible interaction of the EDA type is the π-cation. This interaction mechanism is non-covalent in which the surface of a π system interacts with adjacent cations (Dougherty 2013) and can also be described as a non-covalent bond between a monopole (cation) and a quadrupole (π system) (Singh and Das 2015).

Benzene and acetylene are frequent organic units in organic compounds and have the similar characteristic of not having a resulting dipole moment as a result of their molecular symmetry (Green 1974; Kalsi 2000). However, with a high amount of electrons, benzene's π orbitals can host a partial negative charge, which, in contact with a positive counterweight charge associated with the plane of the benzene atoms, results in an electric quadrupole (Kocman et al. 2014; Mallick et al. 2019). The negatively charged region of the quadrupole can then interact favorably with positively charged species. High-charge density cations (ionic potential) can result in a particularly strong effect. To exemplify this effect in soil under normal conditions, the commonly encountered high ionic potential cations are shown in red in Figure 4.14.

In organic compounds such as biochar, the most studied cation–π interactions involve the bond between an aromatic π system and alkaline and alkaline earth metals (Shen et al. 2020). In the interaction process, the cation is centered at the top of the π face, along the six-fold axis of the aromatic ring by van der Waals forces

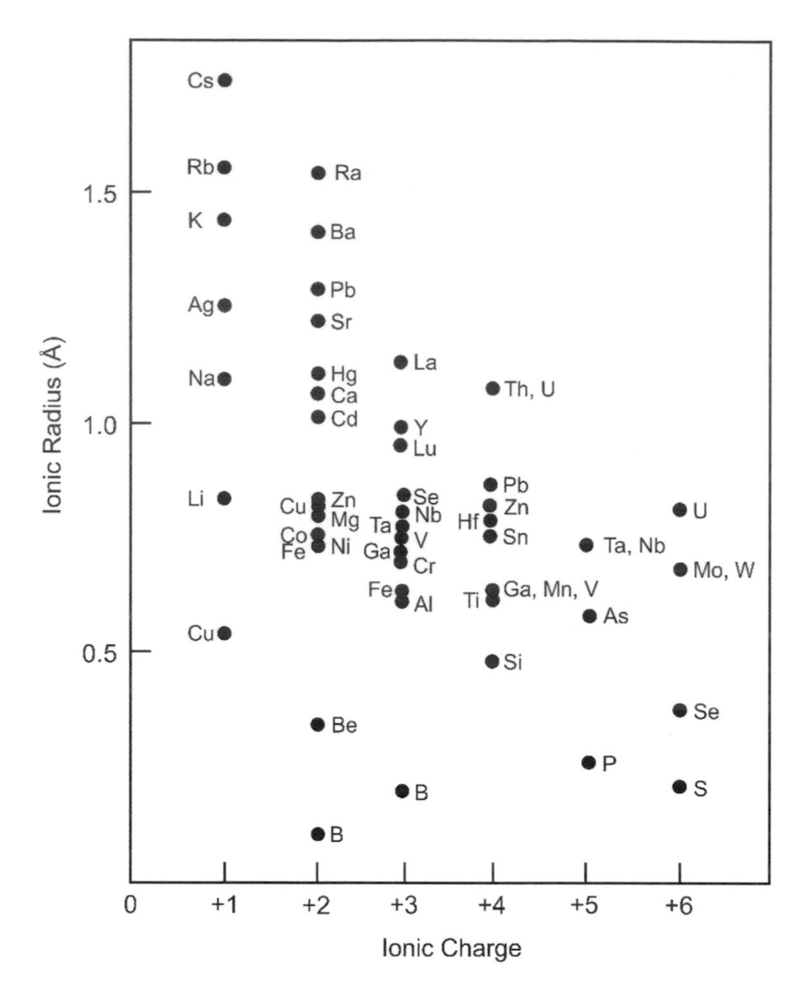

FIGURE 4.14 Diagram of the ionic potential of chemical elements. The main high ionic potential cations found in the soil are shown in red (Feng et al. 2018).

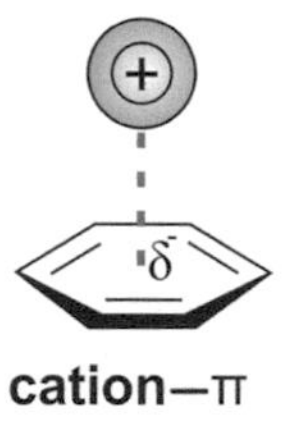

FIGURE 4.15 Spatial positioning of the cation and the π system (Forbes et al. 2017).

(Tsuzuki et al. 2001). Figure 4.15 illustrates the spatial positioning between the π and cation systems.

Some authors have shown that electrostatics dominate interactions in simple systems, and relative binding energies correlate well with electrostatic potential energy

(Mecozzi et al. 1996; Wheeler and Houk 2009). However, electrostatics are not the only component influencing the π–cation bond – for example, the pharmaceutical molecule 1,3,5-trifluorobenzene and the herbicide atrazine can interact with cations despite having a very low resulting dipole moment (Xiao and Pignatello 2015a; Frontera et al. 2011; Garau et al. 2004). This means that even in the absence of electrostatic forces, the charge interaction model ends up being used to predict the relative binding energies since a wide variety of aromatic hydrocarbons exhibit this behavior (Phipps et al. 2015; Wheeler 2013).

In addition to electrostatic attraction, other interaction mechanisms contribute to the occurrence of interaction, including polarization, EDA, and charge transfer. However, these mechanisms are still not well understood, as energy trends do not directly relate to the ability of aromatic hydrocarbons, and cations to take advantage of these effects. For example, if induced dipole were a control effect, aliphatic compounds such as cyclohexane should be good π-cation partners, but they are not (Mecozzi et al. 1996). Even more challenging is to relate the energy transfer of these mechanisms in a complex system such as the soil since there are numerous methodological limitations to isolating and knowing the dominant mechanism in herbicide sorption.

As mentioned earlier, the cation–π interaction is non-covalent and therefore making it fundamentally different from the bond between transition metals and π systems. Transition metals can share electron density with π systems via d orbitals, creating bonds that are highly covalent and cannot be modeled as a cation–π interaction (Dougherty 2013; Neel et al. 2017). This information becomes valuable for future studies that intend to investigate cation–π interaction mechanisms through speciation of unclassified cations, such as transition metals.

4.4.2.3.3 Polar–π Interaction, Hydrophobicity, and Partitioning

The hydrophobicity of biochar is commonly described to be as one of its most favorable properties for sorption, and many studies have noted the importance of hydrophobic effects in immobilizing organic contaminants (Han et al. 2017; Zhang et al. 2014). Typically, the hydrophobicity of biochar is affected by pyrolysis conditions. In general, biochars pyrolyzed at high temperatures have low polarity (low O/C ratio) and therefore demonstrate greater hydrophobic effects (Hassan et al. 2020; Tomczyk et al. 2020).

When the molecule is polar, the formation of a cluster of water on the surface of the biochar is favored (Gámiz et al. 2019; Xiao et al. 2018). The presence of water on the solvent surface enables the formation of a polar–π bond and can make it difficult for the herbicide to access high-energy sorption sites. Furthermore, the presence of water on the surface of the biochar or herbicide favors the partition in the non-pyrolyzed phase of the biochar.

In general, the partitioning mechanism becomes important in biochar as there is a reduction in the specific surface area in the OC content and an increase in the ratio (O+N/C). Partitioning or partitioning between herbicide and biochar occurs in specific structural parts of the biochar that have not been carbonized, or as a result of partial carbonization. As these properties are strongly related to the raw material used in pyrolysis, it is possible to state that the various sorption and partition mechanisms are strongly influenced by the raw material used in the production of biochar (Schreiter et al. 2018).

In practice, the sorption of herbicides and other organic compounds through hydrophobic and polar–π interaction encompasses both partition and hydrophobic sorption (Inyang and Dickenson 2015). However, the distribution of sorption and partition depends on the aging process of the biochar. For example, fresh biochars with low surface oxidation are hydrophobic and can sorb hydrophobic organic compounds by partitioning and hydrophobic adsorption mechanisms (Chen et al. 2008; Fang et al. 2014). However, as biochar ages in the soil, surface oxidation occurs, which reduces its capacity to absorb hydrophobic compounds.

Compared to partitioning, the hydrophobic sorption process takes place on sorbent surfaces with low hydration energy, as a result of direct competition between the water molecules and the nonpolar sorbed molecule (Zhu et al. 2005). An example of this is the sorption of alachlor, metolachlor, and norflurazon in montmorillonite coated with benzyl trimethylammonium (BTMA) (Nir et al. 2000) and cyhalofop in grape wood biochar modified with hydrogen peroxide (Gámiz et al. 2019).

4.4.2.4 Addition and Nucleophilic Substitution

At the end, a particular case of possible interaction between biochar and herbicide is nucleophilic addition and substitution. A nucleophile is a chemical species capable of donating a pair of electrons to form a bond. Generally, any ion or molecule with a free electron pair or at least one π bond can act as a nucleophile (Mayr and Patz 1994; Ritchie 1972). Nucleophilic addition is the interaction of a nucleophile with an unsaturated chemical compound (double or triple bond), resulting in the formation of a saturated compound. In the nucleophilic addition reaction, the π bond of the substrate gives rise to two new bonds of a covalent character (Swain and Scott 1953). The nucleophilic addition and substitution mechanism are exemplified in Figure 4.16.

Addition reactions involve an increase in the number of groups attached to the biochar and therefore a decrease in the degree of unsaturation. This process involves the gain of two groups or atoms at each end of a π bond or the ends of the π system (Kalsi 2000).

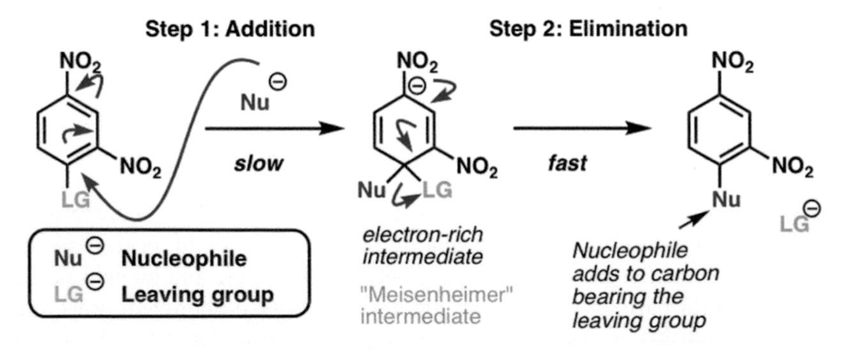

FIGURE 4.16 Mechanism of nucleophilic addition and substitution (Ashenhurst 2021).

Whereas nucleophilic substitution is a chemical reaction in which the nucleophile replaces a functional group of another electron-deficient molecule (Carey and Sundberg 2007). Possibly, in biochar, the main nucleophilic substitution occurs in an aromatic ring, also known as aromatic nucleophilic substitution. Nucleophilic aromatic substitution occurs through the addition of a nucleophile to the aromatic ring followed by the loss of a functional group (Kalsi 2000).

Among the mechanisms of herbicide sorption in biochar, the least-studied mechanism in herbicide is nucleophilic addition/substitution. However, for other organic compounds, such as nematicides, this same mechanism has been widely reported (Qin et al. 2019).

4.5 FACTORS AFFECTING HERBICIDES SORPTION ONTO BIOCHAR

In modern agriculture, herbicide consumption is necessary for weed control and has increased to meet the high food production of the global population (Duke and Dayan 2018; Ighalo et al. 2021). In contrast to this increase in consumption, the possibility of herbicides polluting ecosystems has increased (Schwarzenbach et al. 2010). For this reason, the search for materials for remediation has grown, and the great sorption potential of biochar has stimulated its use (Pignatello and Xing 1996; Beesley et al. 2011; Kookana et al. 2011; Ahmad et al. 2014; Zhang et al. 2013). However, as it is still a fairly new practice, herbicides (chemical structure and mode of action) and varied biochar (sources, ways of obtaining, and properties), as well as studies that are still scarce in terms of behavior in the face of climatic variations (temperature, concentration, competition for activated sites, and pore-blocking) and efficiency, its use on a large scale becomes a careful practice.

The use of biochar is considered to be a sustainable product and a soil quality modifier, and its use has been fundamental to remediate contaminated soils or improve the physicochemical conditions of soils with fertility or water retention problems (Yu et al. 2019; Ahmad et al. 2014). These changes in physicochemical quality can consequently interfere with the sorption of residual herbicides applied to the soil in the PRE. Therefore, understanding the factors that influence the mechanisms of action of these molecules in the presence of biochar is increasingly necessary.

4.5.1 Thermodynamic: Enthalpy, Entropy, and Free Energy (Gibbs)

Sorption is a thermodynamic process – that is, the occurrence of surface interaction or chemical bond formation between the herbicide and the biochar involves the conversion of heat and other forms of energy. There are not many studies, yet, that investigate the thermodynamic conditions of the sorption of these molecules by biochar (Gülen and Aslan 2020; Cusioli et al. 2021). However, among the main state functions (functions that depend only on the initial and final conditions), temperature is the most reported in sorption mechanisms. This is because the practical implications of the sorption mechanisms occur mostly at constant pressure (atmospheric pressure), and, under these conditions, the enthalpy variation (sum of internal energy with the product of system pressure and volume,

$\Delta H = \Delta E + P\Delta V$, where ΔH is the enthalpy, P is the pressure, and ΔV is the volume change) in a system will be a consequence of the internal energy, since there is little change in the volume of the system. The thermodynamic study of sorption can reveal aspects such as the spontaneity of an interaction, energy release/consumption (exo or endothermic), and the influence of the temperature of the environment (favored or unfavorable).

In thermodynamic studies, three state functions are determined: standard enthalpy variation of formation (ΔH°, J/mol), standard entropy variation [ΔS°, J/(mol.K)], and standard free energy variation of Gibbs (ΔG°, J/mol). Simply put, a positive variation of enthalpy indicates energy consumption, in which case the system is said to be endothermic. A negative change indicates the release of energy, usually in the form of heat, in which case the system is said to be exothermic. Endothermic processes are favored by the heating of the surroundings, and exothermic processes are unfavorable. Positive values of entropies indicate spontaneity of the sorption process and negative ones indicate that the process is not spontaneous in the proposed sense. Additionally, a zero value in entropy indicates a sorption equilibrium. It is noteworthy, however, that only the variation of the Gibbs free energy can accurately tell the sense of the spontaneity of the reaction system. Negative values of ΔG° mean that the reaction is spontaneous in the proposed direction, positive the reaction is not spontaneous in the forward direction, but spontaneous in the opposite direction, and null variations indicate equilibrium, which means that there is no global energy variation. Gibbs free energy can also inform about the predominance of physical (–20 to 0 kJ/mol), chemical (–80 to −400 kJ/mol), or both (–20 to −80 kJ/mol) sorption (Jaycock and Parfitt 1981). These functions can be estimated by the linear equation of van't Hoff:

$$\text{In}\, K_e = -\frac{\Delta H^0}{RT} + \frac{\Delta S^0}{R}$$

where K_e = Qe/Ce, R is the gas constant 8.314 J/(mol.K), and T is the absolute temperature (K).

Thermodynamic data from the study of carbonized chestnut pulp (Table 4.3) used in the sorption of the highly toxic herbicide, 2,4-diclophenylacetic acid (2,4-D), revealed to be a spontaneous, exothermic, and predominantly physical process at temperatures of 35, 45, and 55°C (Gülen and Aslan 2020). Furthermore, the increase in temperature reduced the herbicide's sorption capacity, corroborating what was expected because it is an exothermic process (Gülen and Aslan 2020).

The opposite effect of temperature was observed in the sorption of atrazine in biochar derived from the seeds of chemically modified *Moringa oleifera* Lam. – that is, the sorption was disadvantaged by the increase in the analyzed temperatures (25, 35, and 45°C). The thermodynamic data (Table 4.4) was obtained to confirm the pattern observed in the isotherms –that is, a spontaneous, endothermic, and physical and chemical sorption process.

The lack of thermodynamic studies of herbicide sorption in biochar makes it impossible to further explain the effects of environmental variations.

TABLE 4.3
Thermodynamic Results of 2,4-D Sorption in Carbonized Chestnut Pulp (Gülen and Aslan 2020).

Temperature (°C)	ΔG° (kJ mol^{-1})	ΔH° (kJ mol^{-1})	ΔS° (kJ mol^{-1})
35	−9.712	−1.938	0.025
45	−9.965		
55	−10.217		

TABLE 4.4
Thermodynamic Results of Atrazine Sorption on Chemically Modified Biochar Made of *Moringa oleifera* Lam (Cusioli et al. 2021).

Temperature (°C)	Temperature (K)	ΔG° (kJ mol^{-1})	ΔH° (kJ mol^{-1})	ΔS° (kJ mol^{-1})
25	298	−23.449	14.775	0.128
35	308	−24.441		
45	318	−26.075		

4.5.2 Environmental Implications: Soil and Climate

Several biogeochemical interactions can occur when biochar is applied to the soil. One of them is the structural and compositional change with time of residence, a process often called "aging", as already reported (Ahmad et al. 2014). In addition to soil pH, biochar aging and herbicide sorption capacity are especially affected by the presence of natural soil organic matter (NOM) and by oxidation reactions (Uchimiya et al. 2010). Pignatello et al. (2006) demonstrated that NOM can suppress the sorption of organic contaminants by blocking part of the biochar's pores. In terms of biochar–mineral interactions, it is possible that, over time, mineral surfaces may cover the reactive surfaces of the biochar and thus reduce its sorption capacity.

Studies by Ahangar et al. (2008) and Singh and Kookana (2009) based on diuron and insecticide sorption, before and after the removal of paramagnetic materials from carbon surfaces by hydrofluoric acid (HF) treatment (required for nuclear magnetic resonance studies), showed that sorption in natural biochar in soils increases two to three times after treatment with HF. This has been attributed to organomineral interactions that gradually mask the biochar's assortative surface with aging.

Resistance to decomposition and persistence in soil may be an advantageous characteristic of biochar, which, when applied to soil, can have a half-life of thousands of years (Wu et al. 2013). Given these characteristics, studies aimed at soil remediation point to the possible immobilization of contaminants for a longer period due to this recalcitrant effect. However, in agricultural areas, the long persistence of some herbicides in the soil can be a limiting factor for the germination and development of the crop of agronomic interest in rotation/succession, a process known as carryover.

In assessing the stability of biochar in soil, the H/C ratio has been suggested as an indicator (Han et al. 2016). Biochars with H/C ratios less than 0.7 are considered very stable, as aromatic structures are more difficult to degrade. The stability of biochars is influenced not only by the type of raw material and pyrolysis temperature (Gurwick et al. 2013) but also by the interaction of the biochar with the soil under specific climatic conditions (Herath et al. 2015).

Most of the studies that seek to elucidate the mechanisms that govern herbicide sorption in biochar take place under controlled laboratory conditions (temperature between 22 and 25°C) in which the main objective is to evaluate different raw materials and biochar pyrolysis conditions (Table 4.5). The few studies that evaluated the herbicide sorption capacity in biochar in real conditions showed that temperature and humidity are important factors but that they can act indirectly in the retention of biochar and herbicide in the soil, explained in the topic of thermodynamics.

The combination of temperature and humidity (precipitation) are factors that affect the structure and microbial community of the soil (Bundt et al. 2015). The increase in the microbial population and/or the enzymatic activity are the main means responsible for the biodegradation of the herbicide and biochar. Under conditions of high temperature and precipitation, the biodegradation of the herbicide is greater than in regions with low temperatures and/or low precipitation.

In soil, the biodegradation of biochar is less than that of herbicide. However, when the herbicide is sorbed in pyrolyzed biochar at a high temperature (> 500 °C), there is a reduction in hexazinone biodegradation (Fernandes et al. 2021). The reduction in herbicide biodegradation after sorption on biochar produced under these conditions is mainly due to the reduction in the herbicide's bioavailability to microorganisms. On the other hand, in pyrolyzed maize straw and pig manure biochars at low temperature (< 400°C), there may be a stimulus to biodegradation of the thiacloprid insecticide (Zhang, Sun, Ren et al. 2018). The stimulus may occur due to lower porosity and/or release of low molar mass organic compounds, stimulating the enzymatic activity of soil microorganisms (Oni et al. 2019).

Unfortunately, the role of biochar in the biodegradation of herbicides and/or the reduction of bioavailability is still a fairly new issue and not yet consistent, which makes large-scale applications for the remediation of contaminated soils difficult. Furthermore, the behavior of the biochar over the years (tens and thousands) remains uncertain (Lehmann et al. 2011). Some authors suggest minimal microbial actions on biochar over the years (Cui et al. 2013; Xu et al. 2014; Zhu et al. 2016); others such as Spokas et al. (2014) suggest that due to repeated swelling and cracking processes of the graphitic sheets of the biochar, there will be the disintegration of the biochar, resulting in micro-and nano-scale fragments that will be able to percolate in the soil.

4.5.3 Contact Time

Hydrogen bonds, already pointed out in this chapter as a mechanism for the interaction of biochar-herbicides, can increase according to the dynamics of soil, herbicide, and organic matter. The increase in hydrogen bonds is a consequence of the oxidation

TABLE 4.5
Summary of Herbicides Sorbed on Different Biochars and the Main Removal Mechanisms in Soil and Water Cited in the Scientific Literature.

Herbicide	Feedstock	Pyrolysis temperature	Amount of biochar applied	Removal performance (concentration range)[a]	Sorption mechanism	References
Atrazine	Corn cob (*Zea mays*)	350 °C	-	0.09 mg g^{-1} (0.05–30 mg L^{-1})	Hydrophobicity effect, π–π EDA, and the pore-filling	Hao et al. (2013)
		450 °C		0.18 mg g^{-1} (0.05–30 mg L^{-1})		
		550 °C		0.41 mg g^{-1} (0.05–30 mg L^{-1})		
		650 °C		0.73 mg g^{-1} (0.05–30 mg L^{-1})		
	Dairy manure	200 °C	0.2 g	0.02 mg g^{-1} (0–20 mg L^{-1})	Partitioning	Cao et al. (2009)
	Peanut-shell	300 °C	0.2 g	722 mg g^{-1} (0.25–30 mg L^{-1})	hydrophobic partition, π–π EDA, H-bonding, and pore-filling	Wang, Liu et al. (2020)
		450 °C		679 mg g^{-1} (0.25–30 mg L^{-1})		
		600 °C		655 mg g^{-1} (0.25–30 mg L^{-1})		
2,4-Dichlorophenoxyacetic (2,4-D)	Wood chips	600 °C	-	0.72 mg g^{-1} (0.1 mg L^{-1})	Surface sorption	Kearns et al. (2014)
	Corn cob (*Zea mays*)	600 °C	-	32 mg g^{-1} (20–60 mg L^{-1})	H-bonding	Binh and Nguyen (2020)
	Wheat straw	550 °C	0.1 g	42% (15 mg L^{-1})	Hydrophobic bonding	Ćwieląg-Piasecka et al. (2018)
Fluridone	Grass	300 °C	-	10 mg g^{-1} (0.1–10 mg L^{-1})	Partitioning on amorphous-C	Sun et al. (2011)
Glyphosate	*Gliricidia sepium*	700–1000 °C	1 g	44 mg g^{-1} (20 mg L^{-1})	π–π EDA interactions	Mayakaduwa et al. (2016)
Glyphosate	=Rice husk (*Oryza sativa*)	700 °C	0.5 g	123 mg g^{-1} (4–14 mg L^{-1})	Pore-filling, H-bonding, π–π EDA interactions	Herath et al. (2016)
Metolachlor	Wheat straw	550 °C	0.1 g	70% (10 mg L^{-1})	Hydrophobic bonding	Ćwieląg-Piasecka et al. (2018)
	Codonopsis pilosula (Dangshen) and *Angelica sinensis* (Danggui)	300 °C	0.05–0.1 g	0.32 mg g^{-1} (10–100 mg L^{-1})	Pore-filling and polar interaction	Wei et al. (2020)
		500 °C		0.32 mg g^{-1} (10–100 mg L^{-1})		
		750 °C		1.21 mg g^{-1} (10–100 mg L^{-1})		

(Continued)

TABLE 4.5 (Continued)

Herbicide	Feedstock	Pyrolysis temperature	Amount of biochar applied	Removal performance (concentration range)[a]	Sorption mechanism	References
	Rice husk (*Oryza sativa*)	300 °C 500 °C 750 °C	0.05–0.1 g	148 mg g^{-1} (10–100 mg L^{-1}) 222 mg g^{-1} (10–100 mg L^{-1}) 682 mg g^{-1} (10–100 mg L^{-1})	H-bonding and pore-filling	Wei et al. (2017)
Metribuzin	Switchgrass (*Panicum virgatum*)	425 °C	0.2 g	223 mg g^{-1} (40–100 mg L^{-1})	H-bonding and electrostatic interaction	Essandoh et al. (2017)
Nicosulfuron	Peanut-shell	300 °C 450 °C 600 °C	0.2 g	253 mg g^{-1} (0.5–50 mg L^{-1}) 370 mg g^{-1} (0.5–50 mg L^{-1}) 628 mg g^{-1} (0.5–50 mg L^{-1})	hydrophobic partition, π-π EDA, H-bonding, and pore-filling	Wang, Liu et al. (2020)
Paraquat	Swine manure	400 °C	0.1–0.3 g	4.8 mg g^{-1} (0.5–6 mg L^{-1})	Ion exchange	Tsai and Chen (2013)
4-Chloro-2-methylphenoxyacetic acid (MPCA)	Wheat straw	550 °C	0.1 g	38% (10 mg L^{-1})	Hydrophobic bonding	Ćwieląg-Piasecka et al. (2018)

[a] Herbicide concentration used in the study. – there is no information in the article.

of the biochar surface – that is, with oxidation, there is an increase in the oxygen atoms in the structure/surface of the sorbent, and, consequently, there are greater chances of the occurrence of binding of hydrogen (Wang, O'Connor et al. 2020).

Few studies under field conditions have discussed the role of biochar aging in hydrogen bonds. Franz et al. (2000) reported an increase in the sorption of various aromatic compounds in activated biochar after artificially increasing the oxygen content, and this increase occurred in specific parts of the structure (cyclohexane).

Due to its high contribution to overall sorption and strong binding strength, hydrogen bonding is likely to effectively immobilize polar organic contaminants. However, the increase in the abundance of functional groups containing oxygen on the biochar surface over time may promote hydrogen bonds both with the contaminant and with the water molecules, which determines the net effect of these interactions for a certain contaminant.

The interactions between biochar and clay mineral surfaces are similar to those between natural soil organic matter and clay mineral surfaces, especially for biochars with high mineral content (ash content). In the soil, interactions will depend on the type of clay (2:1, 1:1) and the distribution of functional groups in the clay (siloxane, OH) and organic matter (C=O, COOH, CN, C–O), furthermore, the polarity of these compounds and the composition and concentration of cations and anions in the soil solution (Kleber et al. 2007).

Spokas et al. (2009) and Wang et al. (2010) demonstrated that the addition of biochar in agricultural soils increases the sorption and decreases the movement of terbuthylazine, acetochlor, and atrazine in soils, but the knowledge of the organo-mineral interactions of biochar in soils and the consequences of aging biochars for interactions with herbicides are still uncertain. Kookana (2010) proposed that it is possible over time that mineral particles from the soil can cover the reactive surface of the biochar, thus reducing the sorption capacity of biochars for organic compounds such as herbicides. Yan and Sheng (2003), in their study with diuron, observed that the sorption of dissolved organic matter from the soil during the ash surface aging reduced diuron sorption by 50–60%.

4.5.4 Hysteresis

With the increase in the contact time of the biochar with herbicide and with the possible changes in the interaction mechanisms, there is a possibility of hysteresis. Hysteresis is the phenomenon in which part of the sorbed herbicide returns to the solution, and the desorption rate (usually lower) of the colloid molecule is different from the sorption rate (Mendes, de Sousa et al. 2021); that is, hysteresis represents the difference between sorption and desorption. Among the mechanisms studied in this chapter, the trapping of organic compounds in micropores or pore deformation mechanisms are the main causes of desorption hysteresis in biochars (Braida et al. 2003; Loganathan et al. 2009; Zhang et al. 2010).

Understanding the sorption and desorption processes of herbicides in the soil is highly relevant to predict their bioavailability, mobility, degradation, and possibility of contaminating other environmental compartments. The factors that determine the sorption and desorption processes of herbicides in the soil are the physical, chemical, and biological properties of the soil, the environmental conditions, as well as the physicochemical properties of the herbicide (Cha et al. 2016). Mendes et al. (2019) evaluated the sorption and desorption kinetics of diclosulam and pendimethalin in soil remedied with natural biochar and found that the hysteresis index (H) for both herbicides was close to 1, showing that the desorption process is reversible and that there was a difference between the sorption and desorption kinetics of the two herbicides.

Martin et al. (2012) observed marked changes in atrazine and diuron hysteresis as paper mill sludge (PM) and chicken litter (PL) biochars aged pyrolyzed at 550°C. The authors found that the soils recently corrected with biochars showed a two (PM) to five (PL) times increase in herbicide sorption compared to the uncorrected soil. Sorption was reversible in uncorrected soil, but sorption–desorption hysteresis was evident in soil corrected with fresh biochars.

The study of the behavior of herbicides in soils is already consolidated in the scientific community, and the use of biochar as a sorption mechanism has been increasing in use and importance. The biochar–soil–herbicide interaction alters the availability of the herbicide in the soil solution and is mainly governed by the pyrolysis conditions and the type of raw material used in the production of the biochar.

The methodologies that involve the study of sorption mechanisms are still scarce in the literature, and few studies have evaluated the sorption mechanisms of herbicide in biochar (Table 4.5). And the few studies that evaluated the sorption mechanisms showed that there is no standardization of methodology for the study of the mechanism in the same herbicide.

Due to experimental conditions, such as biochar: soil ratio and dose of herbicide used, some of the authors demonstrated that there was no consensus when compared to balanced batch studies proposed by the Organization for Economic Co-operation and Development (OECD) and the United States Environmental Protection Agency (EPA) (Mendes, de Sousa et al. 2021). Even in situations where the purpose of the use is similar, there is no consensus on the method of evaluation or pre-evaluations on dose or mechanism effect based on the residue of the herbicide in the soil or biochar pyrolysis conditions. For this reason, inferring the extent of occurrence of hysteresis for different types of herbicides and biochar is a great field to be studied.

4.6 CONCLUDING REMARKS

Biochar has significant effects on agricultural soils, in the environment, and herbicides' behavior. The properties that stimulate interest in the industry and research concerning biochar are its porous structure and chemical characteristics. In general, biochars of high specific surface area, microporosity, and aromaticity exhibit exceptional herbicides sorption ability.

When in contact with soil, the herbicide can be sorbed by a wide range of physicochemical mechanisms in which its molecules may become unavailable to weed control. The behavior of herbicides in the soil is driven by their physicochemical properties (K_{ow}, K_d, DT_{50}, pK_a/pK_b, VP, and S_w) combined with the chemical, physical, and biological attributes of the soil and its constituents such as biochar, besides the climatic conditions.

Due to the wide structural chemical diversity of herbicides and biochar, as well as the numerous processes of pyrolysis, the possibilities of interaction among these molecules are varied, resulting in differentiated interaction mechanisms that have not yet been fully elucidated. The nature of these interactions can be exclusively physical (pore-filling, Coulombic force, van der Waals, hydrogen bond), chemical (protonation, electron donor–acceptor, π interactions, addition, and nucleophilic substitution interaction), or both and result in the process of sorption. Under field conditions, the herbicide–biochar interaction mechanisms may vary according to contact time (aging) and hysteresis and other factors.

This chapter is the basis for understanding the effect of biochar amendments on herbicide sorption–desorption, leaching, degradation, remediation in contaminated environments, and weed control efficacy.

REFERENCES

Abiven, S., M. W. Schmidt, and J. Lehmann. 2014. Biochar by design. *Nature Geoscience* 7(5):326–327.

Ahangar, A. G., R. J. Smernik, R. S. Kookana, and D. J. Chittleborough. 2008. Separating the effects of organic matter – mineral interactions and organic matter chemistry on the sorption of diuron and phenanthrene. *Chemosphere* 72(6):886–890.

Ahmad, M., A. U. Rajapaksha, J. E. Lim, M. Zhang, N. Bolan, D. Mohan, M. Vithanage, S. S. Lee, and Y. S. Ok. 2014. Biochar as a sorbent for contaminant management in soil and water: A review. *Chemosphere* 99:19–33.

Ahn, S., D. Werner, H. K. Karapanagioti, D. R. McGlothlin, R. N. Zare, and R. G. Luthy. 2005. Phenanthrene and pyrene sorption and intraparticle diffusion in polyoxymethylene, coke, and activated carbon. *Environmental Science & Technology* 39(17):6516–6526.

Alam, M. S., and D. S. Alessi. 2019. Modeling the surface chemistry of biochars. In *Biochar from Biomass and Waste*, ed. Y. S. Ok, D. C. W. Tsang, N. Bolan, and J. M. Novak, 59–72. Amsterdam: Elsevier.

Amondham, W., P. Parkpian, C. Polprasert, R. D. Delaune, and A. Jugsujinda. 2006. Paraquat adsorption, degradation, and remobilization in tropical soils of Thailand. *Journal of Environmental Science and Health, Part B* 41:485–507.

Ashenhust, J. 2021. Nucleofilic aromatic substitution (2) – The benzyne mechanism. www.masterorganicchemistry.com/2018/09/17/nucleophilic-aromatic-substitution-2-benzyne/.

Banerjee, A., A. Saha, and B. K. Saha. 2019. Understanding the behavior of π-π interactions in crystal structures in light of geometry corrected statistical analysis: Similarities and differences with the theoretical models. *Crystal Growth and Design* 19(4):2245–2252.

Beesley, L., E. Moreno-Jiménez, J. L. Gomez-Eyles, E. Harris, B. Robinson, and T. Sizmur. 2011. A review of biochars' potential role in the remediation, revegetation and restoration of contaminated soils. *Environmental Pollution* 159(12):3269–3282.

Binh, Q. A., and H. H. Nguyen. 2020. Investigation the isotherm and kinetics of adsorption mechanism of herbicide 2,4-dichlorophenoxyacetic acid (2,4-D) on corn cob biochar. *Bioresource Technology Reports* 11:100520.

Biniak, S., A. Swiatkowski, and M. Pakula. 2001. Electrochemical studies of phenomena at active carbon-eletrolyte solution interfaces. In *Chemistry and Physics of Carbon: A series of advances*, ed. L. R. Radovic, 126–216. New York: Marcel Dekker.

Bolan, N. S., D. C. Adriano, A. Kunhikrishnan, T. James, R. McDowell, and N. Senesi. 2011. Dissolved organic matter: Biogeochemistry, dynamics, and environmental significance in soils, ed. D. L. Sparks. *Advances in Agronomy* 110:1–75.

Braida, W. J., J. J. Pignatello, Y. Lu, P. I. Ravikovitch, A. V. Neimark, and B. Xing. 2003. Sorption hysteresis of benzene in charcoal particles. *Environmental Science and Technology* 37(2):409–417.

Brassard, P., S. Godbout, V. Lévesque, J. H. Palacios, V. Raghavan, A. Ahmed, R. Hogue, T. Jeanne, and M. Verma. 2019. Biochar for soil amendment. In *Char and Carbon Materials Derived from Biomass*, ed. M. Jeguirin, and L. Limousy, 109–146. Amsterdam: Elsevier.

Brown, R. A., A. K. Kercher, T. H. Nguyen, D. C. Nagle, and W. P. Ball. 2006. Production and characterization of synthetic wood chars for use as surrogates for natural sorbents. *Organic Geochemistry* 37:321–333.

Bundt, A. C., L. A. Avila, A. Pivetta, D. Agostinetto, D. P. Dick, and P. Burauel. 2015. Imidazolinone degradation in soil in response to application history. *Planta Daninha* 33(2):341–349.

Camps-Arbestain, M., J. E. Amonette, B. Singh, T. Wang, and H. P. Schmidt. 2015. A biochar classification system and associated test methods. In *Biochar for Environmental Management: Science, Technology and Implementation*, ed. J. Lehmann, and S. Joseph, 165–194. New York: Routledge.

Cantrell, K. B., P. G. Hunt, M. Uchimiya, J. M. Novak, and K. S. Ro. 2012. Impact of pyrolysis temperature and manure source on physicochemical characteristics of biochar. *Bioresource Technology* 107:419–428.

Cao, X., L. Ma, B. Gao, and W. Harris. 2009. Dairy-manure derived biochar effectively sorbs lead and atrazine. *Environmental Science & Technology* 43(9):3285–3291.

Carey, F. A., and R. J. Sundberg. 2007. *Advanced Organic Chemistry: Part A: Structure and Mechanisms*. New York: Springer.

Cha, J. S., S. H. Park, S. C. Jung, C. Ryu, J. K. Jeon, M. C. Shin, and Y. K. Park. 2016. Production and utilization of biochar: A review. *Journal of Industrial and Engineering Chemistry* 40(1):1–15.

Chan, K. Y., and Z. Xu. 2009. Biochar: Nutrient properties and their enhancement. In *Biochar for Environmental Management: Science, Technology and Implementation*, ed. J. Lehmann, and S. Joseph, 99–116. New York: Routledge.

Chang, R. 2010. Forças Intermoleculares, Líquidos e Sólidos. In *Química Geral: Conceitos essenciais*. Translated by M. J. F. Rebelo, 4th. edition. Porto Alegre: McGraw Hill.

Chen, B., D. Zhou, and L. Zhu. 2008. Transitional adsorption and partition of nonpolar and polar aromatic contaminants by biochars of pine needles with different pyrolytic temperatures. *Environmental Science & Technology* 42(14):5137–5143.

Chen, Z., B. Chen, D. Zhou, and W. Chen. 2012. Bisolute sorption and thermodynamic behavior of organic pollutants to biomass-derived biochars at two pyrolytic temperatures. *Environmental Science & Technology* 46(22):12476–12483.

Cheng, C. H., J. Lehmann, and M. H. Engelhard. 2008. Natural oxidation of black carbon in soils: Changes in molecular form and surface charge along a climosequence. *Geochimica et Cosmochimica Acta* 72:1598–1610.

Chi, Y., G. Zhang, Y. Xiang, D. Cai, and Z. Wu. 2017. Fabrication of a temperature-controlled-release herbicide using a nanocomposite. *Sustainable Chemistry & Engineering* 5(6):4969–4975.

Chokejaroenrat, C., A. Watcharenwong, C. Sakulthaew, and A. Rittirat. 2020. Immobilization of atrazine using oxidized lignite amendments in agricultural soils. *Water, Air, & Soil Pollution* 231:1–15.

Chun, Y., G. Sheng, C. T. Chiou, and B. Xing. 2004. Compositions and sorptive properties of crop residue-derived chars. *Environmental Science & Technology* 38(17):4649–4655.

Colombini, M. P., R. Fuoco, S. Giannarelli, L. Pospıšil, and R. Trskova. 1998. Protonation and degradation reactions of s-triazine herbicides. *Microchemical Journal* 59(2):239–245.

Cui, L., J. Yan, Y. Yang, L. Li, G. Quan, C. Ding, T. Chen, Q. Fu, and A. Chang. 2013. Influence of biochar on microbial activities of heavy metals contaminated paddy fields. *BioResources* 8(4):5536–5548.

Cumming, H., and C. Rucker. 2017. Octanol-water partition coefficient measurement by a simple H-1 NMR method. *ACS Omega* 2:6244–6249.

Cusioli, L. F., C. D. O. Bezerra, H. B. Quesada, A. T. Alves Baptista, L. Nishi, M. F. Vieira, and R. Bergamasco. 2021. Modified *Moringa oleifera* Lam. Seed husks as low-cost biosorbent for atrazine removal. *Environmental Technology* 42(7):1092–1103.

Ćwieląg-Piasecka, I., A. Medyńska-Juraszek, M. Jerzykiewicz, M. Dębicka, J. Bekier, E. Jamroz, and D. Kawałko. 2018. Humic acid and biochar as specific sorbents of pesticides. *Journal of Soils and Sediments* 18(8):2692–2702.

Dela Piccolla, C., D. Hesterberg, T. Muraoka, and E. H. Novotny. 2021. Optimizing pyrolysis conditions for recycling pig bones into phosphate fertilizer. *Waste Management* 131:249–257.

Dougherty, D. A. 2013. The cation-π interaction. *Accounts of Chemical Research* 46(4):885–893.

Duke, S. O., and F. E. Dayan. 2018. *Herbicides*. Chichester: eLS. John Wiley & Sons, Ltd. DOI: 10.1002/9780470015902.a0025264.

EPA – Environmental Protect Agency. 2021. Herbicides. www.epa.gov/caddis-vol2/caddis-volume-2-sources-stressors-responses-herbicides (accessed October 13, 2021).

Essandoh, M., D. Wolgemuth, C. U. Pittman, D. Mohan, and T. Mlsna. 2017. Adsorption of metribuzin from aqueous solution using magnetic and nonmagnetic sustainable low-cost biochar adsorbents. *Environmental Science and Pollution Research* 24(5):4577–4590.

Fang, Q., B. Chen, Y. Lin, and Y. Guan. 2014. Aromatic and hydrophobic surfaces of wood-derived biochar enhance perchlorate adsorption via hydrogen bonding to oxygen-containing organic groups. *Environmental Science & Technology* 48(1):279–288.

Feng, C., C. Aldrich, J. J. Eksteen, and D. W. M. Arrigan. 2018. Removal of arsenic from gold processing circuits by use of novel magnetic nanoparticles. *Canadian Metallurgical Quarterly* 57(4):399–404.

Fernandes, B. C. C., K. F. Mendes, V. L. Tornisielo, T. M. S. Teófilo, V. Takeshita, P. S. F. Chagas, H. A. Lins, M. F. Souza, and D. V. Silva. 2021. Effect of pyrolysis temperature on eucalyptus wood residues biochar on availability and transport of hexazinone in soil. *International Journal of Environmental Science and Technology* 1–16.

Finizio, A., M. Vighi, and D. Sandroni. 1997. Determination of n-octanol/water partition coefficient (Kow) of pesticide critical review and comparison of methods. *Chemosphere* 34:131–161.

Forbes, C. R., S. K. Sinha, H. K. Ganguly, S. Bai, G. P. Yap, S. Patel, and N. J. Zondlo. 2017. Insights into thiol – aromatic interactions: A stereoelectronic basis for S – H/π interactions. *Journal of the American Chemical Society* 139(5):1842–1855.

Franz, M., H. A. Arafat, and N. G. Pinto. 2000. Effect of chemical surface heterogeneity on the adsorption mechanism of dissolved aromatics on activated carbon. *Carbon* 38(13):1807–1819.

Frontera, A., D. Quiñonero, and P. M. Deyà. 2011. Cation – π and anion – π interactions. *Wiley Interdisciplinary Reviews: Computational Molecular Science* 1(3):440–459.

Fuscaldo, F., F. Bedmar, and G. Monterubbianesi. 1999. Persistence of atrazine, metribuzin and simazine herbicides in two soils. *Pesquisa Agropecuaria Brasileira* 34(11):2037–2044.

Gámiz, B., K. Hall, K. A. Spokas, and L. Cox. 2019. Understanding activation effects on low-temperature biochar for optimization of herbicide sorption. *Agronomy* 9(10):588.

Garau, C., A. Frontera, D. Quiñonero, P. Ballester, A. Costa, and P. M. Deyà. 2004. Cation– π versus anion– π interactions: Energetic, charge transfer, and aromatic aspects. *The Journal of Physical Chemistry A* 108(43):9423–9427.

Gaskin, J. W., K. C. Das, A. S. Tassistro, L. Sonon, K. Harris, and B. Hawkins. 2009. Characterization of char for agricultural use in the soils of the southeastern United States. In *Amazonian Dark Earths: Wim Sombroek's Vision*, ed. W. I. Woods, W. G. Teixeira, J. Lehmann, C. Steiner, A. M. G. A. WinklerPrins, and L. Rebellato, 433–443. Dordrecht: Springer.

Gevao, B., K. T. Semple, and K. C. Jones. 2000. Bound pesticide residues in soils: A review. *Environmental Pollution* 108:3–14.

Ghanim, B. M., W. Kwapinski, and J. J. Leahy. 2017. Hydrothermal carbonisation of poultry litter: Effects of initial pH on yields and chemical properties of hydrochars. *Bioresource Technology* 238:78–85.

Glaser B., J. Lehmann, and W. Zech. 2002. Ameliorating physical and chemical properties of highly weathered soils in the tropics with charcoal – a review. *Biology and Fertility of Soils* 35:2219–2230.

Godlewska, P., Y. S. Ok, and P. Oleszczuk. 2021. The dark side of black gold: Ecotoxicological aspects of biochar and biochar-amended soils. *Journal of Hazardous Materials* 403:123833.

Goel, A., L. L. McConnell, and A. Torrents. 2007. Determination of vapor pressure-temperature relationships of current – use pesticides and transformation products. *Journal of Environmental Science and Health Part B* 42(4):343–349.

Green, R. D. 1974. *Hydrogen Bonding by C – H Groups*. London: The MacMillan Press Ltd.

Gülen, J., and S. Aslan. 2020. Adsorption of 2, 4-dichlorophenoxyacetic acid from aqueous solution using carbonized chest nut as low-cost adsorbent: Kinetic and thermodynamic. *Zeitschrift für Physikalische Chemie* 234(3):461–484.

Guo, M., Z. He, and S. M. Uchimiya. 2016. Introduction to biochar as an agricultural and environmental amendment. In *Agricultural and Environmental Applications of Biochar: Advances and Barriers*, ed. M. Guo, Z. He, and S. M. Uchimiya, 1–15. Madison: Soil Science Society of America.

Gurwick, N. P., L. A. Moore, C. Kelly, and P. Elias. 2013. A systematic review of biochar research, with a focus on its stability in situ and its promise as a climate mitigation strategy. *PloS One* 8(9):e75932.

Hamer, U., B. Marschner, S. Brodowski, and W. Amelung. 2004. Interactive priming of black carbon and glucose mineralization. *Organic Geochemistry* 35(7):823–830.

Han, L., L. Qian, J. Yan, and M. Chen. 2017. Effects of the biochar aromaticity and molecular structures of the chlorinated organic compounds on the adsorption characteristics. *Environmental Science and Pollution Research* 24(6):5554–5565.

Han, L., K. S. Ro, K. Sun, H. Sun, Z. Wang, J. A. Libra, and B. Xing. 2016. New evidence for high sorption capacity of hydrochar for hydrophobic organic pollutants. *Environmental Science & Technology* 50(24):13274–13282.

Hansen, V., D. Müller-Stöver, J. Ahrenfeldt, J. K. Holm, U. B. Henriksen, and H. Hauggaard-Nielsen. 2015. Gasification biochar as a valuable by-product for carbon sequestration and soil amendment. *Biomass and Bioenergy* 72:300–308.

Hao, F., X. Zhao, W. Ouyang, C. Lin, S. Chen, Y. Shan, and X. Lai. 2013. Molecular structure of corncob-derived biochars and the mechanism of atrazine sorption. *Agronomy Journal* 105(3):773–782.

Harris, P. 1999. On charcoal. *Interdisciplinary Science Reviews* 24(4):301–306.

Hassan, M., Y. Liu, R. Naidu, S. J. Parikh, J. Du, F. Qi, and I. R. Willett. 2020. Influences of feedstock sources and pyrolysis temperature on the properties of biochar and functionality as adsorbents: A meta-analysis. *Science of the Total Environment* 744:140714.

Helmenstine, A. M. 2020. pKa Definition in Chemistry. *ThoughtCo*. www.thoughtco.com/what-is-pka-in-chemistry-605521 (acessed November 13, 2021).

Herath, H. M. S. K., M. Camps-Arbestain, M. J. Hedley, M. U. F. Kirschbaum, T. Wang, and R. Van Hale. 2015. Experimental evidence for sequestering C with biochar by avoidance of CO_2 emissions from original feedstock and protection of native soil organic matter. *Gcb Bioenergy* 7(3):512–526.

Herath, I., P. Kumarathilaka, M. I. Al-Wabel, A. Abduljabbar, M. Ahmad, A. R. Usman, and M. Vithanage. 2016. Mechanistic modeling of glyphosate interaction with rice husk derived engineered biochar. *Microporous and Mesoporous Materials* 225:280–288.

Hohenstein, E. G., and C. D. Sherrill. 2009. Effects of heteroatoms on aromatic π–π interactions: Benzene– pyridine and pyridine dimer. *The Journal of Physical Chemistry A* 113(5):878–886.

Howard, P. H., and D. C. G. Muir. 2010. Identifying new persistent and bioaccumulative organics among chemicals in commerce. *Environmental Science and Technology* 44:2277–2285.

Hutzinger, O., W. D. Jamieson, J. D. MacNeil, and R. W. Frei. 1971. Electron-donor-acceptor complexing reagents for the analysis of pesticides. I. Survey of reagents and instrumental techniques. *Journal of the Association of Official Analytical Chemists* 54(5):1100–1109.

Igalavithana, A. D., Y. S. Ok, A. R. Usman, M. I. Al-Wabel, P. Oleszczuk, and S. S. Lee. 2016. The effects of biochar amendment on soil fertility. In *Agricultural and Environmental Applications of Biochar: Advances and Barriers*, ed. M. Guo, Z. He, and M. Uchimiya, 123–144. Crown: Soil Science Society of America, Inc.

Ighalo, J. O., A. G. Adeniyi, and A. A. Adelodun. 2021. Recent advances on the adsorption of herbicides and pesticides from polluted waters: Performance evaluation via physical attributes. *Journal of Industrial and Engineering Chemistry* 93:117–137.

Inyang, M., and E. Dickenson. 2015. The potential role of biochar in the removal of organic and microbial contaminants from potable and reuse water: A review. *Chemosphere* 134:232–240.

Ioannidou, O., and A. Zabaniotou. 2007. Agricultural residues as precursors for activated carbon production – A review. *Renewable and Sustainable Energy Reviews* 11:1966–2005.

Ippolito, J. A., L. Cui, C. Kammann, N. Wrage-Mönnig, J. M. Estavillo, T. Fuertes-Mendizabal, M. L. Cayuela, G. Sigua, J. Novak, K. Spokas, and N. Borchard. 2020. Feedstock choice, pyrolysis temperature and type influence biochar characteristics: A comprehensive meta-data analysis review. *Biochar* 2:421–438.

Iriarte-Velasco, U., I. Sierra, L. Zudaire, and J. L. Ayastuy. 2016. Preparation of a porous biochar from the acid activation of pork bones. *Food and Bioproducts Processing* 98:341–353.

Jaycock, M. J., and G. D. Parfitt. 1981. *Chemistry of Interfaces*. Onichester: E. Horwood.

Joseph, S., A. L. Cowie, L. Van Zwieten, N. Bolan, A. Budai, W. Buss, M. L. Cayuela, E. R. Graber, J. A. Ippolito, Y. Kuzyakov, Y. Luo, Y. S. Ok, K. N. Palansooriya, J. Shepherd, S. Stephens, Z. H. Weng, and J. Lehmann. 2021. How biochar works, and when it doesn't: A review of mechanisms controlling soil and plant responses to biochar. *GCB Bioenergy* 13(11):1731–1764.

Joseph, S. D., M. Camps-Arbestain, Y. Lin, P. Munroe, C. H. Chia, J. Hook, L. van Zwieten, S. Kimber, A. Cowie, B. P. Singh, J. Lehmann, N. Foidl, R. J. Smernik, and J. E. Amonette. 2010. An investigation into the reactions of biochar in soil. *Soil Research* 48(7):501–515.

Kalsi, P. S. 2000. *Organic Reactions and Their Mechanisms*, 2nd edition. New Delhi: New Age International.

Kearns, J. P., L. S. Wellborn, R. S. Summers, and D. R. U. Knappe. 2014. 2, 4-D adsorption to biochars: Effect of preparation conditions on equilibrium adsorption capacity and comparison with commercial activated carbon literature data. *Water Research* 62:20–28.

Kleber, M., P. Sollins, and R. Sutton. 2007. A conceptual model or organo-mineral interactions in soils: Self-assembly or organic molecular fragments into zonal structures on mineral surfaces. *Biogeochemistry* 85(1):9–24.

Kocman, M., M. Pykal, and P. Jurečka. 2014. Electric quadrupole moment of graphene and its effect on intermolecular interactions. *Physical Chemistry Chemical Physics* 16(7):3144–3152.

Kookana, R. S. 2010. The role of biochar in modifying the environmental fate, bioavailability, and efficacy of pesticides in soils: A review. *Soil Research* 48(7):627–637.

Kookana, R. S., A. K. Sarmah, L. Van Zwieten, E. Krull, and B. Singh. 2011. Biochar application to soil: Agronomic and environmental benefits and unintended consequences. *Advances in Agronomy* 112:103–143.

Lattao, C., X. Cao, J. Mao, K. Schmidt-Rohr, and J. J. Pignatello. 2014. Influence of molecular structure and adsorbent properties on sorption of organic compounds to a temperature series of wood chars. *Environmental Science & Technology* 48(9):4790–4798.

Lehmann, J., M. C. Rillig, J. Thies, C. A. Masiello, W. C. Hockaday, and D. Crowley. 2011. Biochar effects on soil biota – a review. *Soil Biology and Biochemistry* 43(9):1812–1836.

Li, F., X. Liang, C. Niyungeko, T. Sun, F. Liu, and Y. Arai. 2019. Effects of biochar amendments on soil phosphorus transformation in agricultural soils. *Advances in Agronomy* 158:131–172.

Liu, B., L. Dong, Q. Yu, X. Li, F. Wu, Z. Tan, and S. Luo. 2016. Thermodynamic study on the protonation reactions of glyphosate in aqueous solution: Potentiometry, calorimetry and NMR spectroscopy. *The Journal of Physical Chemistry B* 120(9):2132–2137.

Loganathan, V. A., Y. Feng, G. D. Sheng, and T. P. Clement. 2009. Crop-residue-derived char influences sorption, desorption and bioavailability of atrazine in soils. *Soil Science Society of America Journal* 73(3):967–974.

Lopez-Capel, E., K. Zwart, S. Shackley, R. Postma, J. Stenstrom, D. P. Rasse, A. Budai, and B. Glaser. 2016. Biochar properties. In *Biochar in European Soils and Agriculture: Science and Practice*, ed. S. Shackley, G. Ruysschaert, K. Zwart, and B. Glaser, 41–72. Abington: Routledge, Taylor & Francis Group.

Mackay, D. 1991. *Multimedia Environmental Models: The Fugacity Approach*. Chelsea, MI: Lewis Publishers.

Mallick, S., L. Cao, X. Chen, J. Zhou, Y. Qin, G. Y. Wang, Y. Y. Wu, M. Meng, G. Y. Zhu, Y. N. Tan, T. Cheng, and C. Y. Liu. 2019. Mediation of electron transfer by quadrupolar interactions: The constitutional, electronic, and energetic complementarities in supramolecular chemistry. *Iscience* 22:269–287. doi.org/10.1016/j.isci.2019.11.020.

Manyà, J. J. 2012. Pyrolysis for biochar purposes: A review to establish current knowledge gaps and research needs. *Environmental Science & Technology* 46(15):7939–7954.

March, J. 1992. Localized chemical bonding. In *Advanced Organic Chemistry: Reactions, Mechanisms and Structure*, 4th edition, 8–10. New York: John Wiley and Sons Ltd.

Martin, S. M., R. S. Kookana, L. V. Zwieten, and E. Krull. 2012. Marked changes in herbicide sorption-desorption upon ageing of biochars in soil. *Journal of Hazardous Materials* 231(232):70–78.

Mašek, O., F. Ronsse, and D. Dickinson. 2016. Biochar production and feedstock. In *Biochar in European Soils and Agriculture: Science and Practice*, ed. S. Shackley, G. Ruysschaert, K. Zwart, and B. Glaser, 17–40. Abington: Routledge, Taylor & Francis Group.

Mayakaduwa, S. S., P. Kumarathilaka, I. Herath, M. Ahmad, M. Al-Wabel, Y. S. Ok, A. Usman, A. Abduljabbar, and M. Vithanage. 2016. Equilibrium and kinetic mechanisms of woody biochar on aqueous glyphosate removal. *Chemosphere* 144:2516–2521.

Mayr, H., and M. Patz. 1994. Scales of nucleophilicity and electrophilicity: A system for ordering polar organic and organometallic reactions. *Angewandte Chemie International Edition* 33(9):938–957.

Mecozzi, S., A. P. West, and D. A. Dougherty. 1996. Cation– π interactions in simple aromatics: Electrostatics provide a predictive tool. *Journal of the American Chemical Society* 118(9):2307–2308.

Melo, T. M., M. Bottlinger, E. Schulz, W. M. Leandro, S. B. de Oliveira, A. M. de Aguiar Filho, A. El-Naggar, N. Bolan, H. Wang, Y. S. Ok, and J. Rinklebe. 2019. Management of biosolids-derived hydrochar (Sewchar): Effect on plant germination, and farmer's acceptance. *Journal of Environmental Management* 237:200–214.

Mendes, K. F., K. C. Mielke, L. H. Barcellos Jr, R. A. de la Cruz, and R. N. de Sousa. 2021. Anaerobic and aerobic degradation studies of herbicides and radiorespirometry of microbial activity in soil. In *Radioisotopes in Weed Research*, ed. K. F. Mendes, 95–125. Boca Raton: CRC Press.

Mendes, K. F., G. P. Olivatto, R. N. de Sousa, L. V. Junqueira, and V. L. 2019. Natural biochar effect on sorption-desorption and mobility of diclosulam and pendimethalin in soil. *Geoderma* 347(1):118–125.

Mendes, K. F., R. N. de Sousa, M. B. Soares, D. G. Viana, and A. J. de Souza. 2021b. Sorption and desorption studies of herbicides in the soil by batch equilibrium and stirred flow methods. In *Radioisotopes in Weed Research*, ed. K. F. Mendes, 61–94. Boca Raton: CRC Press.

Meyer, S., B. Glaser, and P. Quicker. 2011. Technical, economical, and climate-related aspects of biochar production technologies: A literature review. *Environmental Science & Technology* 45(22):9473–9483.

Mia, S., F. A. Dijkstra, and B. Singh. 2017. Long-term aging of biochar: A molecular understanding with agricultural and environmental implications. *Advances in Agronomy* 141:1–51.

Mohan, D., A. Sarswat, Y. S. Ok, and C. U. Pittman Jr. 2014. Organic and inorganic contaminants removal from water with biochar, a renewable, low cost and sustainable adsorbent – a critical review. *Bioresource Technology* 160:191–202.

Mukherjee, A., and R. Lal. 2013. Biochar impacts on soil physical properties and greenhouse gas emissions. *Agronomy* 3(2):313–339.

Mukherjee, A., A. R. Zimmerman, and W. Harris. 2011. Surface chemistry variations among a series of laboratory-produced biochars. *Geoderma* 163:247–255.

Mukome, F. N., and S. J. Parikh. 2015. Chemical, physical, and surface characterization of biochar. In *Biochar Production, Characterization, and Applications*, ed. Y. S. Ok, S. M. Uchimiya, S. X. Chang, and N. Bolan, 67–98. Boca Raton: CRC Press.

Neel, A. J., M. J. Hilton, M. S. Sigman, and F. D. Toste. 2017. Exploiting non-covalent π interactions for catalyst design. *Nature* 543(7647):637–646.

Ney, R. 1998. *Fate and Transport of Organic Chemicals in the Environment*, 3rd edition, 370 p. Washington, DC: Government Inst.

Nir, S., T. Undabeytia, D. Yaron-Marcovich, Y. El-Nahhal, T. Polubesova, C. Serban, G. Rytwo, G. Lagaly, and B. Rubin. 2000. Optimization of adsorption of hydrophobic herbicides on montmorillonite preadsorbed by monovalent organic cations: Interaction between phenyl rings. *Environmental Science & Technology* 34(7):1269–1274.

Oliveira Jr, R. S., W. C. Koskinen, and F. A. Ferreira. 2000. Sorption and leaching potential of herbicides on Brazilian soils. *Weed Research* 41:97–110.

Oni, B. A., O. Oziegbe, and O. O. Olawole. 2019. Significance of biochar application to the environment and economy. *Annals of Agricultural Sciences* 64(2):222–236.

Ouse, D. G., J. M. Gifford, J. Schleier III, D. D. Simpson, H. H. Tank, C. J. Jennings, S. P. Annangudi, P. Valverde-Garcia, and R. A. Masters. 2018. A new approach to quantify herbicide volatility. *Weed Technology* 32(6):691–697.

Pesticide Properties Database (PPDB). 2022. Footprint: Creating tools for pesticide risk assessment and management in Europe. https://sitem.herts.ac.uk/aeru/ppdb/en/index.htm (Accessed January 29, 2022).

Peterson, S. C., M. A. Jackson, and M. Appell. 2013. Biochar: Sustainable and versatile. In *Advances in Applied Nanotechnology for Agriculture*, ed. B. Park, and M. Appell, 193–205. Washington: American Chemical Society.

Phipps, M. J., T. Fox, C. S. Tautermann, and C. K. Skylaris. 2015. Energy decomposition analysis approaches and their evaluation on prototypical protein – drug interaction patterns. *Chemical society reviews* 44(10):3177–3211.

Pignatello, J. J. 2013. Adsorption of dissolved organic compounds by black carbon. In *Molecular Environmental Soil Science*, ed. J. Xu, and D. L. Sparks, 359–385. Dordrecht: Springer.

Pignatello, J. J., S. Kwon, and Y. Lu. 2006. Effect of natural organic substances on the surface and adsorptive properties of environmental black carbon (char): Attenuation of surface activity by humic and fulvic acids. *Environmental Science & Technology* 40(24):7757–7763.

Pignatello, J. J., and B. Xing. 1996. Mechanisms of slow sorption of organic chemicals to natural particles. *Environmental Science & Technology* 30(1):1–11.

Qin, J., S. Qian, Q. Chen, L. Chen, L. Yan, and G. Shen. 2019. Cow manure-derived biochar: Its catalytic properties and influential factors. *Journal of Hazardous Materials* 371:381–388.

Ritchie, C. D. 1972. Nucleophilic reactivities toward cations. *Accounts of Chemical Research* 5(10):348–354.

Ruijuan, L., B. Wen, S. Zhang, Z. Pei, and X. Shan. 2009. Influence of organic amendments on the sorption of pentachlorophenol on soils. *Journal of Environmental Sciences* 21:474–480.

Sakulthaew, C., A. Watcharenwong, C. Chokejaroenrat, and A. Rittirat. 2021. Leonardite-derived biochar suitability for effective sorption of herbicides. *Water, Air, & Soil Pollution* 232(2):1–17.

Sander, M., and J. J. Pignatello. 2005. Characterization of charcoal adsorption sites for aromatic compounds: Insights drawn from single-solute and bi-solute competitive experiments. *Environmental Science & Technology* 39(6):1606–1615.

Schimmelpfennig, S., and B. Glaser. 2012. One step forward toward characterization: Some important material properties to distinguish biochars. *Journal of Environmental Quality* 41(4):1001–1013.

Schreiter, I. J., W. Schmidt, and C. Schüth. 2018. Sorption mechanisms of chlorinated hydrocarbons on biochar produced from different feedstocks: Conclusions from single-and bi-solute experiments. *Chemosphere* 203:34–43.

Schwarzenbach, R. P., T. Egli, T. B. Hofstetter, U. Von Gunten, and B. Wehrli. 2010. Global water pollution and human health. *Annual Review of Environment and Resources* 35:109–136.

Schwarzenbach, R. P., P. M. Gschwend, and D. M. Imboden. 1993. *Environmental Organic Chemistry*, 1021 p. New York: John Wiley & Sons.

Scow, K. M. 1993. Effects of sorption-desorption and diffusion processes on the kinetics of biodegradation of organic chemicals in soil. In *Sorption and Degradation Pesticides and Organic Chemicals in Soil*, ed. D. M. Linn, F. H. Carski, M. L. Brusseau, and T. -H. Chang, 73–114. Madison: Soil Science Society of America.

Shen, Z., Y. Zhang, O. McMillan, D. O'Connor, and D. Hou. 2020. The use of biochar for sustainable treatment of contaminated soils. In *Sustainable Remediation of Contaminated Soil and Groundwater*, ed. D. Hou, 119–167. Butterworth-Heinemann: Elsevier.

Sheng, G., Y. Yang, M. Huang, and K. Yang. 2005. Influence of pH on pesticide sorption by soil containing wheat residue-derived char. *Environmental Pollution* 134(3):457–463.

Shows, M. E., and S. V. Olesik. 2000. Extraction of atrazine and its metabolites using supercritical fluids and enhanced-fluidity liquids. *Journal of Chromatographic Science* 38(9):399–408.

Silber, A., I. Levkovitch, and E. R. Graber. 2010. pH-dependent mineral release and surface properties of corn straw biochar: Agronomic implications. *Environmental Science and Technol*ogy 44:9318–9323.

Sing, K. S. 1985. Reporting physisorption data for gas/solid systems with special reference to the determination of surface area and porosity. *Pure and Applied Chemistry* 57(4):603–619.

Singh, B., and A. L. Cowie. 2010. Characterization and evaluation of biochars for their application as soil amendment. *Australian Journal of Soil Research* 48(7):516–525.

Singh, B., M. M. Dolk, Q. Shen, and M. Camps-Arbestain. 2017. Biochar pH, electrical conductivity and liming potential. In *Biochar: A Guide to Analytical Methods*, ed. B. Singh, M. Camps-Arbestain, and J. Lehmann, 23–38. Australia: CSIRO Publishing/Boca Raton: CRC Press, Taylor and Francis Group.

Singh, N., and R. S. Kookana. 2009. Organo-mineral interactions mask the true sorption potential of biochars in soils. *Journal of Environmental Science and Health Part B* 44(3):214–219.

Singh, S. K., and A. Das. 2015. The n→ π* interaction: A rapidly emerging non-covalent interaction. *Physical Chemistry Chemical Physics* 17(15):9596–9612.

Smernik, R. J., R. S. Kookana, and J. O. Skjemstad. 2006. NMR characterization of 13C-benzene sorbed to natural and prepared charcoals. *Environmental Science & Technology* 40(6):1764–1769.

Soares, M. B., F. H. Santos, and L. R. F. Alleoni. 2022. Temporal changes in arsenic and lead pools in a contaminated sediment amended with biochar pyrolyzed at different temperatures. *Chemosphere* 287:132102.

Sohi, S. P. 2012. Carbon storage with benefits. *Science* 338(6110):1034–1035.

Sohi, S. P., E. Krull, E. Lopez-Capel, and R. Bol. 2010. A review of biochar and its use and function in soil. *Advances in Agronomy* 105:47–82.

Song, D., H. Sun, X. Jiang, F. Kong, Z. Qiang, A. Zhang, H. Liu, and J. Qu. 2018. Determination of pKa and the corresponding structures of quinclorac using combined experimental and theoretical approaches. *Journal of Molecular Structure* 1152:53–60.

Sparks, D. L. 2019. Fundamentals of soil chemistry. In *Encyclopedia of Water, Science, Technology and Society*, ed. P. A. Maurice, 1553–1564. Hoboken: John & Wiley & Sons.

Spokas, K. A. 2010. Review of the stability of biochar in soils: Predictability of O: C molar ratios. *Carbon Management* 1(2):289–303.

Spokas, K. A., K. B. Cantrell, J. M. Novak, D. W. Archer, J. A. Ippolito, H. P. Collins, A. A. Boateng, I. M. Lima, M. C. Lamb, A. J. McAloon, R. D. Lentz, and K. A. Nichols. 2012. Biochar: A synthesis of its agronomic impact beyond carbon sequestration. *Journal of Environmental Quality* 41(4):973–989.

Spokas, K. A., W. C. Koskinen, J. M. Baker, and D. C. Reicosky. 2009. Impacts of woodchip biochar additions on greenhouse gas production and sorption/degradation of two herbicides in a Minnesota soil. *Chemosphere* 77(4):574–581.

Spokas, K. A., J. M. Novak, C. A. Masiello, M. G. Johnson, E. C. Colosky, J. A. Ippolito, and C. Trigo. 2014. Physical disintegration of biochar: An overlooked process. *Environmental Science & Technology Letters* 1(8):326–332.

Sposito, G. 2008. *The Chemistry of Soils*, 2nd edition. New York: Oxford University Press.

Steiner, C. 2016. Considerations in biochar characterization. In *Agricultural and Environmental Applications of Biochar: Advances and Barriers*, ed. M. Guo, Z. He, and M. Uchimiya, 87–100. Madison: Soil Science Society of America, Inc.

Stenrød, M. 2015. Long-term trends of pesticides in Norwegian agricultural streams and potential future challenges in northern climate. *Acta Agriculturae Scandinavica, Section B – Soil & Plant Science* 65:199–216.

Sun, H., W. C. Hockaday, C. A. Masiello, and K. Zygourakis. 2012. Multiple controls on the chemical and physical structure of biochars. *Industrial and Engineering Chemistry Research* 51(9):1587–1597.

Sun, K., B. Gao, Z. Zhang, G. Zhang, Y. Zhao, and B. Xing. 2010. Sorption of atrazine and phenanthrene by organic matter fractions in soil and sediment. *Environmental Pollution* 158(12):3520–3526.

Sun, K., M. Keiluweit, M. Kleber, Z. Pan, and B. Xing. 2011. Sorption of fluorinated herbicides to plant biomass-derived biochars as a function of molecular structure. *Bioresource Technology* 102(21):897–9903.

Swain, C. G., and C. B. Scott. 1953. Quantitative correlation of relative rates. Comparison of hydroxide ion with other nucleophilic reagents toward alkyl halides, esters, epoxides and acyl halides 1. *Journal of the American Chemical Society* 75(1):141–147.

Tomczyk, A., and P. Sokołowska. 2019. Biochar efficiency in copper removal from haplic soils. *International Journal of Environmental Science and Technology* 16(8):4899–4912.

Tomczyk, A., Z. Sokołowska, and P. Boguta. 2020. Biochar physicochemical properties: Pyrolysis temperature and feedstock kind effects. *Reviews in Environmental Science and Biotechnology* 19(1):191–215.

Trigo, C., K. A. Spokas, L. Cox, and W. C. Koskinen. 2014. Influence of soil biochar aging on sorption of the herbicides MCPA, nicosulfuron, terbuthylazine, indaziflam, and fluoroethyldiaminotriazine. *Journal of Agricultural and Food Chemistry* 62(45):10855–10860.

Tripathi, M., J. N. Sahu, and P. Ganesan. 2016. Effect of process parameters on production of biochar from biomass waste through pyrolysis: A review. *Renewable and Sustainable Energy Reviews* 55:467–481.

Tsai, W. T., and H. R. Chen. 2013. Adsorption kinetics of herbicide paraquat in aqueous solution onto a low-cost adsorbent, swine-manure-derived biochar. *International Journal of Environmental Science and Technology* 10(6):1349–1356.

Tsuzuki, S., M. Yoshida, T. Uchimaru, and M. Mikami. 2001. The origin of the cation/π interaction: The significant importance of the induction in Li+ and Na+ complexes. *Journal of Physical Chemistry A* 105(4):769–773.

Uchimiya, M., I. M. Lima, K. T. Klasson, and L. H. Wartelle. 2010. Contaminant immobilization and nutrient release by biochar soil amendment: Roles of natural organic matter. *Chemosphere* 80(8):935–940.

Uchimiya, M., L. H. Wartelle, K. T. Klasson, C. A. Fortier, and I. M. Lima. 2011. Influence of pyrolysis temperature on biochar property and function as a heavy metal sorbent in soil. *Journal of Agricultural and Food Chemistry* 59(6):2501–2510.

Vassilev, N., E. Martos, G. Mendes, V. Martos, and M. Vassileva. 2013. Biochar of animal origin: A sustainable solution to the global problem of high-grade rock phosphate scarcity?. *Journal of the Science of Food and Agriculture* 93(8):1799–1804.

Villaverde, J. J., P. Sandín-España, J. L. Alonso-Prados, A. M. Lamsabhi, and M. Alcamí. 2018. Pesticide byproducts formation: Theoretical study of the protonation of alloxydim degradation products. *Computational and Theoretical Chemistry* 1143:9–19.

Wang, H., K. Lin, and Z. Hou. 2010. Sorption of the herbicide terbuthylazine in two New Zealand forest soils amended with biosolids and biochars. *Journal of Soils and Sediments* 10:283–289.

Wang, L., D. O'Connor, J. Rinklebe, Y. S. Ok, D. C. Tsang, Z. Shen, and D. Hou. 2020. Biochar aging: Mechanisms, physicochemical changes, assessment, and implications for field applications. *Environmental Science & Technology* 54(23):14797–14814.

Wang, P., X. Liu, B. Yu, X. Wu, J. Xu, F. Dong, and Y. Zheng. 2020. Characterization of peanut-shell biochar and the mechanisms underlying its sorption for atrazine and nicosulfuron in aqueous solution. *Science of the Total Environment* 702:134767.

Weber, J. B., G. G. Wilkerson, H. M. Linker, J. W. Wilcut, R. B. Leidy, S. Senseman, W. W. Witt, M. Barrett, W. K. Vencil, D. R. Shaw, T. C. Muller, D. K. Miller, B. J. Brecke, R. E. Talbert, and T. F. Peeper. 2000. A proposal to standardize soil/solution herbicide distribution coefficients. *Weed Science* 48:75–88.

Weber, J. B., G. G. Wilkerson, and C. F. Reinhardt. 2004. Calculating pesticide sorption coefficients (K_d) using selected soil properties. *Chemosphere* 55: 157–166.

Wei, L., Y. Huang, L. Huang, Y. Li, Q. Huang, G. Xu, K. Müller, H. Wang, Y. S. Ok, and Z. Liu. 2020. The ratio of H/C is a useful parameter to predict adsorption of the herbicide metolachlor to biochars. *Environmental Research* 184:109324.

Wei, L., Y. Huang, Y. Li, L. Huang, N. N. Mar, Q. Huang, and Z. Liu. 2017. Biochar characteristics produced from rice husks and their sorption properties for the acetanilide herbicide metolachlor. *Environmental Science and Pollution Research* 24(5):4552–4561.

West, A. 2018. Intermolecular forces and solvation. In *Interface Science and Technology*, ed. V. Ball, 49–130. Amsterdam: Elsevier.

Wheeler, S. E. 2013. Understanding substituent effects in noncovalent interactions involving aromatic rings. *Accounts of Chemical Research* 46(4):1029–1038.

Wheeler, S. E., and K. N. Houk. 2009. Substituent effects in cation/π interactions and electrostatic potentials above the centers of substituted benzenes are due primarily to through-space effects of the substituents. *Journal of the American Chemical Society* 131(9):3126–3127.

Wu, M., B. Pan, D. Zhang, D. Xiao, H. Li, C. Wang, and P. Ning. 2013. The sorption of organic contaminants on biochars derived from sediments with high organic carbon content. *Chemosphere* 90(2):782–788.

Wu, P., Q. L. Fu, X. D. Zhu, C. Liu, F. Dang, K. Müller, M. Fujii, D. M. Zhou, H. L. Wang, and Y. J. Wang. 2020. Contrasting impacts of pH on the abiotic transformation of hydrochar-derived dissolved organic matter mediated by δ-MnO_2. *Geoderma* 378:114627.

Xiao, F., and J. J. Pignatello. 2015a. π+-π Interactions between (hetero)aromatic amine cations and the graphitic surfaces of pyrogenic carbonaceous materials. *Environmental Science and Technology* 49(2):906–914.

Xiao, F., and J. J. Pignatello. 2015b. Interactions of triazine herbicides with biochar: Steric and electronic effects. *Water Research* 80:179–188.

Xiao, X., and B. Chen. 2016. Interaction mechanisms between biochar and organic pollutants. In *Agricultural and Environmental Applications of Biochar: Advances and Barriers*, ed. M. Guo, Z. He, and M. Uchimiya, 225–257. Madison: Soil Science Society of America, Inc.

Xiao, X., B. Chen, Z. Chen, L. Zhu, and J. L. Schnoor. 2018. Insight into multiple and multilevel structures of biochars and their potential environmental applications: A critical review. *Environmental Science and Technology* 52(9):5027–5047.

Xu, H. J., X. H. Wang, H. Li, H. Y. Yao, J. Q. Su, and Y. G. Zhu. 2014. Biochar impacts soil microbial community composition and nitrogen cycling in an acidic soil planted with rape. *Environmental Science and Technology* 48(16):9391–9399.

Yang, K., and B. Xing. 2010. Adsorption of organic compounds by carbon nanomaterials in aqueous phase: Polanyi theory and its application. *Chemical Reviews* 110(10):5989–6008.

Yang, K., J. Yang, Y. Jiang, W. Wu, and D. Lin. 2016. Correlations and adsorption mechanisms of aromatic compounds on a high heat temperature treated bamboo biochar. *Environmental Pollution* 210:57–64.

Yang, K., Y. Jiang, J. Yang, and D. Lin. 2018. Correlations and adsorption mechanisms of aromatic compounds on biochars produced from various biomass at 700 °C. *Environmental Pollution* 233:64–70.

Yang, W., J. Shang, B. Li, and M. Flury. 2020. Surface and colloid properties of biochar and implications for transport in porous media. *Critical Reviews in Environmental Science and Technology* 50(23):2484–2522.

Yang, Y., and G. Sheng. 2003. Pesticide adsorptivity of aged particulate matter arising from crop residue burns. *Journal of Agricultural and Food Chemistry* 51(1):5047–5051.

Yu, H., W. Zou, J. Chen, H. Chen, Z. Yu, J. Huang, H. Tang, X. Wei, and B. Gao. 2019. Biochar amendment improves crop production in problem soils: A review. *Journal of Environmental Management* 232:8–21.

Yu, X. Y., G. G. Ying, and R. S. Kookana. 2006. Sorption and desorption behaviors of diuron in soils amended with charcoal. *Journal of Agricultural and Food Chemistry* 54(22):8545–8550.

Yuan, J. H., and R. K. Xu. 2011. The amelioration effects of low temperature biochar generated from nine crop residues on an acidic Ultisol. *Soil Use and Management* 27(1):110–115.

Yvari, S., M. Abualqumboz, N. Sapari, H. A. Hata-Suhaimi, N. Z. Nik-Fuaad, and S. Yavari. 2020. Sorption of imazapic and imazapyr herbicides on chitosan-modified biochars. *International Journal of Environmental Science and Technology* 17:3341–3350.

Zhang, H., K. Lin, H. Wang, and J. Gan. 2010. Effect of Pinus radiata derived biochars on soil sorption and desorption of phenanthrene. *Environmental Pollution* 158(9):2821–2825.

Zhang, M., L. Shu, X. Shen, X. Guo, S. Tao, B. Xing, and X. Wang. 2014. Characterization of nitrogen-rich biomaterial-derived biochars and their sorption for aromatic compounds. *Environmental Pollution* 195:84–90.

Zhang, P., H. Sun, L. Min, and C. Ren. 2018. Biochars change the sorption and degradation of thiacloprid in soil: Insights into chemical and biological mechanisms. *Environmental Pollution* 236:158–167.

Zhang, P., H. Sun, C. Ren, L. Min, and H. Zhang. 2018. Sorption mechanisms of neonicotinoids on biochars and the impact of deashing treatments on biochar structure and neonicotinoids sorption. *Environmental Pollution* 234:812–820.

Zhang, X., H. Wang, L. He, K. Lu, A. Sarmah, J. Li, N. S. Bolan, J. Pei, and H. Huang. 2013. Using biochar for remediation of soils contaminated with heavy metals and organic pollutants. *Environmental Science and Pollution Research* 20(12):8472–8483.

Zheng, W., M. Guo, T. Chow, D. N. Bennett, and N. Rajagopalan. 2010. Sorption properties of greenwaste biochar for two triazine pesticides. *Journal of Hazardous Materials* 181(1–3):121–126.

Zhu, D., S. Kwon, and J. J. Pignatello. 2005. Adsorption of single-ring organic compounds to wood charcoals prepared under different thermochemical conditions. *Environmental Science & Technology* 39(11):3990–3998.

Zhu, D., and J. J. Pignatello. 2005. Characterization of aromatic compound sorptive interactions with black carbon (charcoal) assisted by graphite as a model. *Environmental Science and Technology* 39(7):2033–2041.

Zhu, M., H. Su, Y. Bao, J. Li, and G. Su. 2022. Experimental determination of octanol-water partition coefficient (K_{ow}) of 39 liquid crystal monomers (LCMs) by use of the shake-flask method. *Chemosphere* 287:132407.

Zhu, Y. G., J. Q. Su, Z. Cao, K. Xue, J. Quensen, G. X. Guo, Y. F. Yang, J. Zhou, and J. M. Tiedje. 2016. A buried Neolithic paddy soil reveals loss of microbial functional diversity after modern rice cultivation. *Science Bulletin* 61(13):1052–1060.

Zwetsloot, M. J., J. Lehmann, and D. Solomon. 2015. Recycling slaughterhouse waste into fertilizer: How do pyrolysis temperature and biomass additions affect phosphorus availability and chemistry? *Journal of the Science of Food and Agriculture* 95(2):281–288.

5 Effects of Biochars on Sorption and Desorption of Herbicides in Soil

Kamila Cabral Mielke[1], Kassio Ferreira Mendes[1], and Tiago Guimarães[2]*

[1]Department of Agronomy, Federal University of Viçosa, Viçosa, MG, Brazil

[2]Department of Chemistry, Federal University of Viçosa, Viçosa, MG, Brazil

* Corresponding author: kamila.mielke@ufv.br

CONTENTS

5.1 INTRODUCTION

Biochar is a carbonaceous material produced by heating plant waste, solid animal dung, and other organic waste in oxygen-limited atmospheres (Lehmann and Joseph

DOI: 10.1201/9781003202073-5

2009). The production and application of biochar to soil have been suggested for carbon sequestration (Glaser et al. 2009; Woolf et al. 2010), waste management and energy production (Lehmann and Joseph 2009; Ahmad et al. 2014; Oni et al. 2019), increase in cation exchange capacity (CEC), soil pH, water retention and nutrient source, benefiting agricultural production (Lehmann and Joseph 2009).

Biochar has been reported as a potential sorbent immobilizing soil pesticides (Yu et al. 2011; Lou et al. 2011; Cabrera et al. 2014; Huang et al. 2018; Szmigielski et al. 2018; Mendes et al. 2019a), as well as having the potential in immobilizing heavy metals from soil and water (Singanan et al. 2013; Zhou et al. 2017). Among the pesticides used in agriculture, herbicides have the soil as a transitory and/or final destination, after pre- or post-emergence application. Herbicide sorption and desorption processes in soil are the bases for studying environmental behavior and toxicity to living organisms (Rice et al. 2007). The increased sorption of herbicides by biochar decreases their loss through dissipation, decreasing the risk of human exposure and environmental pollution (Tatarková et al. 2013; Yu et al. 2019). Unlike sorption, there is a reduction of desorption when biochar is added to soil and the herbicide hardly returns to the soil solution, and sorption may become irreversible (Mendes et al. 2019a). Reduced desorption decreases the leaching capacity of herbicides in the soil, reducing the availability for weed degradation and control.

The heterogeneous composition of biochars shows that their surfaces can exhibit hydrophilic, hydrophobic, acidic and basic properties, which contribute to their ability to react with soil solution chemicals (Atkinson et al. 2010). The variability in the physical and chemical properties of biochar depends on the raw material and the conditions used during the pyrolysis process (Lehmann and Joseph 2009). These factors alter the specific surface area, polarity, atomic ratio, pH, elemental composition, and thus the sorption capacity of biochars (Ronsse et al. 2013; Ahmad et al. 2014). In order to increase the sorption capacity of biochars, modification processes of these materials are being adopted (Hagemann et al. 2018). Low temperatures (<400°C) generally produce biochar with a potential for soil fertility but have relatively low sorption capacities (Gámiz et al. 2019a), requiring activation methods to improve the sorptive capacity of biochar.

The application rate of biochar to the soil and aging over time are also factors that affect herbicide sorption. Although soil amendment by biochar is a promising technique for herbicide retention, biochar can decrease its sorptive capacity due to the accumulation of soil minerals on the biochar surface and the presence of easily oxidizable organic fractions that block the sites where sorption is most likely to occur (Joseph et al. 2010). In situations where you have an increased application rate of biochar, you may need high doses of the residual herbicide applied in pre-emergence to control the weed seed bank (Yang et al. 2006; Tang et al. 2013). Therefore, it is important to maintain a balance between the sorption capacity and the application rate of biochar, seeking to obtain potentially promising effects on herbicide efficacy.

This chapter will discuss the influence of biochars on herbicide sorption and desorption processes in soil. The different raw material sources, application rates, pyrolysis temperature, soil aging and chemical modifications of biochars, aiming at enhancing the herbicide sorption process, will be discussed in detail. The main objective is to understand the potential of biochar in the sorption and desorption processes of different herbicides, providing knowledge for the effective use of this promising soil amendment in environmental management.

5.2 CHANGES IN THE PHYSICOCHEMICAL PROPERTIES OF THE SOIL AMENDED WITH BIOCHAR

Biochar can be part of a long-term adaptation strategy in soil, as it improves physicochemical properties, mainly increasing porosity and specific surface area (SSA), water storage capacity, decreasing density, organic matter, pH, and CEC (Glaser et al. 2002; Lu et al. 2014; Lehmann et al. 2015; Nelissen et al. 2015). Improving soil properties contributes strongly to increasing the efficiency of nutrient and water use and crop productivity. The effects of biochar on soil physicochemical characteristics depend on several factors, such as feedstock type, pyrolysis condition, application rate, and environmental condition. The effects of biochar-amended soil on some physicochemical properties are reported later in the chapter.

5.2.1 Apparent Soil Density

Generally biochar has an apparent density between 0.05 and 0.57 kg m^{-3}, being lower than soil, thus biochar application is likely to reduce soil density through mixing effect (Blanco-Canqui 2017). A field study showed that the incorporation of 30 Mg ha^{-1} of wood biochar under fallow pasture led to a reduction in soil density from 2.48 to 1.00 Mg m^{-3} (Usowicz et al. 2016). The application of a higher rate (40 Mg ha^{-1}) of wood chip biochar decreased the bulk density of the soil by 16% (Walters and White 2018). The reduction in soil density by biochar is directly related to biochar type, soil type, biochar particle size, and application rate (Zhang et al. 2021).

5.2.2 Soil Porosity

The type of feedstock and production conditions can influence porosity, volume, and pore size, for example, as pyrolysis temperature increases, volatile matter decreases pore diameter and increases porosity, increasing surface area (Chen and Yuan 2011; Gul et al. 2015). After biochar addition, the porosity, permeability, and saturated hydraulic conductivity of the soil also increase (Oguntunde et al. 2004). The effect of biochar application on different soils and the impact under porosity were reported by Duarte et al. (2019). Porosity increased in both soils (sandy and clayey) with the addition of biochar in 0.15–2 mm particles. Smaller biochar particles changed the pore size distribution and increased macro-pores (> 0.50 mm) and mesopores (0.15–0.50 mm) in both soils. The influence of biochar on the physical quality of a clay soil (Vertisol) was evaluated, and total porosity values increased with the addition of biochar to the soil and varied according to the type of feedstock (Sun and Lu 2014). The relative increase in total porosity was 29%, 12%, and 16% for biochar from crop straw, wood, and wastewater sludge, respectively.

Detailed characterization of pore structure in biochars could effectively predict potential impacts on soil amendment and sorption of organic pollutants and heavy metals (Lu and Zong 2018). The addition of biochar not only changes soil porosity, but also causes the reorganization of soil pores, which subsequently changes the distribution of soil pores (Zhang et al. 2021).

5.2.3 Specific Surface Area (SSA) of Soil

SSA of biochar is generally higher than sand and similar to or higher than clay; consequently, it increases the SSA of the soil when added (Downie et al. 2009). The increase in SSA of a clay soil modified with hardwood chip biochar (700°C) was from 130 to 153 $m^2 g^{-1}$, as the biochar application rate increased from 0 to 20 g kg^{-1} (Laird et al. 2010). Liang et al. (2006) reported that Anthrosols had an approximately 4.8 times higher SSA than other soils due to their higher concentrations of biochar. The increased SSA and porosity caused an increase in water retention (Nair et al. 2017).

5.2.4 Water-Holding Capacity of Soil

Soil modification with biochar can influence the water-holding capacity of the soil. Soil hydraulic properties are important for analyzing the movement and transport of solutes (Carrick et al. 2011). The high porosity and SSA of biochar reduce the water permeability resistance of the soil. The amount of water retained by the soils modified with harvest straw biochar was 1.4%, 6.1%, and 18.4% higher in the treatments with 20, 40, and 60 g kg^{-1} of biochar, respectively, compared to the unmodified soil (Sun and Lu 2014). The application of 50 Mg ha^{-1} of biochar derived from herbaceous plant cuttings to a sandy soil increased porosity, but the hydrophobicity of the biochar may have prevented water from penetrating into the soil pores, which limited the impact of biochar on soil water retention (Jeffery et al. 2015). The biochar derived from switchgrass (*Panicum virgatum*) at an application rate of 25% increased the water-holding capacity of sand by 370% when compared to pure sand (Brockhoff et al. 2010).

The influence of biochars on soil water-holding capacity is also related to soil texture and may have a limited impact when the pore volume of the biochar is low or the soil texture is fine, with a greater presence of clay. Wang et al. (2019) studied a sandy (medium-to-coarse texture) and a clayey (fine texture) soil amended with softwood (600–700°C, low SSA) and walnut shell (900°C, high SSA) biochars. The biochar with higher SSA increased the field capacity of the sandy soil; however, the clayey soil was not influenced by the addition of the biochars. The authors suggest that biochars with a high pore volume can increase the field capacity and available water in coarse-textured soil, until the internal pores of the biochar are filled by clay and soil organic matter. A meta-analysis quantified that across all soil textural groups, biochar on average decreased soil bulk density by 9% (Razzaghi et al. 2020). The authors observed that field capacity, wilting point, and available water content increased for coarse-textured soils (by 51%, 47%, and 45%, respectively) compared to fine-textured soil, suggesting that biochar may have greater benefit in coarse-textured soils.

5.2.5 Organic Matter (OM)

The addition of biochar can increase the OM content of the soil, being determined by the amount and stability of the biochar (Glaser et al. 2001; Zygourakis 2017). Biochar promotes the polymerization of small organic molecules through surface catalytic activity to form soil OM, and the macropores can adsorb small organic molecules in the soil (Liang et al. 2010). Application of biochar to soils can maintain OM levels and soil aggregation stability (Kimetu and Lehmann 2010), as biochar is characterized by recalcitrant C from microbial degradation and a surface loaded with organic functional groups.

As biochar is considered a significant source of organic carbon (OC), studies investigate how biochar application affects the dynamics of humic components and soil OC. The increase in soil OC content was observed when applying 20 Mg ha^{-1} of paper fiber sludge biochar, increasing the OM level of the soil (Šimanský et al. 2016). The addition of 20 Mg ha^{-1} of biochar from grain hulls and paper dregs in a 1:1 ratio increased soil OC content by 28.69% compared to soil without biochar addition (Juriga et al. 2018). Wheat straw biochar applied at different application rates (4 and 8 Mg ha^{-1}) influenced the dissolved organic matter (DOM) content in wheat–maize rotation system soils. The higher addition of biochar increased the average dissolved organic carbon (DOC) concentration from 83.99 to 144.27 mg kg^{-1} in soil (Zhang et al. 2017).

5.2.6 Cation Exchange Capacity (CEC)

CEC is an indirect measure of the ability of soils to retain and exchange nutrients. The acidic aromatic carbon on the biochar surface is oxidized to form abundant functional groups (-OH, -COOH), increasing the cation adsorption capacity of the soil (Atkinson et al. 2010). Even the addition of a small amount of biochar can increase the content of cations and alkali nutrients in the soil (Hossain et al. 2010). Soils amended with biochar increased CEC by 4% to 30%, as compared to unamended soils (Laird et al. 2010). The CEC of a weathered soil increased from 7.41 to 10.8 cmol kg^{-1} after being amended with *Leucaena leucocephala* biochar (Jien and Wang 2013). The addition of biochar to highly organic soils may not improve the CEC of the soil due to the high organic matter contents they already have and consequently a high CEC (Schulz and Glaser 2012).

5.2.7 Soil pH

The application of biochar can increase soil pH value and increase base saturation. After pyrolysis, most biochar samples are alkaline due to the presence of basic functional groups in them that retain alkaline components (nitrates, carbonates, hydroxides) and inorganic minerals (Zhang et al. 2021). Rice husk biochar increased acidic soil pH from 3.33 to 3.63 (Wang et al. 2014). Soil pH increased from 7.6 to 7.8 with the application of corn residue biochar produced at 350 and 500°C (Karimi et al. 2020). Increased pH of acidic soils after biochar addition has also been observed by several researchers (Novak et al. 2009; Zhang et al. 2019; Shetty and Prakash 2020). Soil pH increased from 4.59 to 4.86 in soil corrected with ~ 1 Mg ha^{-1} (Nielsen et al. 2018). Thus, biochars with alkaline characteristics can effectively improve the pH of acidic soils (Van Zwieten et al. 2010).

Changes in soil pH can directly affect the behavior of ionizable herbicides. The sorption of these herbicides is strongly influenced by pH, and this dependence is related to the different proportions of ionic and neutral forms of the herbicide at each pH level and their sorption strength (Kah and Brown 2007). Generally, weak acidic herbicides assume their anionic (dissociated) form when soil pH increases, being more available in the soil solution (Green and Hale 2006; Pintado et al. 2011). The addition of rice-straw-derived biochar increased soil pH by 0.36–1.36 units relative to unamended soil. As pH increased, the fraction of dissociated sulfentrazone increased by 10.2–17.4% (Liu et al. 2016). However, these authors observed that the sorption of sulfentrazone in amended soils increased between 1.5 and 25 times compared to unamended soils which decreased bioavailable concentrations and reduced injury levels in *Oryza sativa*. The relationship of changes in soil pH by the addition of biochar and the bioavailability of ionizable herbicides is still little discussed but should be considered since the addition of biochar directly influences the OM content and pH of the soil and can alter the behavior of ionizable herbicides.

5.3 HERBICIDE SORPTION AND DESORPTION IN BIOCHAR-AMENDED SOILS

The first report of biochar's efficiency in pesticide sorption was in the 1960s, when burned sugarcane waste showed a high affinity for applied pesticides (Hilton and Yuen 1963). Yamane and Green (1972) observed an increased sorption of ametryn and atrazine with the application of biochar to soil, indicating that the sorption isotherm exhibited a shift from linear to nonlinear form, showing sorption, rather than absorption (partitioning), as the dominant sorption mechanism in the presence of these carbonized organic materials.

Sorption refers to the ability of the soil to retain the herbicide molecule by decreasing its activity in the soil solution. On the other hand, the return or release of the sorbed substance is called desorption (Mendes et al. 2021a). Sorption–desorption processes have a direct impact on the ecosystem because they influence the availability of organic pollutants in the soil solution and affect plant uptake, microbial degradation, and leaching potential in the soil profile (Jensen et al. 2018). Three different processes occur in biochar sorption, including adsorption, absorption, and precipitation (Chen et al. 2008; Qiu et al. 2009). In these processes, chemical bonds can be formed by the functional groups on the biochar surface with ions or organic compounds, which lead to adsorption from the surface; in addition, clay minerals present in the soil interact with the biochar and influence the sorption–desorption processes (Liu et al. 2018). Both processes ensure that the herbicide is sorbed into the biochar and not available for root uptake or leaching into the soil profile (Figure 5.1). The studies with biochar originating from different feedstocks, and pyrolysis temperatures on herbicide sorption and desorption are presented in Table 5.1.

The sorption capacity of biochar affects the environmental behavior of herbicides in soil, as it decreases the bioavailability of herbicides to the environment (Khorram et al. 2016). However, the effect of sorption is not completely reversible and not all herbicides sorbed in biochar can be desorbed, suggesting that desorption is much more difficult than sorption (Liu et al. 2018). In the studies conducted by Khorram et al. (2016, 2018), fomesafen was shown to have high affinity for biochars derived from different feedstocks (Table 5.1). Desorption was reduced with the addition of the

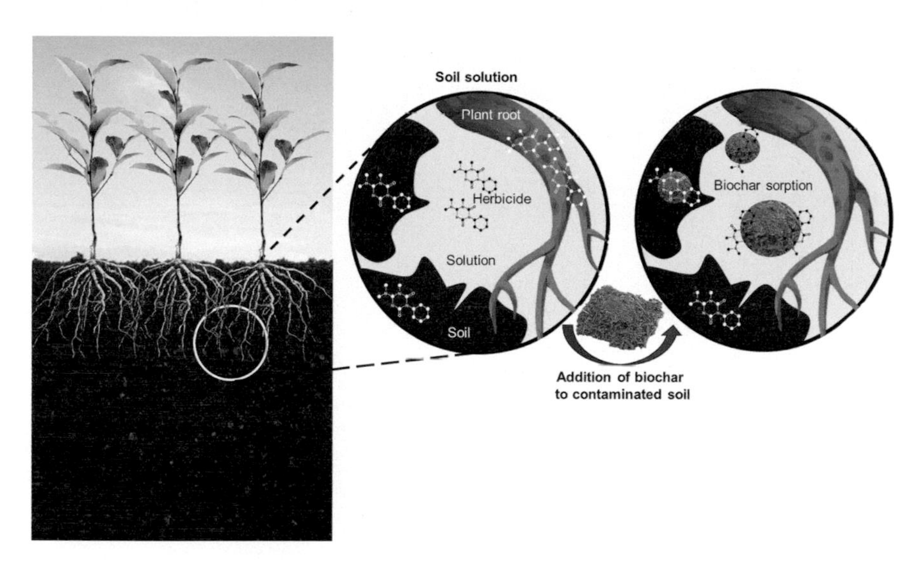

FIGURE 5.1 Herbicide sorption process in biochar-amended soil.

biochars, where, even using relatively low biochar application rate (0.5%), desorption decreased between 9 and 11 times in the different biochars tested compared to the unamended soil.

Biochar typically increases the sorption of herbicides (Table 5.1). 2,4-D was sorbed approximately 100 times more in soil amended with 10% (w/w) biochar regardless of the feedstock source used (Clay et al. 2016). These authors observed that the soil had an influence in enhancing the sorption of biochar. In general, a lower sorption was observed in soil with lower organic matter content (Table 5.1). Metribuzin showed up to a three-fold increase in sorption compared to unamended soil when olive residue biochar was added, reducing the amount of metribuzin leached and the risk of groundwater contamination in soils with low organic matter content (López-Piñeiro et al. 2013).

Increasing herbicide sorption by soil amendment with biochar proved to be an important tool to increase sorption efficiency in situations where the amount of herbicide in the soil needs to be reduced – in this case remediation. However, the application rate of biochar and the source of raw material used directly influence the sorptive capacity of biochar, and it is important to analyze the interaction of these factors for each application site.

5.4 APPLICATION RATE OF BIOCHAR DERIVED FROM DIFFERENT FEEDSTOCKS

Interest in the use of charcoal in agriculture was sparked by studies with "Terra Preta de Índio", anthropogenically created soil in the Amazon, enriched with carbonized organic material (Verheijen et al. 2009). These soils contain high levels of nutrients (mainly P, Ca, and Mn) and organic matter, exhibit higher microbiological activity

TABLE 5.1
Herbicide Sorption and Desorption in Biochar-Amended Soils.

Local (country)	Soil texture (%)				Feedstock	Application rate (%)[b]	Pyrolysis temperature (°C)	Herbicide	Sorption (number of times)[c]	Desorption (number of times)[c]	References
	Sand	Silt	Clay	OM[a]							
China	15.2	73.2	11.5	2.1	Rice husk	0.5	600	Fomesafen	4	-10.5	Khorram et al. (2016)
						1			8	-23	
						2			7	-51	
China	15.2	73.2	11.5	2.1	Rice straw	0.5	500	Fomesafen	3	-10	Khorram et al. (2018)
						1			11	-23.5	
						2			24	-52	
					Softwood	0.5	500		3.5	-9	
						1			11	-23	
						2			25	-47	
					Coconut shell	0.5	550		3.5	-10	
						1			12	-24	
						2			27.5	-55.5	
					Hardwood	0.5	700		4	-12	
						1			13	-29	
						2			30	-73	
					Walnut shell	0.5	400		3.5	-9	
						1			12	-21	
						2			26	-46	
					Bamboo	0.5	800		3.5	-11	
						1			12.5	-23	
						2			28	-61	

Spain	24	47	30	1.3	Hardwood (*Carya tomentosa*)	2	350	Clomazone	2	-2	Gámiz et al. (2017)
							400		1	-1	
							700		11	-7	
					Hardwood (*Carya illinoinensis*)		350		2.5	-2.5	
							400		1.5	-1.5	
							700		8.5	-6.5	
					Hardwood (*Carya tomentosa*)		350	Bispyribac-Sodium	2	-2	
							400		2	-1.5	
							700		7	-14	
					Hardwood (*Carya illinoinensis*)		350		2	-2	
							400		2	-2	
							700		11.5	-7.5	
Malaysia	40	21.5	37.9	0.99	Oil palm empty fruit bunches	1	300	Imazapic	2.5	-2.5	Yavari et al. (2016)
							500		1	-1	
							300	Imazapyr	3	-3.5	
							500		1	-1	
					Rice husk		300	Imazapic	2.5	-2.5	
							500		1	-1	
							300	Imazapyr	2.5	-2.5	
							500		1	-1.5	
Germany	73.3	23.1	3.6	1.7	Beech wood	1.5	550	Imazamox	1	-1	Dechene et al. (2014)
The USA	22	55	23	>2	Mixed sawing	5	500	Atrazine	1	n.a.[d]	Spokas et al. (2009)

(*Continued*)

TABLE 5.1 (Continued)

Local (country)	Soil texture (%)				Feedstock	Application rate (%)[b]	Pyrolysis temperature (°C)	Herbicide	Sorption (number of times)[c]	Desorption (number of times)[c]	References
	Sand	Silt	Clay	OM[a]							
Australia	n.a.	n.a.	n.a.	4.5	Paper mill (sludge)	1	550	Atrazine	2.5	-1	Martin et al. (2012)
					Poultry litter				5	-1	
USA	60	24	16	3.4	Corn straw	1	850	Atrazine	1.5	n.a.	Clay et al. (2016)
						10			4.5		
					Grass (*Panicum vigatum*)	1			1.5		
						10			3.5		
					Pine chips (*Pinus ponderosa*)	1			1.5		
						10			3		
	74	19	8.5	1.6	Corn straw	1			1		
						10			6		
					Grass (*Panicum vigatum*)	1			1		
						10			3		
					Pine chips (*Pinus ponderosa*)	1			4.5		
						10			5		
China	6.7	55.9	7.40	.4	Wheat straw	3	300	Atrazine	1	n.a.	Ren et al. (2018)
							700		5.5		
	19.7	68.4	11.9	.6			300		1.5		

							700		13.5		
China	n.a.	n.a.	.a.	3.2	Sugarcane bagasse	0.2	500	Atrazine	1.5	n.a.	Huang et al. (2018)
						0.5			2		
				2.0		0.2			1.5		
						0.5			2.5		
				3.6		0.2			1.5		
						0.5			2		
China	34	37.8	28.2	0.89	Cassava husk	0.1	750	Atrazine	1.5	-1	Deng et al. (2017)
						0.5			2.5	-1.5	
						1			3.5	-1.5	
						3			4	-2	
						5			5	-2.5	
The USA	22	55	23	>2	Mixed sawing	5	500	Acetochlor	1.5	n.a.	Spokas et al. (2009)
The USA	28	56	16	.4	Soybean residues	10	500	Alachlor	8.5		Mendes et al. (2017)
					Sugarcane bagasse		350		7		
					Wood chips (grape)		500		20		
	10	35	25	3.2	Soybean residues		500		15.5		
					Sugarcane bagasse		350		7	n.a.	
					Wood chips (Grape)		500		26		
	17	56	27	4.9	Soybean residues		500		3.5		
					Sugarcane bagasse		350		8		
					Wood chips (Grape)		500		8		
USA	60	24	16	3.4	Corn straw	1	850	2,4-D	95	n.a.	Clay et al. (2016)

(Continued)

TABLE 5.1 (Continued)

Local (country)	Soil texture (%)				Feedstock	Application rate (%)[b]	Pyrolysis temperature (°C)	Herbicide	Sorption (number of times)[c]	Desorption (number of times)[c]	References
	Sand	Silt	Clay	OM[a]							
						10			150		
					Switchgrass (*Panicum vigatum*)	1			130		
						10			127		
					Pine chips (*Pinus ponderosa*)	1			18		
						10			127		
	74	19	8.5	1.6	Corn straw	1			35		
						10			160		
					Switchgrass (*Panicum vigatum*)	1			50		
						10			53		
					Pine chips (*Pinus ponderosa*)	1			10		
						10			50		
Spain	22	55	23	2.5	Wood chips pellets	10	>500	Aminocyclopyrachlor	25	n.a.	Cabrera et al. (2014)
								Bentazone	0		
					Macadamia nut shells		850	Aminocyclopyrachlor	-2		
								Bentazone	1.5		
					Hardwood		540	Aminocyclopyrachlor	-1		
								Bentazone	2		
Slovakia	29.1	60.1	10.8	1.47	Wheat straw (*Triticum aestivum* L.)	1	300	MCPA	2.5	-1	Tatarková et al. (2013)
The USA	n.a.	n.a.	n.a.	1.2	Charcoal derivative	1	350	Terbuthylazine	3.5	-2.5	Wang et al. (2010)
					Sawdust derivative		700		64	-14.5	

				.0	Charcoal derivative		350		1	-1.4	
					Sawdust derivative		700		2.5	-3	
The USA	22	55	23	2.5	Macadamia nut shells	10	850	Metolachlor	1	n.a.	Trigo et al. (2016)
					Wood chips pellets		550		2		
	74	12	14	0.14	*Eucalyptus* spp.	0.1	500	Isoproturon	1	-1.5	Sopeña et al. (2012)
The UK						1			5	-3.5	
						2			8.5	-9	
Australia	87.8	1.3	8.3	1.4	*Eucalyptus* spp. Derivative	0.1	450	Diuron	1	-2	Yu et al. (2006)
						0.5			6.5	-8.5	
						1			126	-14.5	
						2			26.5	-31	
						5			77	-79	
						0.1	850		6	-7	
						0.2			17.5	-25.5	
						0.5			55	-68.5	
						0.8			85.5	-106.5	
						1			122	-134	
Australia	n.a.	n.a.	n.a.	4.5	Paper mill (sludge)	1	550	Diuron	2	-1	Martin et al. (2012)
					Poultry litter				4.5	-1	
Brazil	62	7.3	30.7	1.26	Biomass of Cerrado vegetation	0.8	450	Diuron	1	-1	Petter et al. (2016)
						1.6			2	-1	

(*Continued*)

TABLE 5.1 (Continued)

Local (country)	Soil texture (%)				Feedstock	Application rate (%)[b]	Pyrolysis temperature (°C)	Herbicide	Sorption (number of times)[c]	Desorption (number of times)[c]	References
	Sand	Silt	Clay	OM[a]							
Costa Rica	13	10	77	n.a.	Pineapple stubble	0.5	300	Diuron	3	n.a.	Chin-Pampillo et al. (2021)
						1			4		
						0.5	600		2		
						1			3		
					Palm oil fiber	0.5	300		4		
						1			6		
						0.5	600		1.5		
						1			1.5		
					Coffee hull	0.5	300		.5		
						1			2		
						0.5	600		1.3		
						1			1		
Russia	3.1	30.4	66.5	n.a.	Woods (*Betula* sp. and *Picea abies*)	1	400	Diuron	1	n.a.	Zhelezova et al. (2017)
						10			10.5		
						20			22		
						30			56.5		
						1		Glyphosate	1		
						10			1		
						20			1		
						30			-1		

	83.7	8.8	7.5	n.a.		1		Diuron	1.5		
						10			23.5		
						20			22		
						30			49		
						1		Glyphosate	-1		
						10			-1		
						20			-1.5		
						30			-2		
Brazil	88.8	3.7	7.5	0.58	Natural Biochar (extracted from the soil)	0.1	n.a.	Diclosulam	1	-1	Mendes et al. (2019b)
						0.5			1.5	-1	
						1			2	-1	
						0.1	n.a.	Pendimethalin	1	-1	
						0.5					
						1					
									1	-1	
									1.5	-1	
Brazil	29.5	19.6	50	4.6	Bonechar (cow bone)	1	n.a.	Aminocyclopyrachlor	1	-1	Mendes et al. (2018)
						5			96	-10	
						10			200	-20	
						100			1000	-100	
						1		Mesotrione	1	-1	
						5			21	-2	
						10			26	-7	
						100			60	-51	

(Continued)

TABLE 5.1 (Continued)

Local (country)	Soil texture (%)				Feedstock	Application rate (%)[b]	Pyrolysis temperature (°C)	Herbicide	Sorption (number of times)[c]	Desorption (number of times)[c]	References
	Sand	Silt	Clay	OM[a]							
Brazil	81.6	3.3	15.1	0.99	Bonechar (Cow bone)	5	n.a.	Hexazinone	173	-3	Mendes et al. (2019c)
						100			3000	-62	
						5		Metribuzin	172	-5	
						100			876	-60	
						5		Quinclorac	1200	-189	
						100			1300	-197	
	28	56	16	2.4	Soybean Residues	10	500	Metolachlor	12.5	-2	Mendes et al. (2020, 2021b)
The USA								Aminocyclopyrachlor	1	-1	
								Indaziflan	11.2	n.a.	
					Sugarcane bagasse		350	Metolachlor	5.5	-1.5	
								Aminocyclopyrachlor	1	-1	
								Indaziflan	5.5	n.a.	
					Wood chips (grape)		500	Metolachlor	18	-4	
								Aminocyclopyrachlor	1	-1	
								Indazilan	18	n.a.	
	17	56	27	2.7	Soybean residues		500	Metolachlor	12	-1.5	
								Aminocyclopyrachlor	1	-1	

								Indaziflan	8.5	n.a.	
					Sugarcane Bagasse		350	Metolachlor	5	-1.5	
								Aminocyclopyrachlor	1	-1	
								Indaziflan	5.5	n.a.	
								Metolachlor	1	-3.5	
					Wood chips (Grape)		500				
								Aminocyclopyrachlor	1	-1	
								Indaziflan	18.5	n.a.	
					Soybean residues		500	Metolachlor	15	-1.5	
	10	65	25	1.9				Aminocyclopyrachlor	1	-1	
								Indaziflan	3.5	n.a.	
					Sugarcane Bagasse		350	Metolachlor	5	-1	
								Aminocyclopyrachlor	1	-1	
								Indaziflan	3	n.a.	
					Wood chips (grape)		500	Metolachlor	10	-2	
								Aminocyclopyrachlor	1	-1	
								Indaziflan	7.5	n.a.	
Spain	43	32	23	0.9	Olive mill waste	2.5	n.a.	Metribuzin	1.5	n.a.	López-Piñeiro et al. (2013)
						5			2.5		
					Olive mill waste plus leaves	2.5			2		
						5			2.5		
	53	32	14	0.6	Olive mill waste	2.5			2		
						5			3		
					Olive mill waste plus leaves	2.5			2		
						5			2.5		
	43	14	42	0.9	Olive mill waste	2.5			1		

(*Continued*)

TABLE 5.1 (Continued)

Local (country)	Soil texture (%)				Feedstock	Application rate (%)[b]	Pyrolysis temperature (°C)	Herbicide	Sorption (number of times)[c]	Desorption (number of times)[c]	References
	Sand	Silt	Clay	OM[a]							
						5			1.5		
					Olive mill waste plus leaves	2.5			1.5		
						5			2		
Brazil	24	36	30	1.18	Wood waste (*Eucalyptus urophylla* x *E. grandi*)	1	650	Hexazinone	1[e]	-1	Fernandes et al. (2021)
							750		1	-1	
							850		1.7	-0.7	
							950		5.3	-0.4	

[a]Organic matter, [b]application rate *vs.* soil mass (w w^{-1}), [c]positive or negative values mean increased or decreased sorption and/or desorption with biochar addition compared to unamended soil, [d]data not available, [e]sorption and desorption values at lower temperature (650°C).

than nearby soils, and also contribute to carbon sequestration (Glaser and Birk 2012). The origin of this kind of soil is the result of the accumulation of plant and animal organic waste that has been subjected to intensive fire use (Sousa et al. 2015). Due to these characteristics, "Terra Preta de Índio" may have potential in the sorption and desorption of pre-emergence-applied herbicides. The sorption and desorption of diuron were studied in two anthropogenic soils in contrast to a sandy Entisols Quartzipsamments soil (Almeida et al. 2020). Sorption of 99% of the herbicide was observed in anthropogenic soil, while in sandy soil it was 60.8%. The desorption of diuron was very low in anthropogenic soils, ranging from 1.36% to 1.70%, and for the sandy soil it was 24%.

Biochars have different physicochemical properties depending on the feedstock used (Janus et al. 2015), production conditions, and nutrient content of the charred materials (Yuan et al. 2019). Biochar can be produced from various carbonaceous materials, including wood, coal, fruit peels, manure, fermentation, agricultural waste, and food processing waste (Table 5.1). Biochar produced from non-woody feedstocks, such as compost and plant residues, has high moisture, high ash content, lower calorific value, low bulk density, nutrient rich, higher pH, and less stable carbon than biochar produced from lignocellulosic feedstocks, such as wood (Mukherjee et al. 2014; Sigua et al. 2015; Jafri et al. 2018). Lignin-based feedstock can be used to produce biochar, being a widely available resource generated by pulp industries (Hu et al. 2018). However, studies using this biochar to immobilize herbicides are still scarce. The quality of biochar produced from lignin-rich waste can be very high, although the costs are also higher compared to other feedstock sources (Gul et al. 2021).

Animal bonechar has high porosity and has the potential to act as a sorbent when added to soil. However, unlike biochar, which is predominantly carbon, bonechar is composed of tricalcium phosphate (hydroxyapatite) with low total carbon content (Mendes et al. 2018). Bonechar showed high sorption potential at 5% application rate for metribuzin, hexazinone, and quinclorac, both herbicides with high leaching capacity in soil [Groundwater Ubiquity Score (GUS) index = 2.96, 4.43, and 6.29, respectively], decreasing the potential for leaching of these herbicides (Mendes et al. 2019c).

Biochar can also be obtained from natural sources originating from anthropogenic soil disturbances, such as vegetation fires, where the oxygen supply is limited (Glaser et al. 2001; Novotny et al. 2009). The physicochemical characteristics of a natural biochar extracted from an Oxisols profile were studied by Mendes et al. (2019b). The carbonaceous material showed a high carbon content (93.28%), low pH value (5.9), irregular surface, high porosity, with pores of varying sizes. Application of natural biochar slightly increased the percentage sorbed of diclosulam and pendimethalin in soil (Mendes et al. 2019b). However, the variation in sorption was more representative for diclosulam compared to pendimethalin due to diclosulam's physicochemical properties of low sorption coefficient and octanol-water (PPDB 2021). The authors evidenced the low sorption capacity of diclosulam (K_d = 0.22 L kg^{-1}) in relation to pendimethalin (K_d = 25.55 L kg^{-1}) in the unamended soil, showing that the addition of natural biochar directly influenced the sorption and desorption of diclosulam due to the higher availability of the herbicide in the soil solution.

Diuron showed high sorption dependent on the type of feedstock used in biochar production (Table 5.1). Although diuron is a non-ionizable (neutral) herbicide in soil solution, it can exhibit polarity and be influenced by soil physicochemical properties, clay minerals, and OM (Rocha et al. 2013). The higher sorption capacity of the soil after the application of biochar from phyto-physiognomy of the Cerrado Brazilian was attributed to the direct effect of biochar on the OM of the soil and consequently on the sorption coefficient and attributed to the effect of the charges generated by the oxidation of aromatic structures in the biochar (Petter et al. 2016). The application rate of 1% biochar from *Eucalyptus* ssp. derivatives increased the sorption of diuron by 125 times and decreased the desorption by 14 times compared to unamended soil (Yu et al. 2006). However, the sorption of diuron in soils modified with biochar from paper mill sludge and poultry manure showed 2 and 4.5 times higher sorption, respectively, than the unamended soil (Martin et al. 2012). The pineapple plant residue biochar showed higher potential in the sorption of diuron than the oil palm fiber and coffee husk biochars (Chin-Pampillo et al. 2021).

Aminocyclopyrachlor and glyphosate behaved differently from the other herbicides (Table 5.1). Soils modified with macadamia nut shell and wood biochars showed lower sorption of aminocyclopyrachlor compared to unamended soil, being attributed to the competition between dissolved organic carbon of the biochars and the herbicides for sorption sites in the soil (Cabrera et al. 2014). Mendes et al. (2020) found similar results (Table 5.1) in which the highest percentage of aminocyclopyrachlor sorbed was in the unamended soil (10.83%). These authors associated the low sorption with the physicochemical characteristics of the herbicide. Aminocyclopyrachlor is a weak acid ($pK_a = 4.6$) (PPDB 2021) and assumes its dissociated form when the soil pH is above the pK_a of the herbicide. The authors observed that pH is greater than pK_a in all the soils studied (pH from 6.0 to 6.7), which may have favored the low sorption of the herbicide in the soil. The high solubility in water (S_w = 3130 mg L^{-1} at 20°C) (PPDB 2021) and low Freundlich sorption coefficient in soil (K_f = 1.13 $\mu mol^{(1-1/n)}\ L^{1/n}\ kg^{-1}$) (Mendes et al. 2019d) of aminocyclopyrachlor also ensure stability in aqueous solutions, decreasing the sorption capacity of the soil, even when amended with biochar. However, bonechar residues showed potential in sorbing aminocyclopyrachlor and mesotrione (Mendes et al. 2018). The 10% application rate of bonechar to soil provided more than 93% sorption for mesotrione and 89% of aminocyclopyrachlor with very low desorption (<10%). Both herbicides were leachable in unamended soil; however, the application of bonechar effectively reduced the leaching of both herbicides such that 100% of the herbicides remained in the surface layer without reaching the deeper layers of the soil profile.

In general, the application rate of biochar directly influences the sorption and desorption of herbicides (Table 5.1). Application rates of 1% and 10% gave a low sorption of glyphosate on clayey soil. However, application rates of 20% and 30% on clay soil and all application rates (1%, 10%, 20%, and 30%) on sandy soil decreased glyphosate sorption compared to unmodified soil (Zhelezova et al. 2017). Sorption of atrazine was tested on different feedstocks, showing affinity for the different biochars (Table 5.1). However, it is observed that the variation in the application rate of biochar has a direct influence on the sorption and desorption of the herbicide. Varying the application rate from 0.1% to 5% increased sorption of atrazine by 5 times and decreased desorption by 2.5 times in the two biochars studied (Deng et al. 2017). For fomesafen, varying the application rate from 0.5% to 2% increased sorption by approximately 25 times and decreased desorption by 5 times regardless of the feedstock used (Khorram et al. 2018).

Each herbicide presents different physical–chemical characteristics, just as the characteristics of biochars depend on the raw material and the production process. Besides the herbicide and biochar physical–chemical properties, the rate of application to the soil is also an important factor, that is, both factors should be analyzed for the adoption of this technique.

5.5 PRODUCTION TEMPERATURE AND IMPACTS ON HERBICIDE SORPTION AND DESORPTION

The technologies that enable biomass recycling for bioenergy production are pyrolysis and hydrothermal carbonization that give rise, respectively, to biochar and hydrochar as solid by-products (Lofredo et al. 2019). Biochar is prepared at temperatures of 200–800°C under limited oxygen conditions (Lehmann and Joseph 2009), while hydrochar is obtained at temperatures of 180–250°C under high pressure (about 2 MPa) in the presence of water (Hitzl et al. 2015). In contrast to pyrolysis, this process does not require an energy-intensive pre-drying step (Mumme et al. 2011). Both materials, especially biochar, are stable and carbon-rich, and their properties are affected by the feedstock and production conditions (Lehmann and Joseph 2009; Hitzl et al. 2015). Although biochars are considered more promising for agricultural soils with respect to climate change mitigation, hydrochars have shown promise for soil water retention (Abel et al. 2013; Gascó et al. 2018).

The sorption of isoproturon was evaluated on soils amended with 0.5% and 5% biochar and hydrochar respectively, produced by different feedstocks (Eibisch et al. 2015). These authors reported lower bioavailability of isoproturon in soils with biochar than with hydrochar, confirmed by the higher amount of non-extractable residues of isoproturon at application rates of 0.5% and 5% (biochar: 18.6–29.6% *vs* 24.2–32.4%) and (hydrochar 4.9–7.7% *vs* 2.7–3.6%). In water, biochar showed a higher potential in the sorption of metribuzin and atrazine compared to hydrochar (Lofredo et al. 2019; Liu et al. 2019; Netto et al. 2021). Studies should be undertaken to optimize the carbonization conditions for hydrochar to increase its sorption capacity (e.g., by increasing pore volumes and SSA), ensuring adequate herbicide degradation under natural conditions (Eibisch et al. 2015).

The pyrolysis temperature of biochar influences sorption, and, in general, efficiency is increased at higher temperatures (Shaaban et al. 2018). The expansion of surface area by maximizing microporosity is the main reason for the increased sorption capacity of carbonized materials at high temperatures (Shinogi and Kanri 2003). Typically, biochars produced at relatively high pyrolysis temperatures (>500°C) are effective in sorption of organic contaminants, increasing SSA, microporosity, and hydrophobicity.

Pine wood biochar produced at 700°C showed higher sorption capacity than those produced at 350°C for terbuthylazine in soil, and this is due to the higher porosity and surface area responsible for sorption of the herbicide (Wang et al. 2010). The pyrolysis temperatures (650, 750, 850, and 950°C) were shown to have an effect on the properties of biochar derived from eucalyptus wood. The pyrolysis temperature of 950°C increased the ratios (nitrogen + oxygen), carbon, and ash and produced a biochar with higher sorption and lower desorption for hexazinone. The pyrolysis temperature of 650°C produced an aliphatic material, with lower sorption and higher desorption of hexazinone (Fernandes et al. 2021). The temperature of 350°C for different raw materials had no effect on glyphosate sorption (<0.5% of sorption) compared to unamended soil; however, with increasing temperature (>500°C), it was possible to sorb up to 68% of glyphosate from the soil solution. However, very high temperatures can cause pore deformation (Abe et al. 1998). Therefore, a maximum temperature value should be considered based on the type of raw material (Yavari et al. 2015). The trend is an increase in the SSA of biochars at temperatures between 800°C and 900°C, at which point the SSA begins to reduce due to structural distortion of the biochar (Downie et al. 2009).

The surface group chemistry is an effective parameter in the sorption efficiency of biochar and is mainly controlled by the pyrolysis temperature in the production process. For example, sorption of polar organic compounds is increased not only by increasing the SSA of the biochars, but also by the polarity of the biochar (Yavari et al. 2015). Increasing the functionality of biochar by decreasing production temperatures down to 150°C would be an effective way to improve sorption of more polar organic contaminants rather than increasing the SSA of biochar by increasing the temperature (Sun et al. 2011). Increasing pyrolysis temperature to 750°C resulted in an increase in SSA of biochar produced from rice husk while a temperature of 300°C increased the number of surface functional groups, especially Si–O–Si surface bonds, which explained the role of rice husk biochar produced at lower temperatures in the sorption of metolachlor in soil (Wei et al. 2017). Empty fruit bunch biochars showed an increased sorption of imazapyr and imazapic at lower pyrolysis temperature (300°C) than at higher temperature (700°C) (Yavari et al. 2016). The authors observed that the lower conversion temperature (300°C) increased the polar surface functional groups of the materials on their outer surfaces.

Production variables, including feedstock type and pyrolysis conditions, provide a wide diversity of biochars with different physicochemical characteristics. The scientific literature reports that higher pyrolysis temperatures increase the sorption of biochar; however, its impacts on different feedstocks are different, since biomasses have varied composition and structure. Therefore, the evaluation of different feedstocks under the different pyrolytic parameters and the optimization of the condition for each feedstock can adjust the characteristics of biochars and enhance their sorption ability.

5.6 FRESH AND AGED BIOCHAR IN HERBICIDE SORPTION AND DESORPTION

Aging of biochars is the main factor responsible for the post-pyrolysis changes in the sorption properties of biochars (Laird et al. 2009). Biochar undergoes physical and chemical changes over time in soils, as a result of which its pore structure and particle size are corrected, influencing its recalcitrant and pollutant sorption capacity

(Nocentini et al. 2010). Aging of biochar in fields is complex and includes several processes, such as abiotic oxidation, carbonate dissolution, chemisorption, interaction of soil constituents, physical disintegration, microbial degradation, macrofauna activity, water erosion, freeze–thaw cycle, and dilatation–retraction (wet–dry) (Wang et al. 2020). Oxidation and hydrolysis are the most effective reactions in the aging process, occurring very rapidly during the first few months of soil incubation (Yavari et al. 2016). During oxidation, an increase in oxygenated functional groups such as -OH, -C = O, and -COOH may occur on its surface and consequently result in an increased CEC value (Liang et al. 2006; Cheng et al. 2006).

Due to their strong sorption ability, biochars can sorb soil minerals and dissolved organic matter that block their micropores to saturation of the active sites and change their surface properties (Kookana 2010). On the other hand, erosion, dissolution, and transport by water flows in the soil also contribute to biochar aging (de la Rosa et al. 2018). Once volatile organic compounds and ash in the blocked pores are flushed out by flows, biochar porosity and SSA can be increased (Suliman et al. 2016).

Aging for 2 years of a macadamia nut biochar increased SSA and improved sorption of indaziflam and fluoroethyldiaminotriazine; however, it decreased sorption of MCPA (Trigo et al. 2014). The sorption–desorption of glyphosate in two tropical soils modified with eucalyptus-derived biochar aged for 30 days had no effect on the sorption–desorption of glyphosate compared to unmodified soils (Junqueira et al. 2019). Kumari et al. (2016) reported higher CEC and SSA in soils modified with wood biochar after 7–19 months, resulting in higher sorption of glyphosate. The sorption of three highly persistent and ionizable herbicides (two anionic: imazamox and picloram) and a weak base (terbuthylazine) in fresh and field-aged biochar for 6 months was studied (Gámiz et al. 2019b). The field-aged biochar immobilized all three herbicides (>85%) than the fresh biochar (<16%). The 10-year aging of charcoal residue biochar was evaluated for sorption and desorption of sulfometuron-methyl in soil (Alvarez et al. 2021). These authors observed that the sorption was approximately 16% (80 Mg ha^{-1} of biochar) and 11% (40 Mg ha^{-1} of biochar) larger than the unamended soil. The amount of herbicide desorbed into the soil solution was approximately 5% (40 Mg ha^{-1} of biochar) and 7% (80 Mg ha^{-1} of biochar) less than the unamended soil.

While adding fresh biochar to soil generally increases herbicide sorption, aging can also reduce herbicide sorption (Hale et al. 2011; Martin et al. 2012). For example, the sorption of atrazine in soil modified with biochar from paper mill sludge and chicken litter for 32 months was similar to that in unamended soil. Aging reduced the sorption capacity of diuron from 47% to 68% (Martin et al. 2012). The sorption of atrazine decreased sharply when the pig manure biochar was aged for 3 months; however, it was still 56.3% higher than in the amended soil (Ren et al. 2016). Fomesafen was less sorbed by rice husk biochar aging for 6 months; consequently, it showed higher desorption, leaching, and bioavailable fraction of the herbicide in soil (Khorram et al. 2018). The studies demonstrate the temporal variability of the sorption capacity of biochars due to soil exposure, which alters the efficacy and bioavailability of herbicides applied to the soil. However, it can be observed that the increase or decrease of sorption–desorption and degradation is also related to the physicochemical characteristics of the herbicide molecule. Thus, aging may limit the usefulness of biochar application for remediation purposes over time.

The degradation of biochar particles during the weathering process changes its mass and its effect on herbicide remediation (Mendes et al. 2018). The biochar mass loss was approximately 40% over 5 years of aging, regardless of the amount applied (30, 60, or 90 Mg ha^{-1}) (Dong et al. 2017). Biochar degradation allows sorbed herbicides to be released into the environment. On a molecular scale, once aliphatic C breaks down, disconnected and intact aromatic portions can break down through soil-constituent interactions and microbial decomposition (Shi et al. 2015). During the incubation period, there was a reduction in the organic carbon content of 80.4 mg L^{-1} of fresh biochar for 31.6 mg L^{-1} of biochar aged for 1 year, which may be due to natural elimination or degradation (Trigo et al. 2014). These chemical and structural changes that influence the sorptive capacity of aged biochar vary according to the characteristic of each carbonaceous material, as well as the incubation time of the biochar, pyrolysis temperature, and environmental conditions.

5.7 BIOCHAR MODIFIED IN HERBICIDE SORPTION AND DESORPTION

Modification refers to the activation of the original biochar by physical and chemical methods (Yang et al. 2019). Chemical and physical activation are the most commonly used for biochar modification. Physical activation is also known as thermal or dry activation (Chen et al. 2011). In this physical activation process, the precursor will be pyrolyzed or carbonized (<800°C) and then activated using steam or CO_2 (Ioannidou and Zabaniotou 2007). This means that there are two main steps in physical activation: carbonization and activation. Chemical activation includes oxidation, sulfonation, amination, impregnation with nanoparticles, and others. Some of the widely used activating agents are classified as acids (HNO_3, H_3PO_4), bases (KOH, NaOH), oxidants (H_2O_2) and sulfonation agents (SO_3H and $ZnCl_2$) (Anto et al. 2021). Oxidation is one of the most studied methods in activation. Chemical oxidation refers to the oxidation of the biochar surface to increase oxygen-containing functional groups such as -OH and -COOH, increasing hydrophilicity. At the same time, the pore size and structure of the biochar would be changed, and finally its sorption capacity for the herbicide would be increased (Yang et al. 2019).

Biochar can have its physicochemical characteristics improved by activation methods. Some studies have shown that the modified biochar has a better effect on the sorption of herbicides than the raw biochar in soil; however, most of the studies showed satisfactory results for sorption of herbicides from water, which promotes further research using this technique for the sorption of herbicides from soil (Table 5.2). The effects of activation on biochars can be the result of a variety of mechanisms, such as changes in ash/mineral ratio, oxygen (O) functionality, aliphatic carbon, SSA, pH, total pore volume π–π interactions or combinations thereof (Gámiz et al. 2019a; Anto et al. 2021). Cyhalofop had improved sorption on biochar activated with H_2O_2 due to the decrease in pH and the availability of O (Gámiz et al. 2019a). The authors observed that ash and C labile (aliphatic groups) had no effect on the sorption of the herbicide. The biochar activated with Nano-MgO increased hydrophilicity thus improving hydrogen bonding and electrostatic attraction between biochar and atrazine and consequently leading to greater sorption (Cao et al. 2021).

TABLE 5.2
Effects of Biochar Activation on Herbicide Sorption.

Local (country)	Feedstock	Pyrolysis temperature (°C)	Substrate	Activation method	Effects of modification on biochar property	Results on herbicide sorption in relation to control	References
Spain	Grape wood pruning waste	350	Water	Hydrogen peroxide (H_2O_2)	Decrease in biochar pH, ash content and carbon content. Biochar oxygen content and surface area increased after activation	Cyhalofop increased sorption from 6.3% to 35.4%, while clomazone did not significantly increase (65.0% to 70.3%)	Gámiz et al. (2019a)
Korea	Ground coffee residues	800	Water	Sodium hydroxide (NaOH)	Increased the specific surface area and total pore volume	Significant increases in the removal rates were found for the herbicides alachlor (from 6.0% to 56.6%), simazine (from 3.1 to 47.4%) and diuron (from 4.7% to 80.6%)	Lee et al. (2021)
China	Fallen leaves	500	Water	Magnesium oxide (Nano-MgO) precursor	Increased the hydrophilicity	The adsorption capacity of biochar was increased by 1.99–5.71 times to atrazine	Cao et al. (2021)
Vietnam	Corn cob biomass	600	Water	Hydrofluoric acid (HF)	Improved the biochar-specific surface area and total pore volume	Removal of 60% of 2,4-D from the water. The biochar had a maximum sorption potential of 37.4 mg g^{-1}	Binh and Nguyen (2020)
China	Wheat straws	500	Soil	Hydrochloric acid (HCl)	[a]n.a.	Decreased the concentrations of fomesafen in wheat roots and shoots in 80%	Meng et al. (2019)

(*Continued*)

TABLE 5.2 (Continued)

Local (country)	Feedstock	Pyrolysis temperature (°C)	Substrate	Activation method	Effects of modification on biochar property	Results on herbicide sorption in relation to control	References
Malaysia	Oil palm empty fruit bunches	300	Soil	Chitosan modification	Cation exchange capacities significantly increased	The capacity of soil for imazapic sorption enhanced from 50% to 73% and from 8% to 84% to imazapyr	Yavari et al. (2020)
	Rice husk biomasses					The capacity of soil for imazapic sorption enhanced from 25% to 36% and from 36% to 67% to imazapyr	
China	Corn straw	500	Water	Sodium sulfide (Na_2S)	H/C ratio slightly decreased, while the O/C and (O + N)/C ratio increased	Atrazine increased by 46.39%	Tan et al. (2016)
				Potassium hydroxide (KOH)		Atrazine increased by 38.66%	
The USA	Cherry pits, *Jatropha presscake* waste pellets, chopped bamboo pieces, sugarcane bagasse pellets, pine forestry waste pellets, and pecan shells	850	Water	Updraft gasifiers under conditions of simultaneous co-pyrolysis thermal air activation	Increase 2.5 times more mesopore SSA (from 40 m^2 g to 110 m^2 g)	2,4-D sorption was more than 10 times greater in biochar-amended soil	Kearns et al. (2019)

[a]Data not available.

The SSA and pore volume of biochar can be improved with biochar activation. Increases in SSA and pore volume from 6.7 to 278.2 $m^2 g^{-1}$ and from 24,4 to 154.2 $mm^3 g^{-1}$, respectively, were reported when pomelo peel biochar was activated by KOH (Zhao et al. 2018). SSA is the main parameter for estimating the sorption capacity of biochar. If a high SSA exposes on the biochar surface, it brings more interfaces to sorb the contaminant (Tan et al. 2016). The SSA and pore volume of NaOH-activated coffee husk biochar were approximately 106 and 21 times, respectively, larger than those of non-activated biochar, improving the sorption of alachlor, simazine and diuron (Lee et al. 2021). Characterization of biochar from empty fruit bunches of palm oil and rice husk showed that the activation of biochar with chitosan decreased its total surface areas. However, its SSA increased, notably improving the sorption capacity of imazapic and imazapyr compared to unmodified soil and soil modified with unactivated biochars (Yavari et al. 2020). Thus, these modifications ensure that the biochar expresses its greatest potential in the sorption of herbicides, and further studies with different activation methods and raw materials under field conditions are important, seeking greater efficiency of herbicide sorption.

5.8 CONCLUDING REMARKS

Biochars are obtained from various raw materials, which are produced under different pyrolysis conditions, but present heterogeneous physicochemical characteristics that interfere with the efficiency in the removal of herbicides from the soil. Biochar aging affects herbicide sorption in soil over time, and herbicide application may need to be adjusted until the biochars lose their herbicide-deactivating capacity. Thus, these relationships should be further studied due to the frequent use of biochar in agriculture, which can present a high sorption capacity of pre-emergence applied herbicides and reduce the efficiency of weed control.

Biochars have great potential in increasing sorption and reducing desorption of herbicides from soil, these processes being important when one intends to use this technique in the removal of herbicides from soil. The sorption process can be improved when the biochar activation technique is used. These processes ensure greater sorptive capacity of biochar for water and soil herbicides and can improve the effectiveness of remediation programs.

ACKNOWLEDGMENTS

The authors are grateful to the "Coordenação de Aperfeiçoamento de Pessoal de Nível Superior" (CAPES – 88887.479265/2020–00) and "Fundação de Amparo à Pesquisa do Estado de Minas Gerais" (FAPEMIG – APQ-01378–21) for financial support.

REFERENCES

Abe, I., S. Iwasaki, Y. Iwata, H. Kominami and Y. Kera. 1998. Relationship between the production method and the adsorption property of charcoal. *Tanso* 185:277–284.

Abel, S., A. Peters, S. Trinks, H. Schonsky, M. Facklam and G. Wessolek. 2013. Impact of biochar and hydrochar addition on water retention and water repellency of sandy soil. *Geoderma* 202:183–191.

Ahmad, M., A.U. Rajapaksha, J.E. Lim, et al. 2014. Biochar as a sorbent for contaminant management in soil and water: A review. *Chemosphere* 99:19–33.

Almeida, C.S., K.F. Mendes, L.V. Junqueira, F.G. Alonso, G.M. Chitonila and V.L. Tornisielo. 2020. Diuron sorption, desorption and degradation in anthropogenic soils compared to sandy soil. *Planta Daninha* 38:e020217146.

Alvarez, D.O., K.F. Mendes, M. Tosi, L.F. Souza, et al. 2021. Sorption-desorption and biodegradation of sulfometuron-methyl and its effects on the bacterial communities in Amazonian soils amended with aged biochar. *Ecotoxicology and Environmental Safety* 207:111222.

Anto, S., M.P. Sudhakar, T.S. Ahamed, et al. 2021. Activation strategies for biochar to use as an efficient catalyst in various applications. *Fuel* 285:119205.

Atkinson, C.J., J.D. Fitzgerald and N.A. Hipps. 2010. Potential mechanisms for achieving agricultural benefits from biochar application to temperate soils: A review. *Plant Soil* 337:1–18.

Binh, Q.A. and H.H. Nguyen. 2020. Investigation the isotherm and kinetics of adsorption mechanism of herbicide 2, 4-dichlorophenoxyacetic acid (2,4-D) on corn cob biochar. *Bioresource Technology Reports* 11:100520.

Blanco-Canqui, H. 2017. Biochar and soil physical properties. *Soil Science Society of America Journal* 8:687–711.

Brockhoff, S.R., N.E. Christians, R.J. Killorn, R. Horton and D.D. Davis. 2010. Physical and mineral-nutrition properties of sand-based turfgrass root zones amended with biochar. *Agronomy Journal* 102:1627–1631.

Cabrera, A., L. Cox, K. Spokas, M.C. Hermosín, J. Cornejo and W.C. Koskinen. 2014. Influence of biochar amendments on the sorption-desorption of aminocyclopyrachlor, bentazone and pyraclostrobin pesticides to an agricultural soil. *Science of the Total Environment* 470:438–443.

Carrick, S., G. Buchan, P. Almond and N. Smith. 2011. Atypical early-time infiltration into a structured soil near field capacity: The dynamic interplay between sorptivity, hydrophobicity, and air encapsulation. *Geoderma* 160:579–589.

Cao, Y., S. Jiang, Y. Zhang, J. Xu, L. Qiu and L. Wang. 2021. Investigation into adsorption characteristics and mechanism of atrazine on Nano-MgO modified fallen leaf biochar. *Journal of Environmental Chemical Engineering* 9:105727.

Chen, B. and M. Yuan. 2011. Enhanced sorption of polycyclic aromatic hydrocarbons by soil amended with biochar. *Journal of Soils and Sediments* 11:62–71.

Chen, Y.N., L.Y. Chai and Y.D. Shu. 2008. Study of arsenic (V) adsorption on bone char from aqueous solution. *Journal of Hazardous Materials* 160:168–172.

Chen, Y.N., Y. Zhu, Z. Wang, Y. Li, et al. 2011. Application studies of activated carbon derived from rice husks produced by chemical-thermal process-a review. *Advances in Colloid and Interface Science* 163:39–52.

Cheng, C.H., J. Lehmann, J.E. Thies, S.D. Burton and M.H. Engelhard. 2006. Oxidation of black carbon by biotic and abiotic process. *Organic Geochemistry* 37:1477–1488.

Chin-Pampillo, J.S., M. Perez-Villanueva, M. Massis-Mora, et al. 2021. Amendments with pyrolyzed agrowastes change bromacil and diuron's sorption and persistence in a tropical soil without modifying their environmental risk. *Science of The Total Environment* 772:145515.

Clay, S. A., K.K. Krack, S.A. Bruggeman, S. Papiernik and T.E. Schumacher. 2016. Maize, switchgrass, and ponderosa pine biochar added to soil increased herbicide sorption and decreased herbicide efficacy. *Journal of Environmental Science and Health – Part B* 51:497–507.

Dechene, A., I. Rosendahl, V. Laabs and W. Amelung. 2014. Sorption of polar herbicides and herbicide metabolites by biochar-amended soil. *Chemosphere* 109: 180–186.

Deng, H., D. Feng, J.X. He, F.Z. Li, H.M. Yu and C.J. Ge. 2017. Influence of biochar amendments to soil on the mobility of atrazine using sorption-desorption and soil thin-layer chromatography. *Ecological Engineering* 99:381–390.

De la Rosa, J.M., M. Rosado, M. Paneque, A.Z. Miller and H. Knicker. 2018. Effects of aging under field conditions on biochar structure and composition: Implications for biochar stability in soils. *Science of the Total Environment* 613–614:969–976.

Dong, X., G. Li, Q. Lin and X. Zhao. 2017. Quantity and quality changes of biochar aged for 5 years in soil under field conditions. *Catena* 159:136–143.

Downie, A., A. Crosky and P. Munroe. 2009. Physical properties of biochar. In *Biochar for Environmental Management: Science and Technology*, ed. J. Lemman, and S. Joseph, 13–32. London: Earthscan.

Duarte, S.J., B. Glaser and C.E.P. Cerri. 2019. Effect of biochar particle size on physical, hydrological and chemical properties of loamy and sandy tropical soils. *Agronomy* 9:1–15.

Eibisch, N., R. Schroll, R. Fuß, R. Mikutta, M. Helfrich and H. Flessa. 2015. Pyrochars and hydrochars differently alter the sorption of the herbicide isoproturon in an agricultural soil. *Chemosphere* 119:155–162.

Fernandes, B.C.C., K.F. Mendes, V.L. Tornisielo, et al. 2021. Effect of pyrolysis temperature on eucalyptus wood residues biochar on availability and transport of hexazinone in soil. *International Journal of Environmental Science and Technology* 1–16.

Gascó, G., J. Paz-Ferreiro, M.L. Álvarez, A. Saa and A. Méndez. 2018. Biochars and hydrochars prepared by pyrolysis and hydrothermal carbonisation of pig manure. *Waste Management* 79:395–403.

Gámiz, B., K. Hall, K.A. Spokas and L. Cox. 2019a. Understanding activation effects on low-temperature biochar for optimization of herbicide sorption. *Agronomy* 9:588.

Gámiz, B., P. Velarde, K.A. Spokas, R. Celis and L. Cox. 2019b. Changes in sorption and bioavailability of herbicides in soil amended with fresh and aged biochar. *Geoderma* 337:341–349.

Gámiz, B., P. Velarde, K.A. Spokas, M.C. Hermosín and L. Cox. 2017. Biochar soil additions affect herbicide fate: Importance of application timing and feedstock species. *Journal of Agricultural and Food Chemistry* 65:3109–3117.

Gul, E., K.A.B Alrawashdeh, O. Masek, et al. 2021. Production and use of biochar from lignin and lignin-rich residues (such as digestate and olive stones) for wastewater treatment. *Journal of Analytical and Applied Pyrolysis* 158:105263.

Gul, S., J.K. Whalen, B.W. Thomas, V. Sachdeva and H. Deng. 2015. Physico-chemical properties and microbial responses in biochar-amended soils: Mechanisms and future directions. *Agriculture, Ecosystems & Environment* 206:46–59.

Glaser, B., L. Haumaier, G. Guggenberger and W. Zech. 2001. The "Terra Preta" phenomenon: A model for sustainable agriculture in the humid tropics. *Naturwissenschaften* 88:37–41.

Glaser, B., J. Lehmann and W. Zech. 2002. Ameliorating physical and chemical properties of highly weathered soils in the tropics with charcoal – a review. *Biology and Fertility of Soils* 35:219–230.

Glaser, B., and J.J. Birk. 2012. State of the scientific knowledge on properties and genesis of anthropogenic dark earths in central Amazonia (Terra Preta de Índio). *Geochimica et Cosmochimica acta* 82:39–51.

Glaser, B., M. Parr, C. Braun and G. Kopolo. 2009. Biochar is carbon negative. *Nature Geoscience* 2:2–2.

Green, J. and T. Hale. 2006. Increasing the biological activity of weak acid herbicides by increasing and decreasing the pH of the spray mixture. In *Pesticide Formulations and Delivery Systems*, ed. M. Salyani, and G. Lindner, 62–71. West Conshohocken: ASTM International.

Hagemann, N., K. Spokas, H.P. Schmidt, R. Kägi, M.A. Böhler and T.D. Bucheli. 2018. Activated carbon, biochar and charcoal: Linkages and synergies across pyrogenic carbon's ABCs. *Water* 10:182.

Hale, S.E., K. Hanley, J. Lehmann, A.R. Zimmerman and G. Cornelissen. 2011. Effects of chemical, biological and physical aging as well as soil addition on pyrene sorption on activated carbon and biochar. *Environmental Science & Technology* 45:10445–10453.

Hilton, H.W. and Q.H. Yuen. 1963. Soil adsorption of herbicides, adsorption of several pre-emergence herbicides by Hawaiian sugar cane soils. *Journal of Agricultural and Food Chemistry* 11:230–234.

Hitzl, M., A. Corma, F. Pomares and M. Renz. 2015. The hydrothermal carbonization (HTC) plant as a decentral biorefinery for wet biomass. *Catalysis Today* 257:154–159.

Hossain, M.K., V. Strezov, K.Y. Chan and P.F. Nelson. 2010. Agronomic properties of wastewater sludge biochar and bioavailability of metals in production of cherry tomato (*Lycopersicon esculentum*). *Chemosphere* 78:1167–1171.

Huang, H., C. Zhang, P. Zhang, et al. 2018. Effects of biochar amendment on the sorption and degradation of atrazine in different soils. *Soil and Sediment Contamination* 27:643–657.

Hu, J., Q. Zhang and D.J. Lee. 2018. Kraft lignin biorefinery: A perspective. *Bioresource Technology* 247:1181–1183.

Ioannidou, O. and A. Zabaniotou. 2007. Agricultural residues as precursors for activated carbon production – a review. *Renewable and Sustainable Energy Reviews* 11:1966–2005.

Jafri, N., W.Y. Wong, V. Doshi, L.W. Yoon and K.H. Cheah. 2018. A review on production and characterization of biochars for application in direct carbon fuel cells. *Process Safety and Environmental Protection* 118:152–166.

Janus, A., A. Pelfrêne, S. Heymans, C. Deboffe, F. Douay and C. Waterlot. 2015. Elaboration, characteristics and advantages of biochars for the management of contaminated soils with a specific overview on Miscanthus biochars. *Journal of Environmental Management* 162:275–289.

Jeffery, S., M.B.J. Meinders, C.R. Stoof, et al. 2015. Biochar application does not improve the soil hydrological function of a sandy soil. *Geoderma* 251:47–54.

Jensen, L.C., J.R. Becerra and M. Escudey. 2018. Impact of physical/chemical properties of volcanic ash-derived soils on mechanisms involved during sorption of ionisable and non-ionisable herbicides. In *Advanced sorption process applications*, ed. S. Edebali, 1–23. Turquia: IntechOpen.

Jien, S.H. and C.S. Wang. 2013. Effects of biochar on soil properties and erosion potential in a highly weathered soil. *Catena* 110:225–233.

Joseph, S. D., M. Camps-Arbestain, Y. Lin, et al. 2010. An investigation into the reactions of biochar in soil. *Soil Research* 48:501–515.

Junqueira, L.V., K.F. Mendes, R.N. Sousa, C.S. Almeida, F.G. Alonso and V.L. Tornisielo. 2019. Sorption-desorption isotherms and biodegradation of glyphosate in two tropical soils aged with eucalyptus biochar. *Archives of Agronomy and Soil Science* 66:1–17.

Juriga, M., V. Šimanský, J. Horák, et al. 2018. The effect of different rates of biochar and biochar in combination with N fertilizer on the parameters of soil organic matter and soil structure. *Journal of Ecological Engineering* 19:153–161.

Kah, M. and C.D. Brown. 2007. Prediction of the adsorption of ionizable pesticides in soils. *Journal of Agricultural and Food Chemistry* 55:2312–2322.

Karimi, A., A. Moezzi, M. Chorom and N. Enayatizamir. 2020. Application of biochar changed the status of nutrients and biological activity in a calcareous soil. *Journal of Soil Science and Plant Nutrition* 20:450–459.

Kearns, J.P., K.K. Shimabuku, D.R.U. Knappe and R.S. Summers. 2019. High temperature co-pyrolysis thermal air activation enhances biochar adsorption of herbicides from surface water. *Environmental Engineering Science* 36:710–723.

Khorram, M.S., D. Lin, Q. Zhang, Y. Zheng, H. Fang and Y. Yu. 2016. Effects of aging process on adsorption – desorption and bioavailability of fomesafen in an agricultural soil amended with rice hull biochar. *Journal of Environmental Sciences* 56:180–191.

Khorram, M.S., A.K. Sarmah and Y. Yu. 2018. The effects of biochar properties on fomesafen adsorption-desorption capacity of biochar-amended soil. *Water, Air, and Soil Pollution* 229:1-13.

Kimetu, J.M. and J. Lehmann. 2010. Stability and stabilisation of biochar and green manure in soil with different organic carbon contents. *Australian Journal of Soil Research* 48:577–585.

Kookana, R.S. 2010. The role of biochar in modifying the environmental fate, bioavailability, and efficacy of pesticides in soils: A review. *Soil Research* 48:627–637.

Kumari, K.G.I.D., P. Moldrup, M. Paradelo, L. Elsgaard and L.W. Jonge. 2016. Soil properties control glyphosate sorption in soils amended with birch wood biochar. *Water, Air & Soil Pollution* 227:1–12.

Laird, D.A., R.C. Brown, J.E. Amonette and J. Lehmann. 2009. Review of the pyrolysis platform for coproducing bio-oil and biochar. *Biofuels, Bioproducts and Biorefining* 3:547–562.

Laird, D.A., P. Fleming, D.D. Davis, R. Horton, B. Wang and D.L. Karlen. 2010. Impact of biochar amendments on the quality of a typical midwestern agricultural soil. *Geoderma* 158:443–449.

Lehmann, J. and S. Joseph. 2009. *Biochar for environmental management: Science and technology*. London: Earthscan.

Lehmann, J., Y. Kuzyakov, G. Pan, et al. 2015. Biochars and the plant-soil interface. *Plant Soil* 395:1–5.

Lee, Y. G., J. Shin, J. Kwak, S. Kim, C. Son, K.H. Cho and K. Chon. 2021. Effects of NaOH activation on adsorptive removal of herbicides by biochars prepared from ground coffee residues. *Energies* 14:1297.

Liang, B., J. Lehmann, S.P. Sohi, et al. 2010. Black carbon affects the cycling of non-black carbon in soil. *Organic Geochemistry* 41:206–213.

Liang, B., J. Lehmann, D. Solomon, et al. 2006. Black carbon increases cation exchange capacity in soils. *Soil Science Society of America Journal* 70:1719–1730.

Liu, K., B. Yu, K. Luo, X. Liu and L. Bai. 2016. Reduced sulfentrazone phytotoxicity through increased adsorption and anionic species in biochar-amended soils. *Environmental Science and Pollution Research* 23:9956–9963.

Liu, Y., L. Lonappan, S.K. Brar and S. Yang. 2018. Impact of biochar amendment in agricultural soils on the sorption, desorption, and degradation of pesticides: A review. *Science of the Total Environment* 645:60–70.

Liu, Y., S.P. Sohi, F. Jing and J. Chen. 2019. Oxidative ageing induces change in the functionality of biochar and hydrochar: Mechanistic insights from sorption of atrazine. *Environmental Pollution* 249:1002–1010.

Loffredo, E., M. Parlavecchia, G. Perri and R. Gattullo. 2019. Comparative assessment of metribuzin sorption efficiency of biochar, hydrochar and vermicompost. *Journal of Environmental Science and Health, Part B* 54:728–735.

López-Piñeiro, A., D. Peña, A. Albarrán, D. Becerra and J. Sánchez-Llerena. 2013. Sorption, leaching and persistence of metribuzin in Mediterranean soils amended with olive mill waste of different degrees of organic matter maturity. *Journal of Environmental Management* 122:76–84.

Lou, L., B. Wu, L. Wang, et al. 2011. Sorption and ecotoxicity of pentachlorophenol polluted sediment amended with rice-straw derived biochar. *Bioresource Technology* 102:4036–4041.

Lu, W., W. Ding, J. Zhang, et al. 2014. Biochar suppressed the decomposition of organic carbon in a cultivated sandy loam soil: A negative priming effect. *Soil Biology and Biochemistry* 76:12–21.

Lu, S. and Y. Zong. 2018. Pore structure and environmental serves of biochars derived from different feedstocks and pyrolysis conditions. *Environmental Science and Pollution Research* 25: 30401–30409.

Martin, S.M., R.S. Kookana, L. Van Zwieten and E. Krull. 2012. Marked changes in herbicide sorption – desorption upon ageing of biochars in soil. *Journal of Hazardous Materials* 231–232: 70–78.

Mendes, K.F., F.G. Alonso, T.B. Mertens, M.H. Inoue, M.G. Oliveira and V.L. Tornisielo. 2019d. Aminocyclopyrachlor and mesotrione sorption – desorption in municipal sewage sludge-amended soil. *Bragantia* 78:131–140.

Mendes, K.F., K.E. Hall, K.A. Spokas, W.C. Koskinen and V.L. Tornisielo. 2017. Evaluating agricultural management effects on alachlor availability: Tillage, green manure, and biochar. *Agronomy* 7: e64.

Mendes, K.F., K.E. Hall, V. Takeshita, M.L. Rossi and V.L. Tornisielo. 2018. Animal bonechar increases sorption and decreases leaching potential of aminocyclopyrachlor and mesotrione in a tropical soil. *Geoderma* 316: 11–18.

Mendes, K.F., A.F.D. Júnior, V. Takeshita, A.P.J. Régo and V.L. Tornisielo. 2019a. Effect of biochar amendments on the sorption and desorption herbicides in agricultural soil. In *Advanced sorption process applications*, ed. S. Edebali, 1–25. London: IntechOpen.

Mendes, K.F., G.P. Olivatto, R.N. Sousa, L.V. Junqueira and V.L. Tornisielo. 2019b. Natural biochar effect on sorption – desorption and mobility of diclosulam and pendimethalin in soil. *Geoderma* 347:118–125.

Mendes, K.F., M.B. Soares, R.N. Sousa, K.C. Mielke, M.G.D.S. Brochado and V.L. Tornisielo. 2021b. Indaziflam sorption – desorption and its three metabolites from biochars-and their raw feedstock-amended agricultural soils using radiometric technique. *Journal of Environmental Science and Health, Part B* 1–10.

Mendes, K.F., R.N. Sousa, M.O. Goulard and V.L. Tornisielo. 2020. Role of raw feedstock and biochar amendments on sorption-desorption and leaching potential of three ^{3}H- and ^{14}C-labelled pesticides in soils. *Journal of Radioanalytical and Nuclear Chemistry* 1:1–14.

Mendes, K.F., R.N. Sousa, M.B. Soares, D.G. Viana and A.J. Souza. 2021a. Sorption and desorption studies of herbicides in the soil by batch equilibrium and stirred flow methods. In *Radioisotopes in weed research*, ed. K.F. Mendes, 17–61. Boca Raton: CRC Press.

Mendes, K.F., R.N. Sousa, V. Takeshita, F.G. Alonso, A.P.J. Régo and V.L. Tornisielo. 2019c. Cow bone char as a sorbent to increase sorption and decrease mobility of hexazinone, metribuzin, and quinclorac in soil. *Geoderma* 343:40–49.

Meng, L., T. Sun, M. Li, M. Saleem, Q. Zhang and C. Wang. 2019. Soil-applied biochar increases microbial diversity and wheat plant performance under herbicide fomesafen stress. *Ecotoxicology and Environmental Safety* 171:75–83.

Mukherjee, A., R. Lal and A.R. Zimmerman. 2014. Effects of biochar and other amendments on the physical properties and greenhouse gas emissions of an artificially degraded soil. *Science of the Total Environment* 487:26–36.

Mumme, J., L. Eckervogt, J. Pielert, M. Diakité, F. Rupp and J. Kern. 2011. Hydrothermal carbonization of anaerobically digested maize silage. *Bioresource technology* 102:9255–9260.

Nair, V.D., P.K. Ramachandran Nair, B. Dari, A.M. Freitas, N. Chatterjee and F.M. Pinheiro. 2017. Biochar in the agroecosystem-climate-change-sustainability nexus. *Frontiers in Plant Science* 8:2051.

Nelissen, V., G. Ruysschaert, D. Manka'Abusi, et al. 2015. Impact of a woody biochar on properties of a sandy loam soil and spring barley during a two-year field experiment. *European Journal of Agronomy* 62:65–78.

Netto, M.S., J. Georgin, D.S. Franco, et al. 2021. Effective adsorptive removal of atrazine herbicide in river waters by a novel hydrochar derived from *Prunus serrulata* bark. *Environmental Science and Pollution Research*, 1–14.

Nielsen, S., S. Joseph, J. Ye, C. Chia, P. Munroe, L.V. Zwieten and T. Thomas. 2018. Crop-season and residual effects of sequentially applied mineral enhanced biochar and N fertiliser on crop yield, soil chemistry and microbial communities. *Agriculture, Ecosystems & Environment* 255:52–61.

Nocentini, C., B. Guenet, E. Di Mattia, G. Certini, G. Bardoux and C. Rumpel. 2010. Charcoal mineralisation potential of microbial inocula from burned and unburned forest soil with and without substrate addition. *Soil Biology and Biochemistry* 42:1472–1478.

Novak, J.M., W.J. Busscher, D.L. Laird, M. Ahmedna, D.W. Watts and M.A.S Niandou. 2009. Impact of Biochar Amendment on Fertility of a Southeastern Coastal Plain Soil. *Soil Science* 174:105–112.

Novotny, E.H., M.H.B. Hayes, B.E. Madari, et al. 2009. Lessons from the Terra Preta de Índio of the Amazon region for the utilisation of charcoal for soil amendment. *Journal of the Brazilian Chemical Society* 20:1003–1010.

Oguntunde, P.G., M. Fosu, A.E. Ajayi and N. Van De Giesen. 2004. Effects of charcoal production on maize yield, chemical properties and texture of soil. *Biology and Fertility of Soil* 39: 295–299.

Oni, B.A., O. Oziegbe and O.O. Olawole. 2019. Significance of biochar application to the environment and economy. *Annals of Agricultural Sciences* 64:222–236.

Petter, F.A., T.S. Ferreira, A.P. Sinhorin, L.B.D. Lima, L.A.D. Morais and L.P. Pacheco. 2016. Sorption and desorption of diuron in Oxisol under biochar application. *Bragantia* 75:487–496.

Pintado, S., M.R. Montoya and J.M.R. Mellado. 2011. Imidazolinone herbicides in strongly acidic media: Speciation and electroreduction. *Comptes Rendus Chimie* 14:957–962.

PPDB. Pesticide Properties Database. 2021. Footprint: Creating tools for pesticide risk assessment and management in Europe. https://sitem.herts.ac.uk/aeru/ppdb/en/index.htm (Accessed August 10, 2021).

Qiu, Y., X Xiao, H. Cheng, Z. Zhou and G.D. Sheng. 2009. Influence of environmental factors on pesticide adsorption by black carbon: pH and model dissolved organic matter. *Environmental Science and Technology* 43:4973–4978.

Razzaghi, F., P.B. Obour and E. Arthur. 2020. Does biochar improve soil water retention? A systematic review and meta-analysis. *Geoderma* 361:114055.

Ren, X., H. Sun, F. Wang and F. Cao. 2016. The changes in biochar properties and sorption capacities after being cultured with wheat for 3 months. *Chemosphere* 144:2257–2263.

Ren, X., X. Yuan and H. Sun. 2018. Dynamic changes in atrazine and phenanthrene sorption behaviors during the aging of biochar in soils. *Environmental Science and Pollution Research* 25:81–90.

Rice, P.J., P.J. Rice, E.L. Arthur and A.C. Barefoot. 2007. Advances in pesticide environmental fate and exposure assessments. *Journal of Agricultural and Food Chemistry* 55:5367–5376.

Rocha, P.R.R., A.T. Faria, G.S.D. Silva, et al. 2013. Half-life of diuron in soils with different physical and chemical attributes. *Ciência Rural* 43:1961–1966.

Ronsse, F., S. Van Hecke, D. Dickison and W. Prins. 2013. Production and characterization of slow pyrolysis biochar: Influence of feedstock type and pyrolysis conditions. *GCB Bioenergy* 5:104–115.

Sigua, G.C., K.C. Stone, P.G. Hunt, K.B. Cantrell and J.M. Bovak. 2015. Increasing biomass of winter wheat using sorghum biochars. *Agronomy for Sustainable Development* 35:739–748.

Šimanský, V., J. Horák, D. Igaz, et al. 2016. How dose of biochar and biochar with nitrogen can improve the parameters of soil organic matter and soil structure? *Biologia* 71:989–995.

Singanan, M. and E. Peters. 2013. Removal of toxic heavy metals from synthetic wastewater using a novel biocarbon technology. *Journal of Environmental Chemical Engineering* 1:884–890.

Sopeña, F., K. Semple, S. Sohi and G. Bending. 2012. Assessing the chemical and biological accessibility of the herbicide isoproturon in soil amended with biochar. *Chemosphere* 8:77–83.

Sousa, S.G.A., M.I. Araújo and E.V. Wandelli. 2015. Saberes tradicionais dos povos amazônicos no contexto do processo de transição agroecológica. *AmbientalMente sustentable* 2:1699–1717.

Sun, F. and S. Lu. 2014. Biochars improve aggregate stability, water retention, and pore-space properties of clayey soil. *Journal of Plant Nutrition and Soil Science* 177:26–33.

Sun, K., M. Keiluweit, M. Kleber, Z. Pan and B. Xing. 2011. Sorption of fluorinated herbicides to plant biomass-derived biochars as a function of molecular structure. *Bioresource Technology* 102:9897–9903.

Suliman, W., J.B. Harsh, N.I. Abu-Lail, A.M. Fortuna, I. Dallmeyer and M. Garcia-Perez. 2016. Modification of biochar surface by air oxidation: Role of pyrolysis temperature. *Biomass and Bioenergy* 85:1–11.

Schulz, H., and B. Glaser. 2012. Effects of biochar compared to organic and inorganic fertilizers on soil quality and plant growth in a greenhouse experiment. *Journal of Plant Nutrition and Soil Science* 175:410–422.

Shaaban, M., L. Van Zwieten, S. Bashir, et al. 2018. A concise review of biochar application to agricultural soils to improve soil conditions and fight pollution. *Journal of Environmental Management* 228:429–440.

Shetty, R. and N.B. Prakash. 2020. Effect of different biochars on acid soil and growth parameters of rice plants under aluminium toxicity. *Scientific Reports* 10:1–10.

Shi, K., Y. Xie and Y. Qiu. 2015. Natural oxidation of a temperature series of biochars: Opposite effect on the sorption of aromatic cationic herbicides. *Ecotoxicology and Environmental Safety* 114:102–108.

Shinogi, Y. and Y. Kanri. 2003. Pyrolysis of plant, animal and human waste: Physical and chemical characterization of the pyrolytic products. *Bioresource Technology* 90: 241–247.

Spokas, K.A., W.C. Koskinen, J.M. Baker and D.C. Reicosky. 2009. Impacts of woodchip biochar additions on greenhouse gas production and sorption/degradation of two herbicides in a Minnesota soil. *Chemosphere* 77:574–581.

Szmigielski, A.M., R.D. Hangs and J.J. Schoenau. 2018. Bioavailability of metsulfuron and sulfentrazone herbicides in soil as affected by amendment with two contrasting willow biochars. *Bulletin of Environmental Contamination and Toxicology* 100:298–302.

Tan, G., W. Sun, Y. Xu, H. Wang and N. Xu. 2016. Sorption of mercury (II) and atrazine by biochar, modified biochars and biochar based activated carbon in aqueous solution. *Bioresource Technology* 211:727–735.

Tang, J., W. Zhu, R. Kookana and A. Katayama. 2013. Characteristics of biochar and its application in remediation of contaminated soil. *Journal of Bioscience and Bioengineering* 116:653–659.

Tataková, V., E. Hiller and M. Vaculík. 2013. Impact of wheat straw biochar addition to soil on the sorption, leaching, dissipation of the herbicide (4-chloro-2-methylphenoxy) acetic acid and the growth of sunflower (*Helianthus annuus* L.). *Ecotoxicology and Environmental Safety* 92:215–221.

Trigo, C., K.A. Spokas, L. Cox and W.C. Koskinen. 2014. Influence of soil biochar aging on sorption of the herbicides MCPA, nicosulfuron, terbuthylazine, indaziflam, and fluoroethyldiaminotriazine. *Journal of Agricultural and Food Chemistry* 62:0855–10860.

Trigo, C., K.A. Spokas, K.E. Hall, L. Cox and W.C. Koskinen. 2016. Metolachlor sorption and degradation in soil amended with fresh and aged biochars. *Journal of Agricultural and Food Chemistry* 64:3141–3149.

Usowicz, B., J. Lipiec, M. Lukowski, W. Marczewski and J. Usowicz. 2016. The effect of biochar application on thermal properties and albedo of loess soil under grassland and fallow. *Soil and Tillage Research* 164:45–51.

Van Zwieten, L., S. Kimber, S. Morris, et al. 2010. Effects of biochar from slow pyrolysis of papermill waste on agronomic performance and soil fertility. *Plant and Soil* 327:235–246.

Verheijen, F., S. Jeffery, A.C. Bastos, M. Van der Velde and I. Diafas. 2009. Biochar application to soils: A critical scientific review of effects on soil properties, processes and functions. *Office for the Official Publications of the European Communities* 24099:162.

Walters, R.D. and J.G. White. 2018. Biochar in situ decreased bulk density and improved soil-water relations and indicators in Southeastern US Coastal plain ultisols. *Soil Science* 183:99–111.

Wang, D., C. Li, S.J. Parikh and K.M. Scow. 2019. Impact of biochar on water retention of two agricultural soils – A multi-scale analysis. *Geoderma* 340:185–191.

Wang, H., K. Lin, Z. Hou, B. Richardson and J. Gan. 2010. Sorption of the herbicide terbuthylazine in two New Zealand forest soils amended with biosolids and biochars. *Journal of Soils and Sediments* 10:283–289.

Wang, L., D. O'Connor, J. Rinklebe, et al. 2020. Biochar aging: Mechanisms, physicochemical changes, assessment, and implications for field applications. *Environmental Science & Technology* 54:14797–14814.

Wang, Y., R. Yin and R. Liu. 2014. Characterization of biochar from fast pyrolysis and its effect on chemical properties of the tea garden soil. *Journal of Analytical and Applied Pyrolysis* 110: 375–381.

Wei, L., Y. Huang, Y. Li, L. Huang, N. Mar, Q. Huang and Z. Liu. 2017. Biochar characteristics produced from rice husks and their sorption properties for the acetanilide herbicide metolachlor. *Environmental Science and Pollution Research* 24:4552–4561.

Woolf, D., J.E. Amonette, F.A. Street-Perrott, J. Lehmann and S. Joseph. 2010. Sustainable biochar to mitigate global climate change. *Nature communications* 1:1–9.

Yamane, V. K. and R.R. Green. 1972. Adsorption of ametryne and atrazine on an oxisol, montmorillonite, and charcoal in relation to pH and solubility effects. *Soil Science Society of America Journal* 36:58–64.

Yang, X., S. Zhang, M. Ju and L. Liu. 2019. Preparation and modification of biochar materials and their application in soil remediation. *Applied Sciences* 9:1365.

Yang, Y., G. Sheng and M. Huang. 2006. Bioavailability of diuron in soil containing wheat-straw-derived char. *Science of the Total Environment* 354:170–178.

Yavari, S., M. Abualqumboz, N. Sapari, H.A. Hata-Suhaimi and N.Z. Nik-Fuaad. 2020. Sorption of imazapic and imazapyr herbicides on chitosan-modified biochars. *International Journal of Environmental Science and Technology* 17:3341–3350.

Yavari, S., A. Malakahmad and N.B. Sapari. 2015. Biochar efficiency in pesticides sorption as a function of production variables-a review. *Environmental Science and Pollution Research* 22:13824–13841.

Yavari, S., A. Malakahmad, N.B. Sapari and S. Yavari. 2016. Sorption-desorption mechanisms of imazapic and imazapyr herbicides on biochars produced from agricultural wastes. *Journal of Environmental Chemical Engineering* 4:3981–3989.

Yu, H., W. Zou, J. Chen, H. Chen, Z. Yu, J. Huang, H. Tang, X. Wei and B. Gao. 2019. Biochar amendment improves crop production in problem soils: A review. *Journal of Environmental Management* 232:8–21.

Yu, X.Y., C.L. Mu, C. Gu, C. Liu and X.J. Liu. 2011. Impact of woodchip biochar amendment on the sorption and dissipation of pesticide acetamiprid in agricultural soils. *Chemosphere* 85:1284–1289.

Yu, X.Y., G.G Ying and R.S. Kookana. 2006. Sorption and desorption behaviors of diuron in soils amended with charcoal. *Journal of Agricultural and Food Chemistry* 54: 8545–8550.

Yuan, P., J. Wang, Y. Pan, B. Shen and C. Wu. 2019. Review of biochar for the management of contaminated soil: Preparation, application and prospect. *Science of the Total Environment* 659:473–490.

Zhang, A., X. Zhou, M. Li and H. Wu. 2017. Impacts of biochar addition on soil dissolved organic matter characteristics in a wheat-maize rotation system in Loess Plateau of China. *Chemosphere* 186:986–993.

Zhang, M., M. Riaz, L. Zhang, Z. El-Desouki and C. Jiang 2019. Biochar induces changes to basic soil properties and bacterial communities of different soils to varying degrees at 25 mm rainfall: More effective on acidic soils. *Frontiers in Microbiology* 10:1321.

Zhang, Y., J. Wang and Y. Feng. 2021. The effects of biochar addition on soil physicochemical properties: A review. *Catena* 202:105284.

Zhao, C., P. Lv, L. Yang, S. Xing, W. Luo and Z. Wang. 2018. Biodiesel synthesis over biochar-based catalyst from biomass waste pomelo peel. *Energy Convers Manage* 160:477–485.

Zhelezova, A., H. Cederlund and J. Stenström. 2017. Effect of biochar amendment and ageing on adsorption and degradation of two herbicides. *Water, Air, and Soil Pollution* 228:216.

Zhou, D., D. Liu, F. Gao, M. Li and X. Luo. 2017. Effects of biochar-derived sewage sludge on heavy metal adsorption and immobilization in soils. *International Journal of Environmental Research and Public Health* 14:681.

Zygourakis, K. 2017. Biochar soil amendments for increased crop yields: How to design a "designer" biochar. *AIChE Journal* 63:5425–5437.

6 Effect of Biochar on Soil Herbicide Transport

Daniel Valadão Silva[1], Tatiane Severo Silva[1], Bruno Caio Chaves Fernandes[1], Taliane Maria da Silva Teófilo[1], Francisca Daniele da Silva[1], and Kassio Ferreira Mendes[2]*

[1]Plant Science Center, Department of Agronomic and Forestry Sciences, Federal Rural University of the Semi-Arid, Mossoró, Brazil.

[2]Department of Agronomy, Federal University of Viçosa, Viçosa, Brazil.

* Corresponding author: daniel.valadao@ufersa.edu.br

CONTENTS

DOI: 10.1201/9781003202073-6

6.1 INTRODUCTION

The use of herbicides is a common practice for plant control in agriculture and an essential component of the production cost of crops. In several countries, chemical control is the method most used by producers. This preference is due to the greater efficiency of herbicides in controlling weeds, combined with lower cost and faster control compared to other methods. However, this efficiency depends on several factors related to the characteristics of the herbicide itself and the interactions that occur with environmental components, such as the soil.

The transport of herbicides is one of the processes directly related to the distribution and dissipation of molecules of these products in the soil. Consequently, this transport is one of the crucial factors playing a role in the efficiency of controlling weed seeds and propagules present in the soil and the potential of herbicides to cause environmental contamination. Upon contact with the soil, the herbicide is subject to interactions with the soil's chemical, physical, and organic components. In this way, agricultural practices that modify soil attributes can also alter the transport of herbicides.

The growing use of biochar in agriculture has naturally promoted to changes in the soil's physical, chemical, and biological attributes. Consequently, it has generated the search for information on the behavior of herbicides in the soil in this new scenario. In this chapter, we seek to elucidate the transport processes of herbicides in the soil, emphasizing the effects of different types of biochar on the leaching, runoff, and volatilization of herbicides. Also, the primary methodologies used to study the implications of the information already known on the subject in the control of weeds and the potential for environmental contamination of herbicides are also discussed.

6.2 TRANSPORTATION OF HERBICIDES ON THE SOIL

The transport of herbicides in the soil can take place both horizontally and vertically. In horizontal transport, herbicides are transported due to runoff/runin. The downward vertical transport of herbicides in the soil profile is known as leaching. These transport routes are primary surface and groundwater pollution sources and can cause serious environmental issues (Sun et al. 2012; Zhang et al. 2018). In this topic, we discuss the transport routes of herbicides in soil and present some more recent studies and discoveries about the process.

6.2.1 RUNOFF/RUNIN

Runoff and runin occur when the total soil infiltration capacity is exceeded (Larsbo et al. 2016). The first occurs on the surface, and the second occurs on the subsurface of the ground. In these processes, herbicides are transported in dissolved or particulate forms across the surface and subsurface of the soil (Tang et al. 2012; Vymazal and Březinová 2015). Herbicides lost in agricultural production fields via runoff and runin can contaminate water bodies and damage local biota (Abdi et al. 2021). In addition, transporting herbicides across agricultural fields can reduce the effectiveness of weed control, increasing the cost of crop management or resulting in crop failure due to reduced yield and quality.

The transport of herbicides in the soil can contribute to the contamination of water bodies. Battaglin et al. (2003), evaluating herbicide concentrations in surface waters in the midwestern United States and using a simple two-component mixture model, indicated that runoff was the source responsible for 90% of herbicide transport to the streams evaluated. Herbicides are transported both dissolved in the soil solution and sorbed to soil colloids (along with soil particles/sediments) (Bento et al. 2017). Several factors influence the herbicide loss process via runin and runoff. The main ones include herbicide and soil properties (Chen et al. 2019), soil management (Nachimuthu et al. 2016), topography (Tayeb et al. 2017), frequency and intensity of irrigation, and the time between application and first rain event and rain conditions (Lupi et al. 2019).

Herbicides applied in pre-planting or pre-emergence (directly to the soil) have higher runoff and runin (Duhan et al. 2020). This factor has increased value for products that have residual activity (remain longer in the soil) and for those that need rain or irrigation water to be activated (movement of the herbicide along with the soil profile until reaching the weed seeds in germination) (Priess et al. 2020). The timing and intensity of these events can cause off-target movement reducing residual activity and weed control effectiveness. Leon et al. (2016), evaluating the lateral movement of pre-emergence herbicides applied to 14% slope golf course grass, reported that all tested herbicides (pronamide, simazine, indaziflam, amicarbazone, dithiopyr, and prodiamine) were moved out of the treated area, regardless of whether or not these herbicides were incorporated with irrigation water before the rain event.

Herbicides applied in post-emergence (direct contact with the plants) also can be moved out of the application area by transport. The transport happens because, in addition to reaching the target plants, some molecules can get the soil, prone to other transport events. Although less of a concern than pre-emergence herbicides, Melland et al. (2016) observed that more than 50% of the glyphosate, 2,4-D, fluoroxypyr, atrazine, and diuron herbicides applied located in sugarcane cultivation were transported in the dissolved phase of the soil. Herbicides commonly applied in post-emergence have already been detected in river waters in several countries, such as the United States (glyphosate and atrazine) (Mahler et al. 2017), Brazil (metolachlor, atrazine, diuron, and simazine) (Barizon et al. 2020), Argentina (glyphosate and glufosinate (Andrade et al. 2021), and Australia (glyphosate and aminomethylphosphonic acid – AMPA) (Okada et al. 2020).

Runoff caused by rainwater is considered one of the primary forms of herbicide pollution in surface water bodies (Harrison et al. 2019). In semi-arid regions such as California in the United States, herbicide runoff generated by irrigation events has been reported as a significant cause for transporting these compounds in agricultural areas during the dry season (Zhang and Goh 2015). Vaz et al. (2021) observed that sugarcane straw on the soil surface reduced diuron runoff, but hexazinone losses were not attenuated. According to the authors, hexazinone is less available to runoff due to its low soil sorption and high solubility in water which contribute to fast leaching in the soil profile (Vaz et al. 2021). Potter et al. (2015) showed high runin of the herbicide fluometuron on a slope in the Atlantic Coastal Plain region of the southeast

United States. Most intriguingly, runin was found to increase with conservation tillage practices, but runoff losses were reduced. Therefore, even with conservation crops, all these factors must be considered when planning the management system.

6.2.1.1 Methodologies for Runoff and Runin

In runoff assessment, generally, the area has a slight slope, where metal plates are inserted below the ground, and a collection system collects runoff water with tanks (Carretta et al. 2021). Water and sediment are subsequently evaluated for herbicide quantification by colorimetric and chromatographic techniques (Gonzalez 2018). Another way to assess runoff is by collecting water from rivers close to cultivated agricultural areas that use herbicides to control weeds (Sandin et al. 2018). River water is sampled at different points using collectors, and the samples are taken to the laboratory for detection and quantification of herbicides.

Runin evaluation (subsurface lateral flow) is also carried out on slopes with small slopes, where perforated tile drainage tubes (about 15 cm in diameter) are installed at a certain depth (about 1.2 m below level). Flow from each drain plate is conducted into metal channels using an unperforated drain pipe. Runoff samples are collected from each chute outlet by automated refrigerated samplers, and these samples are further analyzed by chromatographic methods (Potter et al. 2015).

6.2.2 Leaching (Vertical Transport)

In addition to runin and runoff, leaching is another pathway for transporting herbicides in the soil. This process is considered to be mainly responsible for groundwater contamination (Vryzas 2018). Several studies worldwide have investigated the leaching of herbicides in different environments and soils to understand the behavior and verify the potential risk for groundwater contamination. Hence, the next topic will address the leaching of herbicides and which methodologies are being used to determine the leaching of these chemical compounds.

Leaching is the vertical transport of herbicides into the soil profile that depends on the physicochemical properties of the soil and the herbicide and the intensity and frequency of rainfall and irrigation (dos Reis et al. 2017). Regarding weed control, it is interesting to note that herbicides present small leaching in the soil profile to achieve and exert a good weed control efficacy in the soil seed bank (Carneiro et al. 2020). On the other hand, herbicide leaching deeper into the soil can decrease weed control effectiveness and increase the potential for herbicide molecules to reach groundwater (Jhala et al. 2012).

In recent years, research efforts have been directed toward understanding the complex herbicide–soil interactions. The herbicide leaching process has been addressed quite frequently. Researchers discuss the management of herbicide applications in different scenarios, encouraging safer herbicide applications and adopting management practices that reduce the impacts caused by leaching.

6.2.2.1 Leaching Methodologies

The most adopted methodologies to determine herbicide leaching are the method of bioassay in soil columns, polyvinyl chloride (PVC) or glass, and lysimeter. In the bioassay method, the movement of herbicides in the soil columns is measured by evaluating the visible damage on bioindicator species (crops or weeds) sown in longitudinally divided columns. Figure 6.1 represents the steps used to assess the movement of herbicides using bioindicator species.

(1) Columns are filled with soil.
(2) Columns with soil are placed inside a container with water for 24 h for soil saturation.
(3) After resting for a few hours to drain excess water, the herbicide is applied.
(4) After a specific evaluation time (defined by the researcher), the soil columns are cut longitudinally.
(5) In this step, the sowing bioindicator species seeds are carried out at each depth of the soil column.
(6) After a defined period, the movement of the herbicide is measured by visually evaluating the injuries and collecting the above-ground biomass of the bioindicator species.

The steps used for the PVC column method is similar to the chromatographic analysis method. However, bioindicator species are used to assess the movement of the herbicide in the soil. The columns are cut at each depth, and the soil is removed to subsequently go through a process of chemical extraction of the herbicide. After extraction, the herbicide concentration at each depth is quantified by High Performance Liquid Chromatography (HPLC).

In the glass column method, according to the OECD-312 guidelines, "Leaching in Soil Columns" (OECD 2004), the columns are filled with soil, and you can add sorbent materials such as biochar (on the surface or incorporated) to assess herbicide-holding capacity and leaching consequences (Figure 6.2). The soil is also saturated with water and then drained until it approaches field capacity. After draining, the herbicide is applied to the top of the columns. In this method, as in the bioassay and PVC column, rain of approximately 200 mm using 0.01 M $CaCl_2$ solution in distilled water is simulated for 48 hours after herbicide application (Figure 6.2), and the leachate volume is collected periodically for further analysis (Mendes et al. 2017). After a period determined by each researcher, the soil is removed from the columns by air injection and cut in each defined depth. These are dried, homogenized, and weighed for further quantification of the herbicide.

Assessment of leaching using lysimeters can be performed in the laboratory and the field. Compared to laboratory studies, results using lysimeters in the field are closer to actual conditions due to the size of the scale and the natural condition of rainfall (Schuhmann et al. 2016; OECD 2000). Lysimeters are ground monoliths, cylindrical, and open at the top (Schreiber et al. 2018). Lysimeters generally have variable lengths and diameters produced from inert materials such as stainless steel, metals, and PVC (Schreiber et al. 2018; Albers et al. 2020). Lysimeters are filled with undisturbed soil or installed under field conditions, and the herbicide is applied to the bare soil surface. Generally, the leachate is collected by collectors during the evaluation time of each study, and, later, the samples are analyzed for herbicide detection.

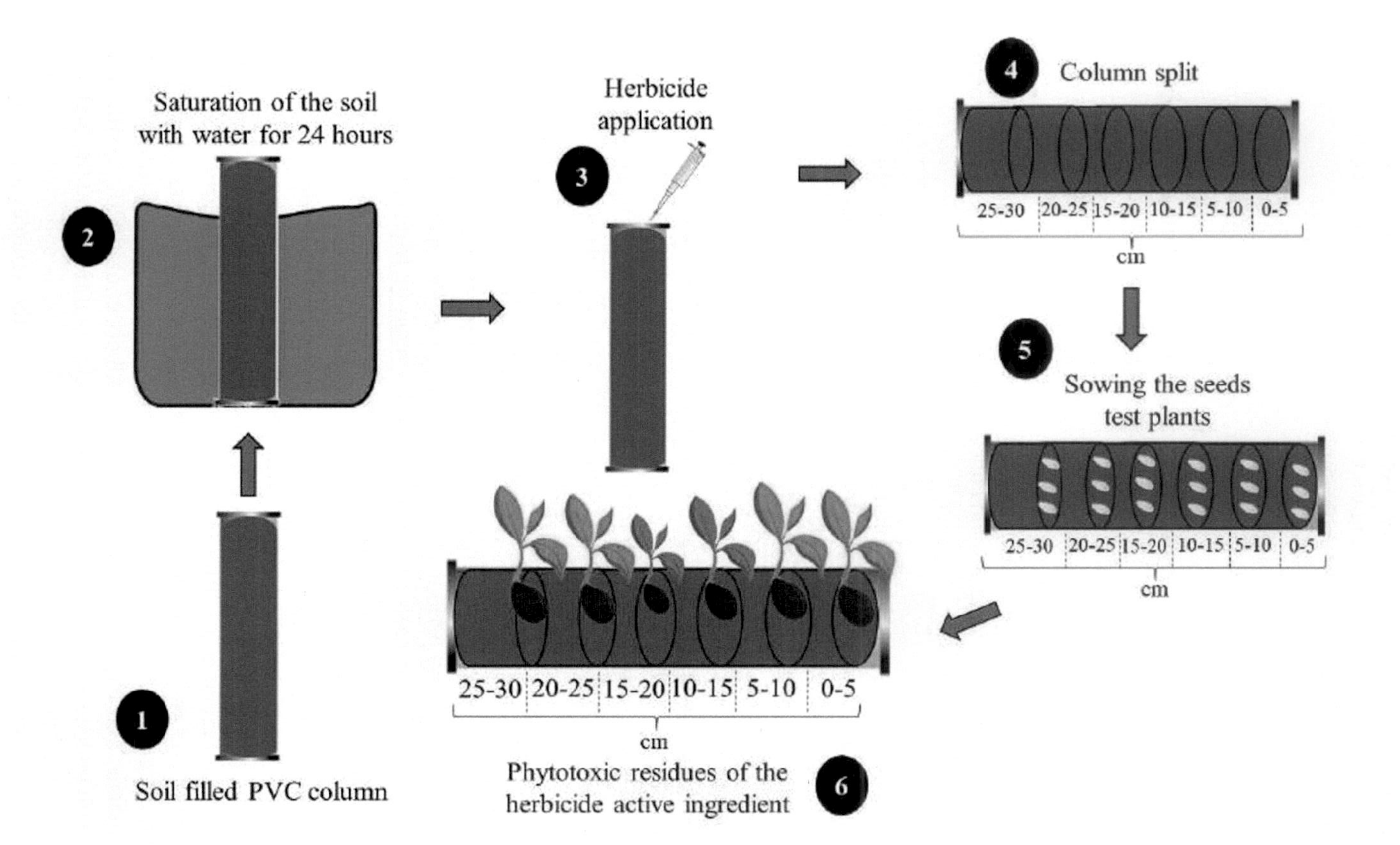

FIGURE 6.1 Representation of the bioassay method to assess herbicide leaching into the soil (adapted from Sekutowski 2011).

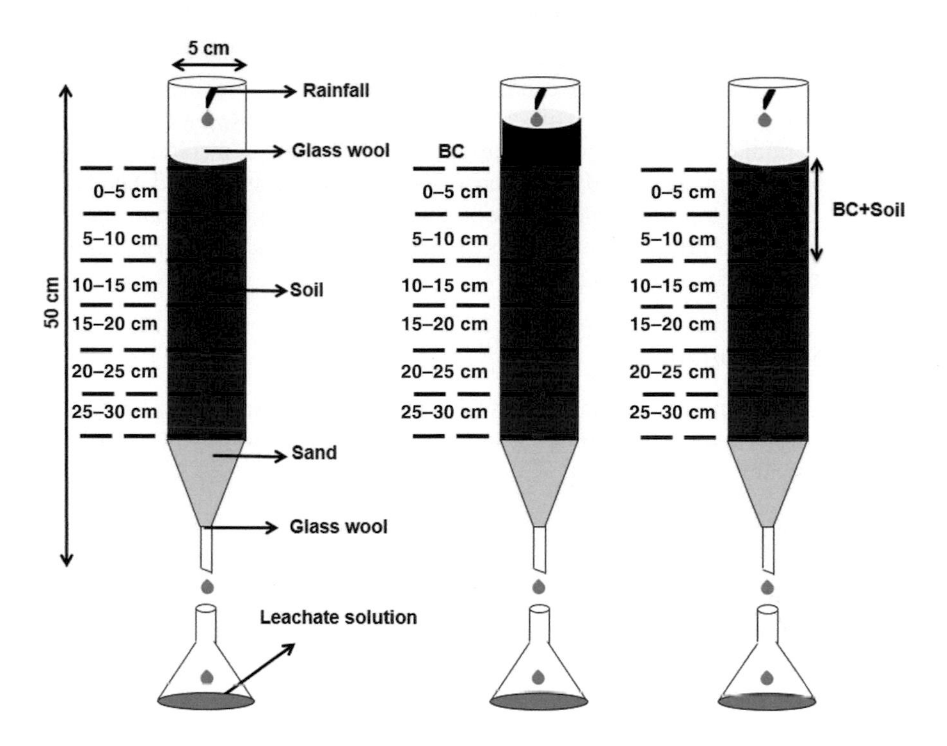

FIGURE 6.2 Method for evaluating the leaching of herbicides through a glass column (adapted from Mendes et al. 2018).

6.2.2.2 Methodology for Ground Mobility

The mobility of herbicides in the soil can be assessed by the method described by the Environmental Protection Agency, "Soil Thin Layer Chromatography" (EPA 1998). According to this methodology, the soil needs to be sieved, ground, and air-dried to reduce the size of the aggregates; after air-drying, they are sieved through a 250-μm sieve. Water is added until the soil is moderately fluid, forming a paste. The paste is spread on a glass plate; the soil layer should be 0.50–0.75 mm. Drying for 24 hours at 25 °C is allowed. A horizontal line of 11.5 cm is drawn above the base through the ground to the glass to stop solvent movement. The herbicide solution 1.5 cm is applied above the base; then, the plate is immersed with the base facing downwards in the chromatographic vat containing water at the height of 0.5 cm. When the water reaches the line made at the top, the plate is removed from the tub. To determine the distance travelled by the herbicide, zone extraction, plaque scanning (in the case of radiolabeled herbicides), or another quantification method can be used.

6.3 FACTORS THAT INFLUENCE THE TRANSPORT OF HERBICIDES

6.3.1 Environmental Factors

The prevailing tropical and subtropical climatic conditions in certain regions can cause variations in the behavior of herbicides in the soil due to rainfall patterns.

The amount, intensity, and frequency of rainfall events can alter the transport of herbicides in soils. Herbicides are exported to groundwater by leaching when rainwater transports its dissolved molecules through the soil profile. Researchers reported that higher rainfall intensity increased atrazine leaching into soil columns (Ouyang et al. 2016), while in the second event, successive rainfall events doubled metolachlor leaching from the top soil (Goldreich et al. 2011). In addition to environmental factors, soil properties are vital components for the fate of herbicides in the soil and will be discussed later in the chapter.

6.3.2 Soil Properties

The main soil properties that affect herbicide leaching include soil composition and pH. Soil composition is formed by the amount of sand, silt and clay in the soil (texture), and organic matter content.

6.3.2.1 Soil Composition

Soil composition affects leaching and herbicide activity due to different soil–herbicide bonds (sorption). Generally, soils with low clay and organic matter content have more potential for leaching due to a decrease in herbicide binding to soil colloids. Increased herbicide leaching can result in less weed control and more significant groundwater contamination.

In general, fine-textured soils with lower organic matter content are less capable of retaining herbicides, and, therefore, they are more leaching. A widely studied alternative is the addition of biochar to the soil to increase holding capacity and reduce herbicide leaching. Mendes et al. (2020) found a reduction in aminocyclopyrachlor and metolachlor leaching in soils treated with biochar produced from sugarcane bagasse, soybean residues, and wood chips. Adding rice husk biochar in columns with sand increased the sorption sites and, consequently, reduced the leaching of the phenylurea group herbicides (monuron, diuron, and linuron) (Dan et al. 2021). Although studies show that leaching is more significant in sandy soils and with lower organic matter content, the transport of herbicides can occur in any soil. Therefore, even clayey soils with higher organic matter content must be monitored. For example, Faria et al. (2018) reported that tebuthiuron leaching reached 50 cm of soil depth, even in soils with higher clay and organic matter content.

6.3.2.2 Soil pH

Soil pH can influence the leaching of herbicides, especially dissociable herbicides (dos Santos et al. 2019) (pH-dependent loads of the medium). Herbicides come in neutral or ionic forms, with the proportion of each form dependent on the soil pH and the acid/basic dissociation constant (pK_a/pK_b) of the herbicides. In addition to altering the shape of the herbicide molecule, changes in pH alter the amount of negative charges in the soil with the sorption capacity of cationic or anionic herbicides (Van der Linden et al. 2009). For example, the solubility and persistence of herbicides from the chemical group of sulfonylureas increase as soil pH increases, and, consequently, the tendency to leach also increases.

Herbicides from the chemical group of sulfonylureas are weak acids with pK_a ranging from 3.0 to 5.2 (Azcarate et al. 2015). Therefore, the neutral form predominates at pH values below the herbicide pK_a, and the anionic form is dominant when the pH is above the pK_a. Generally, neutral and anionic forms of sulfonylureas should predominate, as these pK_a values are within the range of pH values for most cultivated soils (Grey and McCullough 2012). Hence, repulsive interactions with soil particles reduce sorption and increase the leaching potential (Porfiri et al. 2015). As a result of the herbicides' mobility, the occurrence of these compounds in surface and groundwater resources has been reported. Cessna et al. (2010) detected thifensulfuron-methyl in groundwater, with concentrations ranging from 1.2 to 2.5 ng L^{-1}.

As for weak acids, the sorption/desorption and leaching of weak base herbicides such as hexazinone (pK_a 2.2) and atrazine (pK_a 1.7) are also altered by the pH of the soil solution (Guimarães et al. 2018). However, for weak-based herbicides, herbicide adsorption generally increases as the pH value decreases. In the pH range of cultivated soils, hexazinone and atrazine, for example, will exist as neutral species, and therefore the most of the herbicide molecules will not be dissociated. Mandal et al. (2017), characterizing atrazine sorption in biochars with pH ranging from 6.93 to 10.1, observed that atrazine adsorption was low, and the herbicide remained more available in the solution. In alkaline soils it is recommended to apply the lowest recommended dose of hexazinone for pre-emergence application to reduce leaching and consequently groundwater contamination (Dos Santos et al. 2019).

6.3.3 Herbicide Characteristics

The physical and chemical properties of herbicides are crucial elements in the leaching of herbicides into the soil. The main characteristics related to the dynamics of herbicides that influence the leaching are the octanol–water partition constant (Kow), acid/base electrolyte dissociation constant (pK_a and pK_b), and water solubility (S_w).

6.3.3.1 Octanol–Water Partition Constant (K_{ow})

K_{ow} indicates the affinity of the herbicide molecule with the organic and inorganic phases. This property reflects the intensity of the interactions between herbicide molecules with the nonpolar (octanol) or polar (water) phase. K_{ow} classifies herbicides into lipophilic and hydrophilic.

6.3.3.2 Acid/Basic Dissociation Constant (pK_a/pK_b)

A herbicide acid/basic dissociation constant classifies these substances into ionic (pK_a/pK_b non-zero) and non-ionic (pK_a equal to zero). Non-ionic forms, regardless of the pH of the solution, maintain their molecular structure. On the other hand, ionic herbicides can lose or gain protons, depending on the pH of the solution, being classified into weak acids or weak bases, respectively. When the pK_a of a weak acid herbicide or the pK_b of the weak base is equal to the pH of the soil, half the molecules are in ionized form. The increase in soil pH concerning the pK_a value favors that molecules of weak acid herbicides remain predominantly in anionic form, being less attracted by soil colloids. In this same condition, for weak base herbicides, the neutral form prevails.

6.3.3.3 Solubility in Water (S_w)

The S_W of a herbicide in water is the maximum amount of herbicide that can dissolve in a standard amount of pure water under specific temperature and pressure conditions. The herbicide Sw indicates the quantity of its molecules available in the soil solution, S_W being one of the most critical parameters in the dynamics of a herbicide in the environment.

Polar herbicides present high S_W, remaining in high concentrations in the solution, being absorbed by plants and maximizing their bioactivity. In general, depending on the rainfall regime, these herbicides have a high tendency to be transported by leaching or runoff and contaminate water sources (Dechene et al. 2014). On the other hand, nonpolar herbicides are less soluble in water, being strongly attracted by soil particles, reducing their concentration in the solution. Lipophilic herbicides, being less soluble, also tend to be less leached.

6.4 EFFECTS OF BIOCHAR CHARACTERISTICS ON THE TRANSPORTATION OF HERBICIDES

In recent years, the chemical decomposition of organic materials subjected to high temperatures, in the absence of oxygen, has been attracting the attention of several researchers as an alternative to convert biomass from different sources of raw material into high-performance carbon-rich materials, having the physicochemical properties and the effects of their adsorption and repair characteristics under different pyrolysis conditions as one of the main focuses of study (Zhang et al. 2020).

Biochar ability to absorb both organic and inorganic substances has attracted the attention of researchers to different functionalities, such as agricultural inputs, due to its potential to improve soil fertility. Furthermore, several studies suggest that biochar can reduce the transport of herbicides, thus minimizing their environmental impact. This reduction in leaching occurs due to greater herbicide sorption to the biochar, concentrating the retention of molecules in the upper layers of the soil. The source of biomass used to prepare the biochar, the form and rate of application, ageing, and activation are factors that influence the transport of herbicides in the soil.

6.4.1 Feedstock Types and Particle Size

Biochars are developed by the thermal decomposition of various types of biomass, including wood, plant waste, cow bone and animal manure, and their compositions and properties are pretty varied, depending on the raw material and thermal decomposition conditions, which provide changes in its properties, which can substantially impact the ability to sorb herbicides (Wang et al. 2016; Zhang et al. 2011).

A study evaluating three types of rice husk-derived biochars, miscanthus, and long-fiber biochar on atrazine and diuron adsorption found that sorption varied with biochar, herbicide, and soil type (Aldana et al. 2021). The authors found that rice husk biochar had the highest adsorption capacity for all evaluated herbicides. These results are mainly due to the material composition, represented by the lower levels of O/C and H/C and the higher ash content, making it the best sorbent for all herbicides (Aldana et al. 2021).

The incorporation of biochar derived from eucalyptus wood produced at a pyrolysis temperature of 400 °C in soil, regardless of the proportion used, favored the leaching of 2,4-D + picloram and its bioavailability, when compared to soils that received biosolids and aluminum silicate, which increases the environmental risk of groundwater contamination (Barros et al. 2021). These same authors associate the more significant leaching of 2,4-D + picloram with the average particle size of biochar, which, among the other materials used, had the highest average particle diameter among the analyzed residues, which may have contributed to the increased porosity and, consequently, to the increased mobility of these herbicides.

Simazine leaching from high- and low-fertility soils corrected with two hardwood -derived biochars with different particle sizes was evaluated. The presence of biochar decreased the amount of simazine leached from the soil with the biochar application and particle size rate, significantly affecting the loss rate. In high-fertility soil, the presence of large pieces of biochar (>2 mm) tended to have less impact on the amount of simazine leached than fine-grained biochar (<2 mm). Furthermore, the response with large pieces of biochar was highly variable, possibly due to the heterogeneous surface area to piece volume ratio (Jones et al. 2011).

6.4.2 Application Form and Rate

The form and rate of biochar application also influence the leaching of herbicides in the soil profile. Studies evaluating the leaching of aminocyclopyrachlor, hexazinone, metribuzin, and quinclorac herbicides applied on the surface and incorporated into the soil showed that, when applied superficially, there is greater sorption of herbicide molecules in the upper layer of the soil, reaching 100% sorption levels in the first 5-cm depth (Mendes et al. 2018, 2019; Fernandes et al. 2021) (Table 6.1). However, in other studies, where biochar was incorporated into the soil in layers of 0–15 cm, higher leaching rates of fomesafen were observed, with preferential retention of herbicide molecules in layers more significant than 5 cm in-depth, with more accentuated leaching in treatments that received aged biochar, compared with the incorporation of fresh biochar (Khorram et al. 2017) (Table 6.2). This behavior occurs because when the biochar is superimposed on the topsoil, every herbicide applied to the soil column comes into contact with the biochar, with the possibility of more significant adsorption in the upper layer, with a consequent decrease in the amount leached to lower soil layers (Table 6.1).

The presence of biochars in soils has provided good sorption of many herbicides, influencing the mobility of these molecules applied to the soil profile. Several researchers have shown that biochar-corrected soils have prolonged retention and reduced leaching of herbicides, which is environmentally favorable (Jones et al. 2011; Li et al. 2013). The effect of biochar application on the transport of hexazinone, metribuzin, and quinclorac in soil was demonstrated by Mendes et al. (2019). Using TLC plates with soil (Figure 6.3), the authors demonstrated that the most significant reductions in herbicide mobility occurred in treatments when biochar was incorporated into the soil compared to treatments where biochars were applied in the middle or on top of the TLC plate.

TABLE 6.1
Herbicide Leaching in Biochar-Corrected Soils of Different Feedstock and Application Rates.

Herbicide	Biochar		Herbicide distribution by soil depth (%)				Reference
	Feedstock	Rate	0–5 cm		> 5 cm		
		(%) w w^{-1}	Unamended	Biochar-amended soil	Unamended	Biochar-amended soil	
aminocyclopyrachlor	Cow bonechar	1/5	18	100	82	0	Mendes et al. (2018)
Fomesafen	Rice hull (fresh treatment)	0.5	5	8	95	92	Khorram et al. (2017)
		1	5	23	95	77	
		2	5	34	95	66	
	Rice hull (aged 6 months)	0.5	5	5	95	95	Khorram et al. (2017)
		1	5	10	95	85	
		2	5	19	95	81	
Hexazinone	Eucalyptus	1	20	100	80	0	Fernandes et al. (2021)
Hexazinone	Cow bone char	5	7	100	93	0	Mendes et al. (2019)
Mesotrione	Cow bone char	1/5	19	100	81	0	Mendes et al. (2018)
Metribuzin	Cow bone char	5	39	100	61	0	Mendes et al. (2019)
Quinclorac	Cow bone char	5	59	100	41	0	Mendes et al. (2019)

TABLE 6.2
Change in the Amount of Herbicide Leached into the Soil by Adding Biochar Produced from Different Feedstock and Application Rates.

Herbicide	Biochar		Reduction of leached herbicide	Reference
	Feedstock	Rate		
		(%) w w^{-1}	(%)	
Atrazine	Pine chip	0.8	52.0	Delwiche et al. (2014)
Fluometuron	Wood pellets	2	81.0	Cabrera et al. (2011)
Fomesafem	Rice hull (fresh treatment)	0.5	19.6	Khorram et al. (2017)
		1	42.4	
		2	74.7	
	Rice hull (aged 6 months)	0.5	9.50	Khorram et al. (2017)
		1	15.6	
		2	43.7	
Imazapic	Oil palm empty fruit bunch	1	77.6	Yavari et al. (2021)
Imazapic	Rice hull	1	73.3	Yavari et al. (2021)
Imazapyr	Oil palm empty fruit bunch	1	80.3	Yavari et al. (2021)
Imazapyr	Rice hull	1	72.0	Yavari et al. (2021)
MCPA	Wheat straw	1	37.5	Tatarková et al. (2013)
MCPA	Wood pellets	2	36.7	Cabrera et al. (2011)
Mesotrione	Oak wood	2	23.2	Gámiz et al. (2019b)

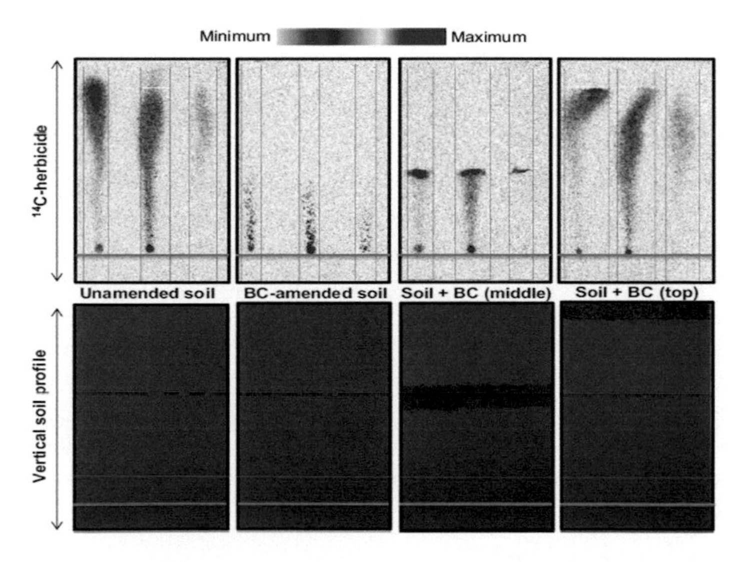

FIGURE 6.3 Mobility of ^{14}C-labeled quinclorac, metribuzin, and hexazinone (left to right, respectively) in four soil treatments in TLC: unamended soil (control), soil amended with biochar, soil + biochar strip (1 cm) in the middle, and soil + biochar band at the top of the soil TLC (adapted from Mendes et al. 2019).

Applying biochar at appropriate rates can reduce herbicide leaching, provide manageable solutions for reducing carbon dioxide emissions directly from the atmosphere through plant uptake, and contribute to the growth of food production. Khorram et al. (2017), evaluating the influence of three application rates of rice-husk-derived biochar (0.5%, 1%, and 2%) on the leaching of fomesafen, observed a direct correlation between the application rate and the reduction in the leaching of fomesafen, obtaining a more significant reduction in the leaching of herbicide molecules (74.7%) when the highest application rate (2%) of rice-husk biochar was used (Table 6.1).

However, the effect of reducing herbicide leaching induced by the application of increasing rates of biochar may be weakened by excessively increasing these rates. The optimal application rate will vary with the type of soil, contaminant to be remedied, and the physical and chemical characteristics of the adsorbent material. Thus, the biochar additive can reduce herbicide leaching and retain upper soil layers when applied at optimal rates. With the retention of the herbicide in the upper layers, less portion of the herbicide will be leached out (Table 6.2). This percentage of herbicide leached into the soil profile will vary, among other factors, with the type of biomass, percentage of application, and pyrolysis temperature used to produce the biochar. Several studies have observed reductions in the ratios of herbicides leached into soils corrected with biochars derived from different biomasses. However, these results are not a general rule. Cabrera et al. (2011) evaluated the sorption and leaching of fluometuron and of 4-chloro-2-methylphenoxyacetic acid (MCPA) applied to soils corrected with six biochars derived from different raw materials. These authors observed that the sorption of both herbicides increased in all corrected soils but decreased in soil corrected with biochar produced from macadamia nut shells produced by the rapid pyrolysis process when compared to unaltered soil. This result can be explained by the smaller surface area of the biochar particles produced under these conditions, which provides a greater competition to herbicide molecules for the sorting sites, thus favoring their mobility in the soil.

6.4.3 Biochar Activation

Researchers have been studying the effectiveness of different biochar activation techniques to enhance the sorption of soil contaminants and thus reduce groundwater pollution through the leaching of these contaminants. Activation techniques are applied to biochars through physical or chemical treatments to maximize sorption capacity and reduce contaminant leaching. Upon physical activation, the biochar's porosity and specific surface area (SSA) are increased through exposure to gaseous agent fluxes CO_2, steam, or burning above 700 °C. Some of the widely used activating agents are classified as acids (HNO_3, H_3PO_4, HCl), bases (KOH, NaOH), and oxidizing agents (H_2O_2) (Anto et al. 2021). Given in subsequent sections are some physical and chemical activation techniques that researchers have adopted to optimize the effectiveness of biochars.

6.4.3.1 Chemical Activation

Chemical activation of biochars using activation treatments such as acids, bases, and oxidative agents can increase the number of functional groups containing oxygen on the biochar surface, increasing the possibility of hydrogen bonding, electron donor–acceptor, π–π interactions, and covalent bonding (Zhou et al. 2017). The alachlor, diuron, and simazine behavior were evaluated using biochars from ground coffee residue without and with NaOH activation and compared to provide more in-depth information about their behavior and adsorption mechanisms. The biochar derived from ground coffee that underwent chemical activation with NaOH could more effectively remove herbicides, presenting itself as a promising alternative to make biochar more practical and effective in removing herbicides in aqueous solutions (Lee et al. 2021).

Studies evaluating the effect of using biochar chemically activated with HCl on atrazine leaching under field conditions, with high soil heterogeneity, demonstrated that the use of 10 t ha^{-1} of HCl acidified biochar before application to the soil reduced 53% peak atrazine concentrations in groundwater relative to the soil without addition of biochar (Delwiche et al. 2014). This result occurred because the acidification process increased the surface area of the biochar, thus exposing it to more atrazine adsorption sites.

Activation of low temperature (350 °C) grape wood-derived biochar with H_2O_2 altered the biochar surface chemistry and bulk composition, improving cyhalofop-butyl sorption. This activation is beneficial from an agronomic and environmental point of view, as weak acid herbicides are anionic, making them particularly susceptible to leaching and groundwater contamination (Gámiz et al. 2019a).

6.4.3.2 Physical Activation

The physical activation of biochar is considered advantageous because it does not involve incorporating additives that often promote impurities, as seen in the case of chemical activation. The carbonization temperature is a factor that significantly impacts the development of SSA and the degree of aromaticity of the biochar. According to Deng et al. (2017), the higher pyrolysis temperature favors the degradation of long-chain aliphatic and aromatic hydrocarbons in the biomass, affecting the development of the specific surface area and the degree of aromaticity of the biochar. Studies evaluating the leaching of hexazinone in soils modified with biochar derived from eucalyptus wood at different pyrolysis temperatures revealed that biochars produced at temperatures above 750 °C had their properties altered, increasing the (N+O)/C ratios and content of ash, providing greater hexazinone sorption and less leaching when compared to unmodified soil (Fernandes et al. 2021).

Another study evaluated the effects of six biochars derived from two types of hardwood from the species *Carya tomentosa* and *Carya illinoinensis* at three different pyrolysis temperatures (350, 500, and 700 °C) on the dissipation and leaching of the herbicides clomazone and sodium bispyribac. The results of this study demonstrated different behaviors, depending on both the biochar and the herbicide characteristics. All biochars suppressed clomazone and bispyribac leaching compared to uncorrected soil; however, the larger surface area of biochars made at 700 °C provided greater sorption capacity for clomazone and bispyribac, reducing leaching and increasing persistence on the ground. In this way, it was evident that biochars produced at a lower pyrolysis temperature (350–500 °C) can be the best route to minimize the environmental impacts of applied pesticides, optimizing effectiveness (Gámiz et al. 2017).

6.4.4 Aged Biochar

Biochar ageing can alter its sorption properties and, consequently, herbicide transport. Therefore, changes in biochar's sorting properties during soil ageing must be considered when planning the use of biochar in agriculture and for soil remediation purposes (Zhelezova et al. 2017). These same authors, evaluating the sorption of glyphosate and diuron in clayey and sandy soil, using fresh biochar and aged for 3.5 months, observed that the addition of aged biochar increased soil pH and reduced the sorption of both herbicides. In contrast, the addition of fresh biochar increased diuron adsorption in clayey and sandy soils and decreased glyphosate sorption in sandy soil. The ageing process of rice-husk-derived biochar and its effect on the leaching of fomesafen in agricultural soil were studied. The results showed that the adsorption capacity of fomesafen by aged biochar was reduced when compared to soil with the addition of fresh biochar, since the sorption coefficient values ranged from 1.9 to 12.4 mg L^{-1} when the soil was corrected with 0.5 to 2% of fresh biochar, and 1.36 to 4.16, 1.13 to 2.78 and 0.95 to 2.31 mg L^{-1} when the soils were corrected with biochars aged for 1, 3 and 6 months of age, respectively. Consequently, a more significant leaching and the bioavailable fraction of fomesafen were observed in treatments that received aged biochars, emphasizing the treatment at 6 months of age (Khorram et al. 2017).

The addition of biochar as an agricultural soil corrector has shown to be a promising method to reduce herbicide mobility in soil, minimizing the risk of contamination of water bodies. However, the studies presented showed that the residence time of biochar in the soil could affect the biochar–herbicide interactions, reducing the sorption capacity of the biochar due to the elimination of the reactive surface, caused by the reduction in the availability of micropores present in the particles-adsorbent material.

6.5 CONCLUDING REMARKS

The use of biochar in agricultural soils represents a promising alternative to be explored, mainly as a sorbent to reduce herbicide transport. This chapter addresses issues involving the use of biochar in soil and its effects on herbicide transport. It is essential to know the physicochemical characteristics of the herbicide and biochars and their interaction with the soil to obtain promising results regarding the use of biochar. The biochar source and the adequate management for production are essential points to get biochar with desired aspects. Because of the complexity of understanding the behavior of biochar in soils, further field studies are suggested so that the use of the material as a reducing agent for herbicide transport can be safely recommended from an environmental and agronomic point of view –this way, reducing the contamination of water resources and keeping the effectiveness of weed control.

REFERENCES

Abdi, D. E., Owen Jr, J. S., Wilson, P. C., Hinz, F. O., Cregg, B., and Fernandez, R. T. 2021. Reducing pesticide transport in surface and subsurface irrigation return flow in specialty crop production. *Agricultural Water Management*, *256*, 107124.

Albers, C. N., Jacobsen, O. S., Bester, K., Jacobsen, C. S., and Carvalho, P. N. 2020. Leaching of herbicidal residues from gravel surfaces – a lysimeter-based study comparing gravels with agricultural topsoil. *Environmental Pollution*, *266*, 115225.

Aldana, G. O., Hazlcrigg, C., Lopez-Capel, E., and Werner, D. 2021. Agrochemical leaching reduction in biochar-amended tropical soils of Belize. *European Journal of Soil Science*, *72*(3), 1243–1255.

Andrade, V. S., Gutierrez, M. F., Regaldo, L., Paira, A. R., Repetti, M. R., and Gagneten, A. M. 2021. Influence of rainfall and seasonal crop practices on nutrient and pesticide runoff from soybean dominated agricultural areas in Pampean streams, Argentina. *Science of The Total Environment*, *788*, 147676.

Anto, S., Sudhakar, M. P., Ahamed, T. S., Samuel, M. S., Mathimani, T., Brindhadevi, K., and Pugazhendhi, A. 2021. Activation strategies for biochar to use as an efficient catalyst in various applications. *Fuel*, *285*, 119205.

Azcarate, M. P., Montoya, J. C., and Koskinen, W. C. 2015. Sorption, desorption and leaching potential of sulfonylurea herbicides in Argentinean soils. *Journal of Environmental Science and Health, Part B*, *50*(4), 229–237.

Barizon, R. R., Figueiredo, R. D. O., de Souza Dutra, D. R. C., Regitano, J. B., and Ferracini, V. L. 2020. Pesticides in the surface waters of the Camanducaia River watershed, Brazil. *Journal of Environmental Science and Health, Part B*, *55*(3), 283–292.

Barros, R. E., Reis, M. M., Montes, W. G., Lopes, É. M. G., Figueiredo, F. F., and Santos, L. D. T. 2021. Impacts of the addition of biochar, biosolid and aluminum silicate on the leaching and bioavailability of 2, 4-D+ picloram in soil. *Environmental Technology & Innovation*, 101682.

Battaglin, W. A., Thurman, E. M., Kalkhoff, S. J., and Porter, S. D. 2003. Herbicides and transformation products in surface waters of the Midwestern United States 1. *JAWRA Journal of the American Water Resources Association*, *39*(4), 743–756.

Bento, C. P., Goossens, D., Rezaei, M., Riksen, M., Mol, H. G., Ritsema, C. J., and Geissen, V. 2017. Glyphosate and AMPA distribution in wind-eroded sediment derived from loess soil. *Environmental Pollution*, *220*, 1079–1089.

Cabrera, A., Cox, L., Spokas, K. A., Celis, R., Hermosín, M. C., Cornejo, J., and Koskinen, W. C. 2011. Comparative sorption and leaching study of the herbicides fluometuron and 4-chloro-2-methylphenoxyacetic acid (MCPA) in a soil amended with biochars and other sorbents. *Journal of Agricultural and Food Chemistry*, *59*(23), 12550–12560.

Carneiro, G. D. O. P., de Freitas Souza, M., Lins, H. A., das Chagas, P. S. F., Silva, T. S., da Silva Teófilo, T. M., Pavão, Q. S., Grangeiro, L. C., and Silva, D. V. 2020. Herbicide mixtures affect adsorption processes in soils under sugarcane cultivation. *Geoderma*, *379*, 114626.

Carretta, L., Tarolli, P., Cardinali, A., Nasta, P., Romano, N., and Masin, R. 2021. Evaluation of runoff and soil erosion under conventional tillage and no-till management: A case study in northeast Italy. *Catena*, *197*, 104972.

Cessna, A. J., Elliott, J. A., and Bailey, J. 2010. Leaching of three sulfonylurea herbicides during sprinkler irrigation. *Journal of Environmental Quality*, *39*(1), 365–374.

Chen, C., Guo, W., and Ngo, H. H. 2019. Pesticides in stormwater runoff – A mini review. *Frontiers of Environmental Science & Engineering*, *13*(5), 1–12.

Dan, Y., Ji, M., Tao, S., Luo, G., Shen, Z., Zhang, Y., and Sang, W. 2021. Impact of rice straw biochar addition on the sorption and leaching of phenylurea herbicides in saturated sand column. *Science of the Total Environment*, *769*, 144536.

Dechene, A., Rosendahl, I., Laabs, V., and Amelung, W. 2014. Sorption of polar herbicides and herbicide metabolites by biochar-amended soil. *Chemosphere*, *109*, 180–186.

Delwiche, K. B., Lehmann, J., and Walter, M. T. 2014. Atrazine leaching from biochar-amended soils. *Chemosphere*, *95*, 346–352.

Deng, H., Feng, D., He, J. X., Li, F. Z., Yu, H. M., and Ge, C. J. 2017. Influence of biochar amendments to soil on the mobility of atrazine using sorption-desorption and soil thin-layer chromatography. *Ecological Engineering*, *99*, 381–390.

Dos Reis, F. C., Tornisielo, V. L., Pimpinato, R. F., Martins, B. A., and Victória Filho, R. 2017. Leaching of diuron, hexazinone, and sulfometuron-methyl applied alone and in mixture in soils with contrasting textures. *Journal of Agricultural and Food Chemistry*, *65*(13), 2645–2650.

Dos Santos, L. O. G., Souza, M. D. F., Das Chagas, P. S. F., Fernandes, B. C. C., Silva, T. S., Dallabona Dombroski, J. L., Souza, C. M. M., and Silva, D. V. 2019. Effect of liming on hexazinone sorption and desorption behavior in various soils. *Archives of Agronomy and Soil Science*, *65*(9), 1183–1195.

Duhan, A., Oliver, D. P., Rashti, M. R., Du, J., and Kookana, R. S. 2020. Organic waste from sugar mills as a potential soil ameliorant to minimize herbicide runoff to the Great Barrier Reef. *Science of The Total Environment*, *713*, 136640.

EPA – Environmental Protection Agency Soil Thin Layer Chromatography. EPA – Fate. 1998. Transport and Transformation Test Guidelines – OPPTS 835.1210 EPA, Washington, DC, USA (6 pp).

Faria, A. T., Souza, M. F., de Jesus Passos, A. B. R., da Silva, A. A., Silva, D. V., Zanuncio, J. C., and Rocha, P. R. R. 2018. Tebuthiuron leaching in three Brazilian soils as affected by soil pH. *Environmental Earth Sciences*, *77*(5), 1–12.

Fernandes, B. C. C., Mendes, K. F., Tornisielo, V. L., Teófilo, T. M. S., Takeshita, V., das Chagas, P. S. F., Lins, H. A., Souza, M. F., and Silva, D. V. 2021. Effect of pyrolysis temperature on eucalyptus wood residues biochar on availability and transport of hexazinone in soil. *International Journal of Environmental Science and Technology*, 1–16.

Gámiz, B., Hall, K., Spokas, K. A., and Cox, L. 2019a. Understanding activation effects on low-temperature biochar for optimization of herbicide sorption. *Agronomy*, *9*(10), 588.

Gámiz, B., Velarde, P., Spokas, K. A., and Cox, L. 2019b. Dynamic effect of fresh and aged biochar on the behavior of the herbicide mesotrione in soils. *Journal of Agricultural and Food Chemistry*, *67*(34), 9450–9459.

Gámiz, B., Velarde, P., Spokas, K. A., Hermosín, M. C., and Cox, L. 2017. Biochar soil additions affect herbicide fate: Importance of application timing and feedstock species. *Journal of Agricultural and Food Chemistry*, *65*(15), 3109–3117.

Goldreich, O., Goldwasser, Y., and Mishael, Y. G. 2011. Effect of soil wetting and drying cycles on metolachlor fate in soil applied as a commercial or controlled-release formulation. *Journal of Agricultural and Food Chemistry*, *59*(2), 645–653.

Gonzalez, J. M. 2018. Runoff and losses of nutrients and herbicides under long-term conservation practices (no-till and crop rotation) in the US Midwest: A variable intensity simulated rainfall approach. *International Soil and Water Conservation Research*, *6*(4), 265–274.

Guimarães, A. C. D., Mendes, K. F., Dos Reis, F. C., Campion, T. F., Christoffoleti, P. J., and Tornisielo, V. L. 2018. Role of soil physicochemical properties in quantifying the fate of diuron, hexazinone, and metribuzin. *Environmental Science and Pollution Research*, *25*(13), 12419–12433.

Grey, T. L., and McCullough, P. E. 2012. Sulfonylurea herbicides' fate in soil: Dissipation, mobility, and other processes. *Weed Technology*, *26*(3), 579–581.

Harrison, S., McAree, C., Mulville, W., and Sullivan, T. 2019. The problem of agricultural 'diffuse' pollution: Getting to the point. *Science of the Total Environment*, *677*, 700–717.

Jhala, A. J., Ramirez, A. H., and Singh, M. 2012. Leaching of indaziflam applied at two rates under different rainfall situations in Florida Candler soil. *Bulletin of Environmental Contamination and Toxicology*, *88*(3), 326–332.

Jones, D. L., Edwards-Jones, G., and Murphy, D. V. 2011. Biochar mediated alterations in herbicide breakdown and leaching in soil. *Soil Biology and Biochemistry*, *43*(4), 804–813.

Khorram, M. S., Lin, D., Zhang, Q., Zheng, Y., Fang, H., and Yu, Y. 2017. Effects of aging process on adsorption – desorption and bioavailability of fomesafen in an agricultural soil amended with rice hull biochar. *Journal of Environmental Sciences*, *56*, 180–191.

Larsbo, M., Sandin, M., Jarvis, N., Etana, A., and Kreuger, J. 2016. Surface runoff of pesticides from a clay loam field in Sweden. *Journal of Environmental Quality*, *45*(4), 1367–1374.

Lee, Y. G., Shin, J., Kwak, J., Kim, S., Son, C., Cho, K. H., and Chon, K. 2021. Effects of NaOH activation on adsorptive removal of herbicides by biochars prepared from ground coffee residues. *Energies*, *14*(5), 1297.

Leon, R. G., Unruh, J. B., and Brecke, B. J. 2016. Relative lateral movement in surface soil of amicarbazone and indaziflam compared with other pre-emergence herbicides for turfgrass. *Weed Technology*, *30*(1), 229–237.

Li, J., Li, Y., Wu, M., Zhang, Z., and Lü, J. 2013. Effectiveness of low-temperature biochar in controlling the release and leaching of herbicides in soil. *Plant and Soil*, *370*(1), 333–344.

Lupi, L., Bedmar, F., Puricelli, M., Marino, D., Aparicio, V. C., Wunderlin, D., and Miglioranza, K. S. 2019. Glyphosate runoff and its occurrence in rainwater and subsurface soil in the nearby area of agricultural fields in Argentina. *Chemosphere*, *225*, 906–914.

Mahler, B. J., Van Metre, P. C., Burley, T. E., Loftin, K. A., Meyer, M. T., and Nowell, L. H. 2017. Similarities and differences in occurrence and temporal fluctuations in glyphosate and atrazine in small Midwestern streams (USA) during the 2013 growing season. *Science of the Total Environment*, *579*, 149–158.

Mandal, A., Singh, N., and Purakayastha, T. J. 2017. Characterization of pesticide sorption behaviour of slow pyrolysis biochars as low cost adsorbent for atrazine and imidacloprid removal. *Science of the Total Environment*, *577*, 376–385.

Melland, A. R., Silburn, D. M., McHugh, A. D., Fillols, E., Rojas-Ponce, S., Baillie, C., and Lewis, S. 2016. Spot spraying reduces herbicide concentrations in runoff. *Journal of Agricultural and Food Chemistry*, *64*(20), 4009–4020.

Mendes, K. F., de Sousa, R. N., Goulart, M. O., and Tornisielo, V. L. 2020. Role of raw feedstock and biochar amendments on sorption-desorption and leaching potential of three ^{3}H- and ^{14}C-labelled pesticides in soils. *Journal of Radioanalytical and Nuclear Chemistry*, *324*(3), 1373–1386.

Mendes, K. F., de Sousa, R. N., Takeshita, V., Alonso, F. G., Régo, A. P. J., and Tornisielo, V. L. 2019. Cow bone char as a sorbent to increase sorption and decrease mobility of hexazinone, metribuzin, and quinclorac in soil. *Geoderma*, *343*, 40–49.

Mendes, K. F., Hall, K. E., Takeshita, V., Rossi, M. L., and Tornisielo, V. L. 2018. Animal bonechar increases sorption and decreases leaching potential of aminocyclopyrachlor and mesotrione in a tropical soil. *Geoderma*, *316*, 11–18.

Mendes, K. F., Martins, B. A. B., Reis, F. C., Dias, A. C. R., and Tornisielo, V. L. 2017. Methodologies to study the behavior of herbicides on plants and the soil using radioisotopes. *Planta Daninha*, *35*.

Nachimuthu, G., Halpin, N. V., and Bell, M. J. 2016. Effect of sugarcane cropping systems on herbicide losses in surface runoff. *Science of The Total Environment*, *557*, 773–784.

OECD – Organisation for Economic Co-Operation and Development. 2000. Guidance document for the performance of out-door monolith lysimeter studies. OECD Environmental Health and Safety Publications, Paris, OECD (Test 106) (25 p).

OECD – Organisation for Economic Co-Operation and Development. 2004. OECD – Organisation for economic co-operation and development OECD guidelines for testing of chemicals. Test number 312: Leaching in soil columns OECD, Paris (15 pp).

Okada, E., Allinson, M., Barral, M. P., Clarke, B., and Allinson, G. 2020. Glyphosate and aminomethylphosphonic acid (AMPA) are commonly found in urban streams and wetlands of Melbourne, Australia. *Water Research*, *168*, 115139.

Ouyang, W., Huang, W., Wei, P., Hao, F., and Yu, Y. 2016. Optimization of typical diffuse herbicide pollution control by soil amendment configurations under four levels of rainfall intensities. *Journal of Environmental Management*, *175*, 1–8.

Porfiri, C., Montoya, J. C., Koskinen, W. C., and Azcarate, M. P. 2015. Adsorption and transport of imazapyr through intact soil columns taken from two soils under two tillage systems. *Geoderma*, *251*, 1–9.

Potter, T. L., Bosch, D. D., and Strickland, T. C. 2015. Tillage impact on herbicide loss by surface runoff and lateral subsurface flow. *Science of the Total Environment*, *530*, 357–366.

Priess, G. L., Norsworthy, J. K., Roberts, T. L., and Gbur, E. E. 2020. Weed control and soybean injury from preplant vs. pre-emergence herbicide applications. *Weed Technology*, *34*(5), 718–726.

Schreiber, F., Scherner, A., Andres, A., Concenço, G., Ceolin, W. C., and Martins, M. B. 2018. Experimental methods to evaluate herbicides behavior in soil. *Revista Brasileira de Herbicidas*, *17*(1), 71–85.

Schuhmann, A., Gans, O., Weiss, S., Fank, J., Klammler, G., Haberhauer, G., and Gerzabek, M. H. 2016. A long-term lysimeter experiment to investigate the environmental dispersion of the herbicide chloridazon and its metabolites – comparison of lysimeter types. *Journal of Soils and Sediments*, *16*(3), 1032–1045.

Sandin, M., Piikki, K., Jarvis, N., Larsbo, M., Bishop, K., and Kreuger, J. 2018. Spatial and temporal patterns of pesticide concentrations in streamflow, drainage and runoff in a small Swedish agricultural catchment. *Science of the Total Environment*, *610*, 623–634.

Sekutowski, T. 2011. Application of bioassays in studies on phytotoxic herbicide residues in the soil environment. In *Herbicides and Environment*, 1st edition, 253–272. Rijeka (CRO): IntechOpen.

Sun, K., Gao, B., Ro, K. S., Novak, J. M., Wang, Z., Herbert, S., and Xing, B. 2012. Assessment of herbicide sorption by biochars and organic matter associated with soil and sediment. *Environmental Pollution*, *163*, 167–173.

Tang, X., Zhu, B., and Katou, H. 2012. A review of rapid transport of pesticides from sloping farmland to surface waters: Processes and mitigation strategies. *Journal of Environmental Sciences*, *24*(3), 351–361.

Tatarková, V., Hiller, E., and Vaculík, M. 2013. Impact of wheat straw biochar addition to soil on the sorption, leaching, dissipation of the herbicide (4-chloro-2-methylphenoxy) acetic acid and the growth of sunflower (Helianthus annuus L.). *Ecotoxicology and Environmental Safety*, *92*, 215–221.

Tayeb, M. A., Ismail, B. S., and Khairiatul-Mardiana, J. 2017. Runoff of the herbicides triclopyr and glufosinate ammonium from oil palm plantation soil. *Environmental Monitoring and Assessment*, *189*(11), 1–8.

Van der Linden, A. M. A., Tiktak, A., Boesten, J. J. T. I., and Leijnse, A. 2009. Influence of pH-dependent sorption and transformation on simulated pesticide leaching. *Science of the Total Environment*, *407*(10), 3415–3420.

Vaz, L. R. L., Barizon, R. R. M., de Souza, A. J., and Regitano, J. B. 2021. Runoff of hexazinone and diuron in green cane systems. *Water, Air, & Soil Pollution*, *232*(3), 1–11.

Vryzas, Z. 2018. Pesticide fate in soil-sediment-water environment in relation to contamination preventing actions. *Current Opinion in Environmental Science & Health*, *4*, 5–9.

Vymazal, J., and Březinová, T. 2015. The use of constructed wetlands for removal of pesticides from agricultural runoff and drainage: A review. *Environment International*, *75*, 11–20.

Wang, Z., Han, L., Sun, K., Jin, J., Ro, K. S., Libra, J. A., Liu, X., and Xing, B. 2016. Sorption of four hydrophobic organic contaminants by biochars derived from maize straw, wood dust and swine manure at different pyrolytic temperatures. *Chemosphere*, *144*, 285–291.

Yavari, S., Kamyab, H., Asadpour, R., Yavari, S., Sapari, N. B., Baloo, L., Manan, T. S. B. A., Ashokkumar, V., and Chelliapan, S. 2021. The fate of imazapyr herbicide in the soil amended with carbon sorbents. *Biomass Conversion and Biorefinery*, 1–9.

Zhang, G., Zhang, Q., Sun, K., Liu, X., Zheng, W., and Zhao, Y. 2011. Sorption of simazine to corn straw biochars prepared at different pyrolytic temperatures. *Environmental Pollution*, *159*(10), 2594–2601.

Zhang, P., Zhang, X., Li, Y., and Han, L. 2020. Influence of pyrolysis temperature on chemical speciation, leaching ability, and environmental risk of heavy metals in biochar derived from cow manure. *Bioresource Technology*, *302*, 122850.

Zhang, X., and Goh, K. S. 2015. Evaluation of three models for simulating pesticide runoff from irrigated agricultural fields. *Journal of Environmental Quality*, *44*(6), 1809–1820.

Zhang, X., Luo, Y., and Goh, K. S. 2018. Modeling spray drift and runoff-related inputs of pesticides to receiving water. *Environmental Pollution*, *234*, 48–58.

Zhelezova, A., Cederlund, H., and Stenström, J. 2017. Effect of biochar amendment and ageing on adsorption and degradation of two herbicides. *Water, Air, & Soil Pollution*, *228*(6), 216.

Zhou, Y., Liu, X., Xiang, Y., Wang, P., Zhang, J., Zhang, F., Wei, J., Luo, L., Lei, M., and Tang, L. 2017. Modification of biochar derived from sawdust and its application in removal of tetracycline and copper from aqueous solution: Adsorption mechanism and modelling. *Bioresource Technology*, *245*, 266–273.

7 Effects of Biochar on Degradation Herbicides in the Soil

Vanessa Takeshita[1], Fernando Sarmento de Oliveira[2], Gustavo Vinícios Munhoz-Garcia[1], Bruno Caio Chaves Fernandes[3], Taliane Maria da Silva Teófilo[3], Kassio Ferreira Mendes[4], and Valdemar Luiz Tornisielo[1]*

[1]Center of Nuclear Energy in Agriculture, University of São Paulo, Brazil
[2]Faculty of Technology - Sertão Central, Technological Education Institute, Brazil
[3]Federal Rural University of the Semi-Arid, Brazil
[4]Departament of Agronomy, Federal University of Viçosa, Brazil
* Corresponding author: vanessatakeshita@usp.br

CONTENTS

DOI: 10.1201/9781003202073-7

7.1 INTRODUCTION

Biochar is the solid product resulting from the pyrolysis of plant biomass under anaerobic or low oxygen conditions at high temperatures (Koufopanos et al. 1989). This product is both found in nature and is artificially produced as well. Characteristics such as porosity, organic carbon (OC) content, surface area, and cation exchange capacity (CEC) vary according to the source material and pyrolysis performance temperature (Guo et al. 2020; Ippolito et al. 2020).

Among the benefits of applying biochar to agricultural soils, increased OC and mineral content, reduced organic matter (OM) mineralization, increased carbon sequestration, and greenhouse gas reductions are noteworthy (García-Jaramillo et al. 2020; Joseph et al. 2020; Xu et al. 2020; Zhang et al. 2020), consequently, reducing environmental agricultural impacts. Biochar, however, may cause unwanted management system effects, such as the retention of pre-emergent herbicides. These herbicides are applied directly to soil, and their behavior is influenced by physical, chemical, and biological soil; intrinsic physicochemical properties; and environmental conditions. Therefore, biochar may alter sorption–desorption herbicide dynamics and local biodegradation, decreasing herbicide availability and effectiveness (Nag et al. 2011; Tatarková et al. 2013; Abujabhah et al. 2016; Khorram et al. 2016; Cha et al. 2016; Mendes et al. 2018).

Biochar can also reduce pesticide and metal soil availability due to biochar pesticide retention and microbial degradation (Li et al. 2020). Microbial development stimulation generally promotes organic contaminant degradation through microorganism action (Guo et al. 2020). These processes, alongside sorption, are responsible for the environmental unavailability of these molecules due to the strong bond formed by pesticide–biochar complexes (Zhang et al. 2005; Yu et al. 2011).

Reduced herbicide availability due to biochar application has been reported for different compounds employed in agricultural activities, such as atrazine, nicosulfuron, imazapic, imazapyr, pyrazosulfuron-ethyl, indaziflam, sulfometuron-methyl, aminocyclopyrachlor, and metolachlor (Spokas et al. 2009; Cabrera et al. 2014; Dechene et al. 2014; García-Jaramillo et al. 2014; Tan et al. 2015; Liu et al. 2020; Voorhees et al. 2020; Wang et al. 2020; Manna et al. 2020; Mendes et al. 2020, 2021; Dan et al. 2021; Obregón-Alvarez et al. 2021). Biochar's versatility and ability to dissipate different herbicides comprise a favorable perspective in its application to the remediation of contaminated areas.

Interactions between biochar, microorganisms, and contaminants are complex and will be addressed throughout this chapter, focusing on the use of biochar in agricultural areas and its relationship with herbicides, especially regarding soil herbicide biodegradation processes.

7.2 SOIL HERBICIDE BIODEGRADATION PROCESSES

Soil herbicide trapping processes, whether through chemical or physical sorption mechanisms, occur concurrently with the dissipation of these products through transport and degradation processes. These processes interact in highly compartmentalized soil compartments, limiting herbicide availability to the plants they are intended to control, as well as to the microorganisms responsible for their degradation (Sims et al. 1991; Sims and Cupples 1999). Figure 7.1 represents the complex soil environment where herbicide degradation processes occur.

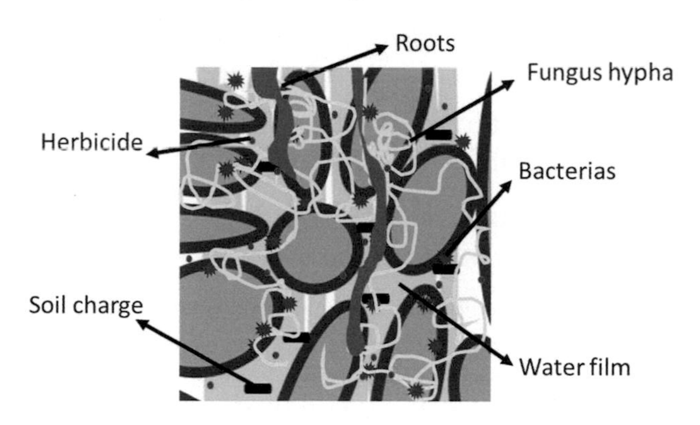

FIGURE 7.1 Schematic representation of the complex soil environment where biological herbicide degradation processes take place.

One of the main soil herbicide degradation processes comprises biodegradation, consisting of biological degradation by microorganisms. During this process, the chemical herbicide structure is altered, and molecules can be transformed into metabolites and even be fully mineralized, mainly through biotic processes (Terzaghi et al. 2018; Hussain et al. 2015). In these natural environments, herbicides and herbicide degradation products (metabolites formed from parental molecules) are transformed directly by microorganisms or as a complement to complete compound mineralization, such CO_2 and H_2O, among others (Ortiz-Hernández et al. 2013). This process involves the complete breakdown of compounds and occurs through the ability of microorganisms to physically and chemically interact with molecules, altering their structural conformations (Ortiz-Hernández et al. 2013).

In addition to the ability of the microbial community to use herbicide molecules, as well as pesticides in general, as a source of energy, OM soil maintenance is paramount to soil biodegradation process. The addition of organic compounds, such as crop residues, manure, sludge, and biochar, has been reported to be able to improve the degradation capacity of herbicides in soil (Cox et al. 2001; Sánchez et al. 2004; Kanissery and Sims 2011; He et al. 2016). Moorman et al. (2001), when assessing the remediation of herbicide-contaminated areas, also indicated that the addition of different organic materials, such as corn stalks and cattle manure, contributed to the stimulation of microorganisms and soil herbicide degradation. However, the same authors indicate that this does not always take place, as even when accounting for general microbial activity, specific populations able to degrade herbicides were not stimulated. Furthermore, the addition of these materials can contribute to soil herbicide unavailability and increased degradation half-life time (DT_{50}) through the formation of bound residues, as observed by Takeshita et al. (2020) for aminocyclopyrachlor in soils containing sugarcane raw feedstock and metabolites. Furthermore, slow and reduced alachlor and sulfometuron-methyl degradation has also been reported when biochar is added to the soil (Mendes et al. 2017; Obregón-Alvarez et al. 2021).

The influence of biochar in this soil herbicide degradation process mediated by microorganisms will be further explored in the following section.

7.3 BIOCHAR FACTORS AFFECTING HERBICIDE BIODEGRADATION

7.3.1 Effects of Biochar Production and Raw Feedstock on Biochar Characteristics

Physicochemical biochar properties, determined by the type of raw material, mechanisms, pyrolysis temperature, and residence time employed in the biomass conversion process of the material used in biochar production, can influence soil herbicide biodegradation processes, further detailed later in the chapter.

7.3.1.1 Raw Feedstock

Biomasses consisting of mainly plant residues, which contain lignin, cellulose, and hemicellulose in their composition, are termed lignocellulosic biomasses (Wang et al. 2017). On the other hand, biomasses composed of animal waste, skin, bones, algae and sewage sludge, among others, which may contain proteins, lipids, saccharides, and inorganic compounds and a small portion of cellulose lignin are termed non-lignocellulosic biomasses (Li and Jiang 2017).

Table 7.1 displays biochar property variations in biochars produced from different biomasses at the same pyrolysis temperature (500°C). Certain characteristics, such as ash content and total and fixed carbon, have been pointed out by Zhao et al. (2013) to be most influenced by the type of employed biomass (Table 7.1.).

The observed biochar differences may occur due to the varied composition of each feedstock, as well as cellulose, hemicellulose, protein, saccharides, and inorganic compound contents, which respond differently to the conversion mechanism. Biomasses richer in lignin, for example, result in biochars displaying greater porous structure, aromaticity, and carbon content (Leng and Huang 2018), while biomasses presenting higher hemicellulose and lignin content result in biochars with a higher amount of functional groups containing oxygen (Hassan et al. 2020). These characteristics directly influence herbicide retention and, consequently, compound availability for environmental dissipation processes.

According to Ali et al. (2019), biochars produced from different feedstock, at the same temperature (500°C) led to changes in some soil microbial community phyla, probably due to the amount and quality of carbon present in each feedstock. Considering that the microbial community plays an important role in herbicide

TABLE 7.1
Physicochemical Properties of Biochars from Different Feedstocks Produced at a 500°C Pyrolysis Process Temperature.

Biochar feedstock	TC (%)	FC (%)	Yield (%)	VM (%)	Ash (%)	pH	CEC (cmol kg^{-1})	SSA (m^2 g^{-1})
Cow manure	43.7	14.7	57.2	17.2	67.5	10.2	149	21.9
Pig manure	42.7	40.2	38.5	11.0	48.4	10.5	82.8	47.4
Shrimp hull	52.1	18.9	33.4	26.6	53.8	10.3	389	13.3
Bone dregs	24.2	10.5	48.7	11.0	77.6	9.57	87.9	113
Wastewater sludge	26.6	20.6	45.9	15.8	61.9	8.82	168	71.6
Waste paper	56.0	16.4	36.6	30.0	53.5	9.88	516	133
Sawdust	75.8	72.0	28.3	17.5	9.94	10.5	41.7	203
Grass	62.1	59.2	27.8	18.9	20.8	10.2	84.0	3.33
Wheat straw	62.9	63.7	29.8	17.6	18.0	10.2	95.5	33.2
Peanut shell	73.7	72.9	32.0	16.0	10.6	10.5	44.5	43.5
Chlorella	39.3	17.4	40.2	29.3	52.6	10.8	562	2.78
Waterweeds	25.6	3.84	58.4	32.4	63.5	10.3	509	3.78

Total carbon: TC, fixed carbon: FC, volatile matter: VM, cation exchange capacity: CEC, and BET-N_2 specific surface area: SSA. *Source*: Adapted from Zhao et al. (2013).

biodegradation processes, these changes can alter the degradation process dynamics of these products in the environment.

7.3.1.2 Production Mechanisms

According to Kumar et al. (2020a), converted feedstock can be categorized as biochars and hydrochars, based on production mechanisms and characteristics (Figure 7.2). The type of mechanism, temperature, and residence time will directly influence biochar/hydrochar physicochemical characteristics, and, consequently, herbicide degradation. The most common processes applied to biochar production are gasification and pyrolysis (slow and fast), and for hydrochar, hydrothermal carbonization is the most common process.

Biochar yields following the gasification process are less than 10%, as the main product of interest in this process is biogas. The feedstock is oxidized in a fluidized bed gasification unit, at about 800°C, with a residence time of 10 to 20 s, at atmospheric pressure or higher (Meyer et al. 2011). As the focus of this process is not biochar production, we will mostly discuss the main product differences obtained through pyrolysis and hydrothermal carbonization.

Pyrolysis is one of the most applied processes used to convert biomass into biochar, classified as slow or fast pyrolysis, consisting of the thermal degradation of dry material in the absence or under reduced oxygen conditions to avoid feedstock combustion. Slow pyrolysis results in higher biochar yields compared to the fast process, from 20% to 40%, at lower heating rates less than 10°C min and usually at

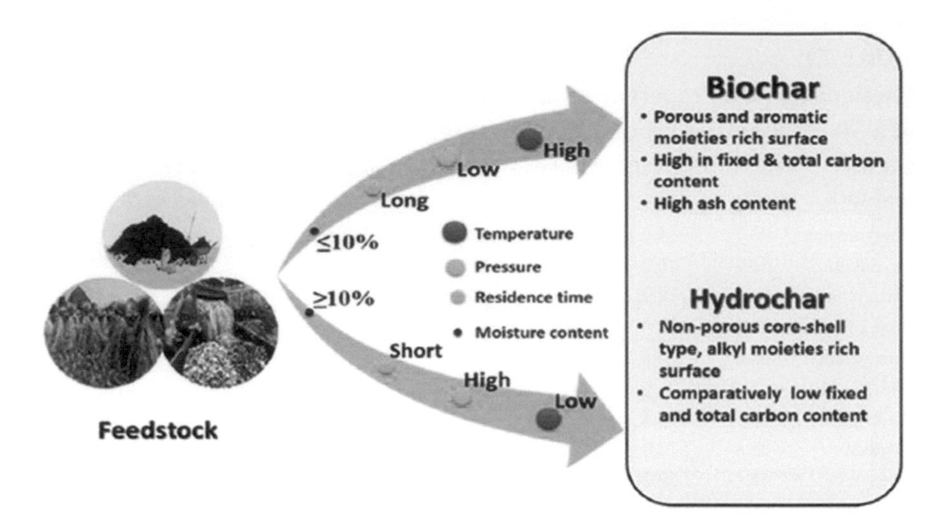

FIGURE 7.2 Differences between biochar and hydrochar according to their production mechanisms and physicochemical characteristics. *Source*: Adapted from Kumar et al. (2020a).

temperatures ranging from 300 to 600 °C and a residence time of hours to days (Mohan et al. 2014). The temperature range of fast pyrolysis is similar to the slow process but with an applied heating rate of 200°C/min and with a residence time of only a few seconds (Kumar et al. 2020a). Rapid pyrolysis produces biochar with a smaller specific surface area (SSA) due to tar pore trapping. Despite the shorter process time, the main products obtained are bio-oil and biogas, with a biochar yield of only 10% to 20%. Other biochar production processes have also been applied, such as instantaneous, vacuum, and microwave methods.

Hydrothermal carbonization is employed to produce hydrochar, developed for the carbonization of feedstock presenting high moisture content at temperatures between 180 and 350°C and at pressures from 2 to 6 MPa. This product is less stable than biochar, due to a higher percentage of labile carbon in its structure and the presence of an alkyl fraction (Mohan et al. 2014), although containing high amounts of functional surface oxygen groups (Figure 7.2). Biochar, on the other hand, exhibits a porous structure, aromatic structure, higher fixed carbon content, and ash content (Figure 7.2). The lower applied temperature, high pressure, and humidity used in hydrochar production are the main processes responsible for physicochemical differences when compared to biochar, even when originating from the same feedstock.

The aforementioned physical–chemical differences due to different production methods alter soil herbicide biodegradation dynamics. Eibisch et al. (2015), when evaluating the effect of soil biochar and hydrochar addition on isoproturon degradation, observed that mineralization was more strongly reduced in the biochar application (81% reduction) compared to hydrochar (56% reduction). These authors also mention that the production method leads to a greater influence when compared to the type of utilized raw feedstock, and that the lower mineralization rate is observed in the biochar application due to greater herbicide sorption, making this compound less available to microorganisms responsible for herbicide soil biodegradation.

7.3.1.3 Pyrolysis Temperature

Physicochemical biochar characteristics are affected by pyrolysis temperature. Increasing temperatures increase fixed carbon content, ash, aromaticity, SSA, zeta potential, while decreasing yield, hydrogen, oxygen, and polarity (Figure 7.3). According to Kumar et al. (2020a), feedstock constituent decomposition (cellulose, hemicellulose, and lignin) result in the formation and transformation of intermediate compounds and a temperature-dependent removal of volatiles.

The reactions involved in the biomass pyrolysis process comprise dehydration, depolymerization, fragmentation, condensation, and conversion of short substituents in the benzene ring (Lian et al. 2017) (Figure 7.4). Increasing pyrolysis temperatures results in biomass constituent degradation, promoting the release of volatile

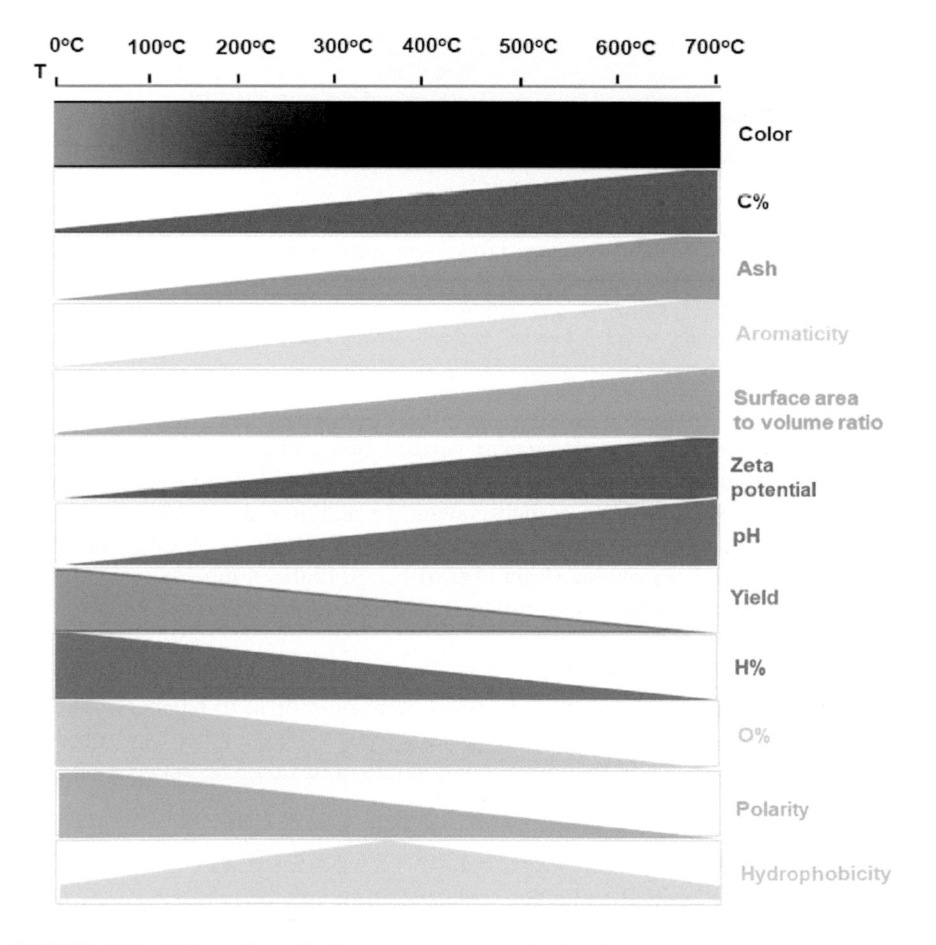

FIGURE 7.3 Effect of pyrolysis temperature on physicochemical biochar characteristics. *Source*: Adapted from Ghodake et al. (2021).

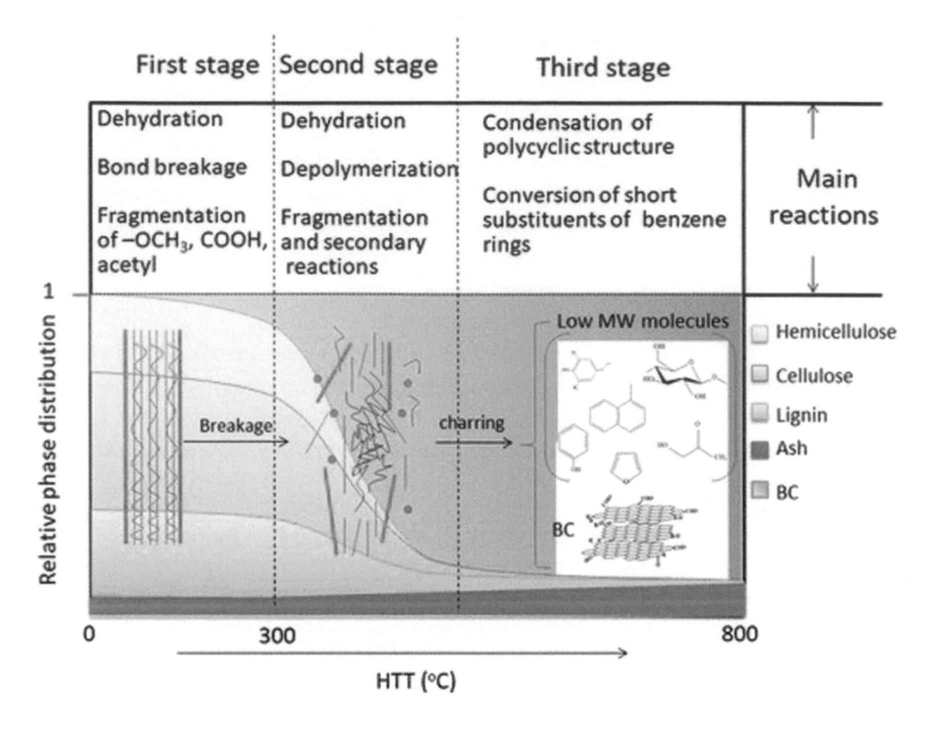

FIGURE 7.4 Processes and reactions involved in increasing heat treatment temperature (HTT) in the production of the biochar (BC). As the temperature increases, a large part of the biomass (hemicellulose, cellulose, and lignin) is converted into molecules of low molecular weight (MW – molecular weight), biochar, and ash. *Source*: Adapted from Lian et al. (2017).

compounds, thereby increasing biochar surface area and porosity while reducing oxygen and hydrogen content. This, in turn, increases aromatization due to structure condensation as well as fixed carbon content. In addition, a pH reduction is also observed due to decreases in acidic functional groups and the accumulation of minerals with increased ash content (Lian et al. 2017). At higher temperatures (~800°C), the biomass is almost completely carbonized, transformed into overlying aromatic carbon sheets and ash (Figure 7.4).

According to Cheng et al. (2017a), simazine degradation is reduced by the addition of wheat straw biochar, with this effect more prominent in biochars produced at lower pyrolysis temperatures. Fernandes et al. (2021), on the other hand, observed no significant difference in the mineralization rate of hexazinone between biochars derived from eucalyptus wood produced at different pyrolysis temperatures.

7.3.1.4 Residence Time

Depending on the applied technique, residence time can comprise seconds (rapid pyrolysis) or hours (slow pyrolysis or hydrothermal carbonization), affecting biomass carbonization degree and, consequently, physicochemical biochar properties. Similar to pyrolysis temperature, the applied residence time also affects final product yields and physicochemical characteristics. According to Kumar et al. (2020b), a residence time of up to 2 h increases the biochar surface area and total pore volume, and residences times above 2 h have the opposite effect on these characteristics.

The influence of residence time on biochar characteristics is dependent on pyrolysis temperature. As reported by Sun et al. (2017), increased residence times promote decreased yields and increased pH at 300°C, which was not observed at 600°C. At higher temperatures, the largest feedstock fraction is found carbonized, thus not significantly altering some produced biochar characteristics.

Physicochemical biochar properties, determined from the choice of feedstock in the applied production processes, are paramount for herbicide biodegradation, but do not act alone. Thus, their relationship with herbicide properties and soil changes should also be assessed.

Herbicide biochar sorption/desorption will determine its availability to microorganisms responsible for biodegradation and may retain the herbicide, reducing its availability and, consequently, its degradation (Kookana 2010; Beesley et al. 2011; Sopeña et al. 2012). As reported by Jones et al. (2011), the strong simazine sorption to biochar reduced its biodegradation due to a lower availability of microbial soil communities. Biochar properties, chemical groupings, porosity, surface area, pore size, pH, zero charge point (PZC) are crucial for herbicide–biochar interactions (Kupryianchyk et al. 2016; Lian et al. 2017).

Before addressing the relationship between biochar properties and reduced herbicide degradation process availability, we emphasize that chemical herbicide characteristics must be favorable for surface bond formation for this biochar interaction to take place. Characteristics such as log K_{ow} (octanol–water partition constant), pK_a (acid dissociation constant), water-solubility (S_w), molecule size, and the ability to carry out hydrogen bonds, as well as the presence of electronegative atoms in the aromatic ring, are determinant of the types of interactions that the herbicide molecules can display with biochar (Mojiri et al. 2020).

In general, more polar herbicides, which dissociate, will exhibit greater affinity with biochars produced at lower temperatures, due to their high polarity, lower zero charge point, and pH (Lian et al. 2017). The predominant bond types with these biochars comprise partition, hydrogen bonding, Coulomb forces, and Lewis acid–base interactions (Li et al. 2013; Lian et al. 2016) (Figure 7.5). On the other hand, more nonpolar herbicides (larger log of K_{ow}), neutral and with the presence of electronegative atoms in the aromatic ring, will result in sorption to biochars produced at higher pyrolysis temperatures due to their high aromaticity and low polarity (Lattao et al. 2014; Tong et al. 2019; Hassan et al. 2020). The predominant bond types in this interaction are hydrophobic interactions, Van der Wall strengths, π–π bonds, and pore-filling (Figure 7.5) (Xiao and Pignatello 2015a; Tong et al. 2019).

Biochar functional groups and aromatic rings will contribute to material polarity. In general, biochars will be polar and hydrophilic when presenting a greater amount of surface functional groups containing oxygen and nitrogen (Hassan et al. 2020), while nonpolar and hydrophobic biochars exhibit reduced functional groups containing O and N and high aromaticity. Polarity can be assessed through O/C (hydrophilicity index) and H/C (aromaticity index) ratios by elemental analyses. According to Wei et al. (2020), the H/C ratio can predict the metolachlor sorption, a

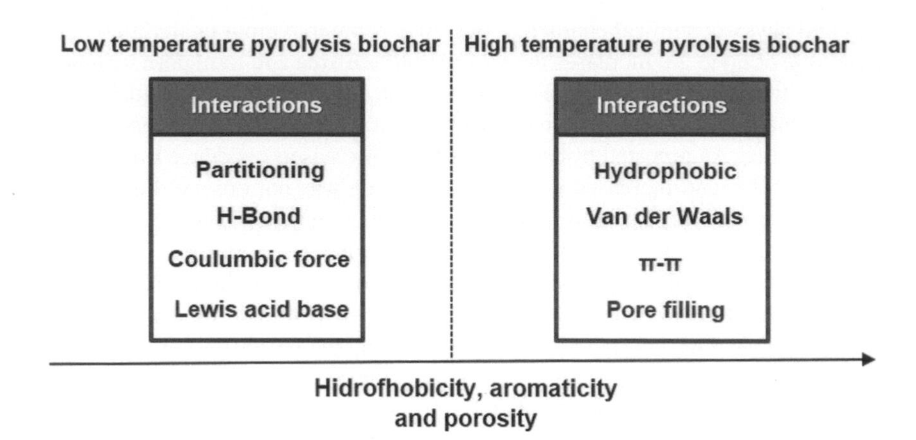

FIGURE 7.5 Types of predominant interactions between herbicides and biochars produced at low/high pyrolysis temperatures.

polar herbicide, to biochar, where the highest sorption was observed in the biochar presenting the lowest H/C ratio. Functional groups directly affect herbicide bonds, altering their availability through hydrogen bonds and Lewis acid–base interactions.

Some of the main biochar characteristics comprise SSA and porous structure. When herbicide chemical properties are favorable to biochar interactions, the greater these characteristics, the greater the herbicide sorption and the lower its medium availability (Graber et al. 2012; Tan et al. 2017; Khalid et al. 2020). Depending on the employed biomass, method, and pyrolysis temperature, biochar pore sizes can vary from nano (<0.9 nm), micro (<2 nm) to macropores (>50 nm) (Shaaban et al. 2018). Biochars produced at high pyrolysis temperatures, in general, exhibit greater SSA and porosity, with increased micropore volume (Lian et al. 2017). Pore size is an important factor to be considered in herbicide biochar retention processes as, depending on the size of the herbicide molecule, it cannot access micropore areas, even if displaying favorable chemical properties concerning biochar interactions (Xiao and Pignatello 2015b; Suo et al. 2018).

The pH values generally exhibit a positive correlation with biochar ash contents. When produced at high temperatures, biochars exhibit higher pH and ash content due to decreased acidic functional groups and original biomass burning. Depending on the amount of biochar, its pH can alter the pH of the medium, affecting ionizable herbicide dissociation, modifying interactions, and consequently, biodegradation mechanism availability. The biochar pH also interferes with herbicide hydrolysis processes, so that it can be catalyzed by acids and/or bases. Zhang et al. (2013), for example, observed that the high pH of biochar produced at 700°C was mainly responsible for the hydrolysis of atrazine (strongly hydrolyzed in strongly acidic or alkaline media).

Ashes can reduce herbicide availability due to increasing herbicide interactions through new bond formation and increased biochar surface polarity, attracting molecules that display an affinity for this compound. In addition, they can also stimulate hydrolysis, as the metal atoms present in the ashes can form herbicide complexes on the biochar surface or in the soil solution (released ions), facilitating nucleophilic water molecule attack (Zhang et al. 2013).

Biochar surface loads can influence herbicide availability in the medium, determined by the individual pK_a of each functional group and, mainly, by the biochar PZC and medium pH (Lian et al. 2017). In general, when the pH of the medium is lower than the PZC, functional group protonation takes place, forming positive surface loads and, thus, reducing the availability of anionic herbicides (Banik et al. 2018). However, when the pH of the medium is greater than the PZC, the predominant biochar surface loads are negative, due to functional group deprotonation, reducing cationic herbicide availability (Banik et al. 2018).

The lower herbicide availability to soil microorganisms due to greater sorption is caused not by an isolated biochar characteristic but by the set of interactions between these characteristics with the chemical properties of the herbicide, establishing a repulsion or affinity balance. As noted by Fernandes et al. (2021), the highest hexazinone sorption was not obtained in the biochar presenting the largest SSA but in the one with the highest polarity index. Biochar herbicide retention cannot be considered as the only interfering factor in the biodegradation and dissipation processes of these products. In this sense, Zhelezova et al. (2017) observed that diuron mineralization was not altered due to increased sorption in biochar-amended soils. Other factors that influence the herbicide biodegradation in biochar-amended soils will be discussed in the next sections.

7.3.2 Modification of Chemical Biochar-Amended Soil Characteristics

Changes in chemical soil properties due to biochar addition directly affect herbicide soil biodegradation, as various chemical soil properties (i.e., pH, electrical conductivity, CEC, OM) are directly and indirectly associated to this process. Thus, understanding the role played by biochar in chemical soil properties can aid in predicting environmental herbicide behavior, making herbicide use more effective and reducing environmental risks. In this context, the most suitable biochar should be selected based on raw material composition, production parameters (mainly pyrolysis temperature), and applied concentration.

The addition of different types of biochars to soil can alter different chemical soil properties, and, in general, their contributions vary due to different physical and chemical properties and type of soil (Chintala et al. 2014).

Soil solution pH alterations comprise one of the most common changes in biochar-amended soils. In general, when biochars, generally alkaline-rich in basic carbonates and organic anions, are added to acidic soil, increases in soil pH and pH-buffering capacity are observed (Lehmann and Joseph 2009; Yuan and Xu 2011; Xu et al. 2012; Chintala et al. 2014; Dai et al. 2014; Zhelezova et al. 2017). Soil pH plays an important role in herbicide molecule soil dissociation (Oliveira Jr. et al. 2001) and in microbial soil community composition and activity (Khorram et al. 2016), which can affect herbicide degradation rates. For example, increasing soil pH can increase the negative surface charge of soil colloids and acid herbicides, mostly in the anionic form, making herbicides more bioavailable in the soil solution due to decreased sorption), thus increasing herbicide biodegradation.

Biochar-amended soils also affect soil electrical conductivity (EC) and CEC, reported to be significantly increased in biochar-amended soils compared to unamended soils (Chintala et al. 2014; Muter et al. 2014). This is generally due to the addition of broad anionic functional groups, acid-neutralizing capacity, and a high concentration of biochar basic cations (Chintala et al. 2014; Jiang et al. 2015). Therefore, it is expected that biochar-amended soils displaying high EC and CEC contain more exchange sites to retain ionic herbicides, which can reduce their bioavailability and, thus, decrease herbicide soil biodegradation.

Biochar soil addition can also affect soil OC content, as the high OC contents observed in biochars can increase OC contents in soil, affecting herbicide degradation by (i) increasing herbicide sorption and decreasing herbicide desorption, resulting in lower herbicide availability for degradation by soil microorganisms and, therefore, lower biodegradation rates (Spokas et al. 2009; Tatarková et al. 2013; Muter et al. 2014) and (ii) increasing microbial activity, probably due to the supply of nutrients, energy, and moisture contained in biochar, resulting in increased microbial herbicide degradation (Zhang et al. 2005).

In general, biochars affect soil herbicide degradation due to their influence on soil characteristics, which are related to herbicide soil sorption and desorption (Liu et al. 2018; Zhang et al. 2018a; Yavari et al. 2019). Soil changes caused by biochar addition generally increase herbicide soil sorption, decreasing herbicide biodegradation bioavailability. Although some studies have indicated that biochar in the soil can also stimulate microbial herbicide degradation (Zhang et al. 2005; Yavari et al. 2019), their soil effects are more likely to reduce herbicide dissipation in agricultural soils, leading to the gradual accumulation of these compounds on the soil surface. However, further research is required to better understand the role of biochar on soil properties and its influence on soil herbicide degradation under specific local conditions.

7.4 BIOCHAR HERBICIDE SOIL BIODEGRADATION EFFECTS

7.4.1 Biochar Interactions with Soil Microorganisms

Soil is a heterogeneous, dynamic, and complex environment that hosts a high diversity of microorganisms such as fungi and bacteria, essential in ecological dynamics and soil functionality maintenance (Cotta 2016; Callieri et al. 2019). These microorganisms act in OM decomposition, biological nitrogen fixation, plant growth promotion, pesticide biodegradation, increased nutrient availability, and water stress tolerance (Bini et al. 2016; Jacoby et al. 2017). Most soil microorganisms are heterotrophic and chemotrophic, requiring OC and nutrients as energy sources for metabolism and growth maintenance (Bini et al. 2016).

The main biochar effects on biological soil attributes comprise increased OC content and nutrient mineralization (Xie et al. 2013; Lu et al. 2014). Following biochar application and increased OC contents, the characteristic environment of local microorganism communities becomes altered, resulting in altered microbial activity (Gul et al. 2015).

Biochar application generally stimulates the establishment of local microbial communities, such as arbuscular mycorrhizal fungi and bacteria (Hammer et al. 2014;

Kolton et al. 2011), due to higher OC and nutrient availability. On the other hand, only a small part of this carbon will be available to microorganisms (Ameloot et al. 2013). Therefore, biochar effects on microorganisms depend not only on increased carbon content, but also on soil conditions and nutrient availability (Anders et al. 2013).

The biochar–microbial community relationship and its effect on pesticide biodegradation are complex processes that depend on both biochar and environmental conditions. Soil microbiota biochar effects are evaluated through soil quality indicators such as microbial diversity, microbial biomass, and enzymatic activity (Ekenler and Tabatabai 2004; Ferreira et al. 2017).

Biochar promotes rapid microorganism growth, increasing microbial biomass, while at the same time specializing the microorganism population in the treated area and reducing microbial community diversity (Khodadad et al. 2011). The observed biomass increase is associated to biochar porosity, which assumes the habitat function for the development of microorganisms, increasing available microbiota survival spaces and providing a favorable development environment. The pores also serve as protective spaces for smaller microorganisms to survive predation by larger microorganisms that are unable to enter these spaces (Verheijen et al. 2010).

The biochar can also affect enzymatic soil activity, comprising enzyme expression due to microorganism action during their development in soil and indirectly reflecting soil ability to stabilize contaminants and decompose OC (Nannipieri et al. 2012; Cui et al. 2013). Biochar can increase, reduce, or not affect enzymatic activity at all, depending on the intrinsic characteristics of the material, the applied amount, the local microbial community, and the pre-established characteristic environment (Gorovtsov et al. 2020).

Enzymes affected by biochar are involved in nutrient cycling, OC decomposition, and herbicide degradation, such as phosphatase, peroxidase, phenoloxidase, ammonia-monooxygenase, β-glucosidase, N-acetyl-glucosaminidase, β-xylosidase, l-leucine aminopeptidase, urease, phosphatase, dehydrogenase, and arylsulfatase (Bailey et al. 2011; Demisie et al. 2014; Sun et al. 2014; Wang et al. 2015; Yang et al. 2015; Li et al. 2021).

Biochar-amended soils may display increased enzymatic activity. Li et al. (2021), for example, detected increases ammonia–monooxygenase activity produced by bacteria. Increased lipase and β-glucosidase, urease, and l-leucine aminopeptidase activities may also occur (Bailey et al. 2011; Demisie et al. 2014; Wang et al. 2015). Decreased enzymatic activity due to biochar use has been reported for phosphatase and dehydrogenase (Wang et al. 2015; Yang et al. 2015), while arylsulfatase was not influenced by the biochar application (Sun et al. 2014). Lower dehydrogenase activity can be explained by the addition of recalcitrant carbon present in biochar, which does not stimulate microbial activity or enzymatic activity (Wu et al. 2013) or as a consequence of the enzyme or substrate sorption process due to high biochar sorption capacity (Bailey et al. 2011).

The amount of biochar added to soil can influence microorganism interactions, although there is no defined dose pattern. The use of high amounts (5% w/w) of corn straw biochar, for example, resulted in the inhibition of the enzymatic activity of extracellular enzymes involved in the carbon and sulfur cycle, impairing nutrient and OC decomposition dynamics, while values 10-fold lower (0.5% w/w) led to an increased activity of these same enzymes (Wang et al. 2015). The fact that higher amounts of biochar reduce enzymatic activity is associated with the increased OC content and, at the same time, decreased OC decomposition that reduces the microorganism energy source and depends on enzymatic activity.

Biochar can also alter microbial diversity levels, including bacteria involved in the nitrogen cycle (Anderson et al. 2011) and those responsible for the OC degradation, such as Actinobacteria, Firmicutes, and Proteobacteria (Khodadad et al. 2011; Zhou et al. 2019). This is associated to the pre-existing microbial community composition. Chen et al. (2015) observed that microbial diversity varies depending on location and not on biochar application when comparing biochar effects in soils from three different regions in China, indicating the role of the local microbial community concerning biochar interaction.

Biochar application may also result in no microbial community effects (Elzobair et al. 2016). With increasing soil porosity and contact surface in biochar-amended soils, soil moisture maintenance depends less on fungal activity, thus not stimulating the development of these microorganisms (Gao et al. 2018).

In sum, it is not possible to state that biochar soil remediation will always promote beneficial microorganism or plant effects. Factors such as the C/N ratio, volatile compounds, and internal biochar colonization are associated with soil–biochar–microorganism interactions, highlighted here:

- *C/N ratio*: Materials with a high C/N ratio can decrease soil nutrient availability and reduce microorganism plant growth (Johnson 1993; Gaur and Adholeya 2000; Wallstedt et al. 2002).
- *Volatile compounds*: When adsorbed on the biochar surface, these compounds can become toxic to microorganisms (Sun et al. 2015) and impair biochar microbial community effectiveness.
- *Biochar internal colonization*: This may be hampered due to decreased OC and nutrient availability inside biochar pores due to biochar sorption capacity (Quilliam et al. 2013).

Microbial community composition and location, herbicides present in soil and biochar, the presence of plants, increased product persistence and desorption processes drive biochar herbicide effects. Here, we describe each factor and how it acts on biochar–microorganism–herbicide complexes:

- *Microbial community composition*: The relationship between sorption, leaching, and herbicide biodegradation is complex. Bending et al. (2007) observed that isoproturon degradation was greater in 40–80-cm deep layers compared to the first centimeters of the soil due to the specialized microorganisms that consume isoproturon molecules. These microorganisms are established in deep soil layers, emphasizing microbial community in pesticide biodegradation, i.e., if the groups of microorganisms present in the treated area are not capable of degrading the applied herbicide, the biochar effect on these groups will not allow for molecule biodegradation.

- *Colonization site*: comprises another example of soil–biochar–microorganism interaction toward herbicide degradation (Tao et al. 2020). Biochar-amended soils exhibit increased phosphorus-solubilizing bacterial activity. In addition, according to the same authors, for example, these bacteria have been reported as colonizing the biochar surface and using atrazine as a development substrate. Atrazine biodegradation was facilitated due to molecule biochar surface stabilization alongside the bacteria.
- *Presence of plants in treated soil*: Plants remove fewer soil contaminants but result in greater herbicide degradation (Cheng et al. 2017b), as they stimulate the microbial rhizosphere community, facilitating herbicide degradation due to the release of enzymes and exudates (Sun et al. 2004).
- *Desorption process*: The pesticide biochar sorption and its slow desorption cause the release of pesticide molecules at low concentrations, stimulating and allowing for the action of. High organic contaminants concentrations, on the other hand, can inhibit microbial communities, as observed for a chlorimuron-ethyl and atrazine combination (Wang et al. 2018).

When soil is treated with biochar, the biochar does not always significantly alter physicochemical soil properties and reduce herbicide bioavailability through the sorption process at first (Meng et al. 2019) or through biodegradation. In this regard, the biotic and abiotic effect balance involved in biochar–microorganism–pesticide complex behavior is quite delicate, comprising the result of several factors, including biochar environment modifications.

In summary, biochar application provides biological soil attribute and the microbial community benefits. It is important to note that no standard microbial community behavior takes place due to biochar application. Their variation is a consequence of the biochar characteristics and its effect on physical, chemical, and biological soil properties, which, in turn, influence microbial community composition and activity, thus altering biochar–soil–microorganism interactions (Pokharel et al. 2020).

7.4.2 Biochar Soil Addition Herbicide Biodegradation Effects

Herbicides applied as pre- and post-emergence weed control come in different forms, namely ionic-acidic, basic, and non-ionic. The behavior of each herbicide form in soil differs, which generally determines its bioavailability (effectiveness) and, thus, soil biodegradation. For example, basic herbicides bind more to OM and are less bioavailable for biodegradation than anionic forms (Oliveira Jr. 2001). Yavari et al. (2019) reported that imazapyr photodegradation and biodegradation rates are lower than imazapic, as the former exhibits greater affinity to biochar compared to the latter. On the other hand, weak acid herbicides applied to agricultural soils, such as indaziflam, remain mainly in anionic form and, thus, are more bioavailable in the soil solution and less sorbed to soil colloids, resulting in quicker biodegradation rates (Mendes, Furtado et al. 2021).

Biochar soil addition can affect applied herbicide bioavailability. In general, the bioavailability of post-emergence herbicides is less affected than that of those applied pre-emergence. In addition, the form of pre-or post-emergence application comprises another important factor when evaluating soil–biochar–herbicide efficacy effects (Gámiz et al. 2017). In a recent study carried out in tropical cow-bonechar-amended soil, decreased indaziflam pre-emergence activity was observed concerning weed control due to high herbicide sorption and unavailability in the soil solution (Mendes, Furtado et al. 2021).

Many other studies have reported biochar effects on herbicide soil degradation (Table 7.2). Based on these studies, biochar has two evident effects: (i) increased herbicide sorption without affecting degradation and (ii) increased herbicide persistence in soil (DT_{50}) due to increased sorption. For example, Junqueira et al. (2020) reported that the biochar addition to tropical soils did not influence glyphosate biodegradation compared to unamended soils. On the other hand, biochar can affect soil organism composition and activity, for example stimulating microbial soil activity community, thus increasing herbicide biodegradation. Wu et al. (2019) confirmed this effect, reporting that the degradation rate of oxyfluorfen in biochar-amended soil is faster compared to unamended soil, as biochar is rich in many nutrients and provides a suitable environment for microorganisms, stimulating microbial soil activity. Furthermore, biochar usually presents several surface functional groups (Qiu et al. 2008), which can react and degrade part of the adsorbed herbicide molecules, thus increasing chemical herbicide degradation.

Si et al. (2011) reported that biochar addition greatly reduced isoproturon biodegradation in different soils. These authors reported that, when the biochar addition

TABLE 7.2
Selected Studies Concerning the Biochar Herbicide Soil Degradation Effects.

Feedstock of biochar	Pyrolysis temperature (°C)	Applied rates (t ha^{-1})	Herbicide	Degradation time half-life (DT_{50})*	Reference
Sawdust	500 °C	24–720	Atrazine and acetochlor	34 days (increase of 256%)	Spokas et al. (2009)
Charcoal waste	300 to 500 °C	40 and 80	Sulfometuron methyl	55 days (increase of 51%)	Alvarez et al. (2021)
Vegetable waste (pineapple, palm oil fiber, and coffee husk)	300 or 600 °C	10 and 20	Bromacil and diuron	300 days for bromacil and 73 days for diuron	Chin-Pampillo et al. (2021)
Empty bunches of palm fruits	300 °C	12	Imazapic	46 days (increase of 33%)	Yavari et al. (2019)
			Imazapyr	53 days (increase of 38%)	
Rice husk	300 °C	12	Imazapic	41 days (increase of 17%)	
			Imazapyr	46 days (increase of 20%)	

*Degradation time half-life of herbicides in treatments with higher rate of biochar (percentage of increase in DT_{50} in biochar-amended soil compared to unamended soil).

rate was increased from 0 to 50 g kg^{-1}, the herbicide DT_{50} increased from 54.6 to 71.4 days in paddy soil, from 16.0 to 136 days in Alfisol, and from 15.2 to 107 days in Vertisol. This clearly demonstrated that biochar soil changes greatly prolongs the isoprouron soil DT_{50}. In another study, the degradation rate of simazine was significantly reduced in biochar soil compared to unamended soil. After 21 days of incubation, only about 10% of the herbicide was mineralized compared to unamended soil, probably due to increased simazine sorption due to the presence of biochar (Jones et al. 2011). Sopeña et al. (2012) observed that amended soils with red gum wood biochar (1% and 2% w/w) increased isoproturon sorption, reducing isoproturon soil biodegradation.

Herbicide soil degradation is a complex dynamic process that depends on chemical and biological processes, as well as on soil and herbicide characteristics. Biochar soil addition can significantly alter biodegradation processes (Haskis et al. 2019). However, the high persistence of herbicides in biochar-amended soils, due to the unavailability of herbicides to the biodegradation process. This can result in low microbial activity and lead to important ecosystem implications, such as increasing herbicide soil residues and decreasing herbicide effectiveness. However, most of biochar herbicide degradation effects are still poorly understood especially in the long term, which may limit the wider application of biochars to soil aiming at agronomic and environmental benefits.

7.4.3 Biochar Herbicide-Contaminated Soil Remediation Effects

When herbicide molecules reach the soil directly and indirectly after application, they are found in the form of extractable residues when they are extractable by organic solvents or in the form of non-extractable residues (when the molecules are strongly soil-bound). Concerning the latter, as reported by Francisco et al. (2018) for aminocyclopyrachlor, a decrease in herbicide unavailability is noted for microorganism degradation. The formation of bound residues during the herbicide transformation process can considerably contribute to long herbicide soil persistence (Pareja et al. 2012). Viti et al. (2021) reported that bound diuron, hexazinone, and metribuzin residues, as well as their metabolites can become bioavailable and mineralized through microorganism action or returned to the soil solution in biochar-unamended soils, depending on soil conditions and interventions, which can negatively affect successive cultures or nontarget organisms. However, the strong retention of molecules in materials such as biochar can contribute to alternative ways to manage contaminated areas.

Persistent herbicides can comprise a threat to both the environment and successive crops. Yavari et al. (2019), aiming to investigate the effects of the biochar application on the degradation of imidazolinone herbicides, reported that the presence of rice husk biochar in the soil provided DT_{50} reductions of 40.7–25.6 days, 46.2–26.6 days, and 43.3–26.5 days for imazapic, imazapyr, and imazapic + imazapyr (Onduty®), respectively. The same authors confirmed that biochar represents an innovative sustainable strategy with the potential to reduce the persistence of imidazolinones and minimize their environmental risks. Jing et al. (2018) also observed much lower herbicide rates in biochar-amended soils, using rice husk biochar for fenoxaprop soil degradation, of over 85% on the 35th day, also reporting estimated half-lives of 56.9 and 1.5 days in control and biochar-amended soils, respectively, and indicating that biochar addition can reduce the risks of soil fenoxaprop-ethyl.

The negative effects of persistent herbicides on soil health have become the target of emerging concerns. Herbicide soil degradation normally takes place through biodegradation, hydrolysis, photolysis, and oxidation, with biodegradation being the main dissipation and decomposition route of most herbicides (Liu et al. 2018). Meng et al. (2019) evaluated the effects of altering fomesafen-contaminated soils using wheat-straw-derived biochar on microbial soil activity, demonstrating that the presence of biochar influenced the balance between ecologically different bacteria (oligotrophic *vs.* copiotrophic), generating relatively more diverse microbial communities essential for plant growth.

The remediation potential of a consortium of degrading bacteria immobilized on biochar was evaluated in soils containing potato and metribuzin by Wahla et al. (2020). The findings indicated that a metribuzin degradation rate of 82% in soils submitted to the addition of isolated degrading bacteria, while soils that received degrading bacteria immobilized on rice-husk-derived biochar presented 96% degradation. This suggests that adding biochar to herbicide-contaminated soils can reduce environmental risks and help improve microbial soil ecology. In another study, flumioxazin degradation behavior in characteristic soils from China containing 0.5% (w/w) corn stem biochar was faster in soil samples not containing biochar (Chen et al. 2021), attributed to the nutrients provided by biochars that act as microbial population stimuli.

The biochar remediation capacity for some herbicides is influenced by pyrolysis temperature. Fernandes et al. (2021) evaluated the effect of different pyrolysis temperatures (650, 750, 850, and 950°C) on the properties of biochar derived from eucalyptus wood and their influence on hexazinone behavior in the soil. The authors verified that a pyrolysis temperature of 950°C increased the (nitrogen + oxygen)/carbon and ash ratios, producing biochar with a higher sorption coefficient and a lower hexazinone desorption coefficient, decreasing hexazinone environmental availability. Gámiz et al. (2019) observed the effects of H_2O_2 activation in biochar derived from grapevine wood produced at a low temperature (350°C), indicating subsequent changes in the removal efficiency of cihalofop and clomazone, enabling an improved organic acid herbicide removal.

Given these findings, biochar is a promising material for the remediation of herbicide-contaminated soils, improving soil quality and increasing microbial stimulation, acting on herbicide molecule biodegradation, reducing the environmental damage caused by these compounds and allowing for the anticipation of the establishment of sensitive crops in succession.

7.4.4 Modified Biochar Used to Enhance Herbicide Soil Biodegradation

The use of ecological alternatives for the remediation of herbicide-contaminated soils seems to be a relevant technique to solve this problem. In this sense, biochar has gained attention due to its high capacity to retain and immobilize organic contaminants in soil due to high porosity, surface area, pH, abundant functional groups, and highly aromatic structure, depending mainly on the raw material utilized in its production and the applied pyrolysis temperature (Liu et al. 2018). These factors make it a promising material to adsorb and, thus, reduce herbicide bioavailability in contaminated soils.

However, original biochar characteristics must be improved in order to obtain greater contaminant removal efficiency and develop wider applications to polluted environments (Zhang et al. 2018b). Thus, biochar improvements concerning pore structure, SSA, and surface functional group enrichment, among others, have become essential to optimize the use of this material.

Biochar modifications enhance herbicide biodegradation processes, stimulating soil microbial flora and molecule breakdown, reducing the risk of environmental contamination. Chemicals are the most frequent biochar modification method, offering greater biochar SSA and porosity development improvements due to higher activation temperatures (Wang and Wang 2019; Takaya et al. 2016). The following sections detail the most common chemical biochar modification methods, namely acid modification, alkaline modification, oxidizing agents, metal salts, and metal oxides.

7.4.4.1 Acid Modification

This method employs acids (sulfuric, hydrochloric, nitric, and phosphoric) to treat biochar and remove impurities, consequently increasing the material surface area and introducing acidic functional groups. Phosphoric acid is one of the most frequently utilized chemical modification activators, acting on the decomposition of lignocellulosic, aliphatic, and aromatic materials. Phosphate and polyphosphate bridges are formed and prevent material shrinkage during pore development, while also comprising more environmentally friendly alternatives than other corrosive and hazardous reagents (Kang et al. 2011; Yang et al. 2011).

Pesticide sorption studies in soils containing biochars chemically modified by washing with low molecular weight organic acids (LMWOA) indicate that this modification promotes particle removal, improving biochar porosity and favoring contaminant retention (Zheng et al. 2019). However, information on LMWOA-modified biochar applied to herbicides is still scarce.

7.4.4.2 Alkaline Modification

The most common alkaline agents used in biochar modification are sodium hydroxide (NaOH) and potassium hydroxide (KOH). This type of treatment promotes biochar porous structure changes, facilitating impurity removal, increasing the surface area, and enriching surface hydroxyl groups (Mahdi et al. 2019).

NaOH activation of biochars produced from coffee residue grounds results in greater alachlor, diuron, and simazine adsorption, with significantly higher removal rates observed with the amended biochar (43.9–81.9%) compared to unamended biochar (2.9–5.5%), due to increases in SSA and pore volume provided by the NaOH activation, which are about 106- and 21-fold higher than unamended biochar (Lee et al. 2021).

7.4.4.3 Oxidizing Agents

The most employed oxidizing agent used to activate biochars is hydrogen peroxide (H_2O_2). The results following oxidizing treatment include increased functional groups containing oxygen, increased surface area, and decreased pH (Lawrinenko et al. 2016). Furthermore, Mao et al. (2012) reported that oxidized coal forms a more stable material and significantly increases soil CEC.

Grape wood biochar prepared at 350°C before and after H_2O_2 activation added to soils contaminated with cyhalofop and clomazone exhibits surface modifications, increasing cyhalofop sorption capacity (Gámiz et al. 2019), although not clomazone. This suggests that H_2O_2 treatments can be useful to sorb organic acid herbicides but do not influence the removal of non-ionizable polar compounds.

7.4.4.4 Salts or Metal Oxides

Metals can be incorporated before or after the feedstock pyrolysis process. The most employed elements are iron, aluminum, manganese, magnesium, and cobalt. This type of modification has the advantage of improving biochar adsorption for a given pollutant, facilitating recycling and increasing biochar catalytic characteristics. For example, iron-modified biochar (FeMBC) was manufactured through the chemical co-precipitation of Fe^{3+} in corn stalk biochar to remove atrazine from soil, with herbicide degradation rates increased by 11.46% compared to unmodified biochar (Tao et al. 2019). Biochar modified by zero-valent iron (BC-nZVI) has also been used as a heterogeneous catalyst, activating peroxymonosulfate (PMS) for atrazine soil removal, with a 10% removal rate in biochar alone in 240 min, while the modified biochar achieved almost 96% removal under optimal reaction conditions (Diao et al. 2021).

7.5 CONCLUDING REMARKS

This chapter discussed aspects involving biochar soil addition and its effects on herbicide biodegradation. The complexity of the herbicide degradation process is directly related to herbicide biodegradation in the presence of biochar. In addition, the greatest biochar effects are not only due to its physicochemical characteristics, which generally contribute to increased absorption of several compounds, including herbicides, but also due to its contributions to increased soil organic carbon and nutrients. In some cases, beneficial soil microbiota effects and herbicide degradation intensification are also observed. However, this does not always take place, resulting in increased persistence of some compounds and decreased effectiveness of pre-emergent herbicides, considering the high biochar retention of these molecules. In this sense, the use of biochar in soil remediation practices can comprise an alternative for herbicide removal from the environment and reduces the environmental risks associated with these contaminants.

The information presented herein contributes to understanding the effect of biochar addition in agricultural systems, which can affect local organism diversity and contaminant molecules. The particularities of each system can determine biochar potential in assisting in the degradation or removal of herbicide molecules. Understanding these aspects can contribute to the adequate use of this OM, as well as correct herbicide management and decreased environmental risks.

REFERENCES

Abujabhah, I. S., Bound, S. A., Doyle, R., and Bowman, J. P. 2016. Effects of biochar and compost amendments on soil physico-chemical properties and the total community within a temperate agricultural soil. *Applied Soil Ecology*, 98:243–253.

Ali, N., Khan, S., Yao, H., and Wang, J. 2019. Biochars reduced the bioaccessibility and (bio) uptake of organochlorine pesticides and changed the microbial community dynamics in agricultural soils. *Chemosphere*, 224, 805–815.

Ameloot, N., Graber, E. R., Verheijen, F. G. A., and Neve, S. De. 2013. Interactions between biochar stability and soil organisms: Review and research needs. *European Journal of Soil Science*, 64(4):379–390.

Anders, E., Watzinger, A., Rempt, F., Kitzler, B., Wimmer, B., Zehetner, F., Stahr, K., Zechmeister-Boltenstern, S., and Soja, G. 2013. Biochar affects the structure rather than the total biomass of microbial communities in temperate soils. *Agricultural and Food Science*, 22(4):404–423.

Anderson, C. R., Condron, L. M., Clough, T. J., Fiers, M., Stewart, A., Hill, R. A., and Sherlock, R. R. 2011. Biochar induced soil microbial community change: Implications for biogeochemical cycling of carbon, nitrogen and phosphorus. *Pedobiologia*, 54(5–6):309–320.

Bailey, V. L., Fansler, S. J., Smith, J. L., and Bolton, H. 2011. Reconciling apparent variability in effects of biochar amendment on soil enzyme activities by assay optimization. *Soil Biology and Biochemistry*, 43(2):296–301.

Banik, C., Lawrinenko, M., Bakshi, S., and Laird, D. A. 2018. Impact of pyrolysis temperature and feedstock on surface charge and functional group chemistry of biochars. *Journal of Environmental Quality*, 47(3):452–461.

Beesley, L., Moreno-Jiménez, E., Gomez-Eyles, J. L., Harris, E., Robinson, B., and Sizmur, T. 2011. A review of biochars' potential role in the remediation, revegetation and restoration of contaminated soils. *Environmental Pollution*, 159(12):3269–3282.

Bending, G. D., and Rodriguez-Cruz, M. S. 2007. Microbial aspects of the interaction between soil depth and biodegradation of the herbicide isoproturon. *Chemosphere*, 66(4):664–671.

Bini, D., Lopez, M. V., and Cardoso, E. J. B. N. 2016. Metabolismo Microbiano. In: *Microbiologia do Solo*, ed. Cardoso, E. J. B. N., and Andreote, F. D. 2ª Edição. Piracicaba, Brazil: ESALQ.

Cabrera, A., Cox, L., Spokas, K., Hermosín, M. C., Cornejo, J., and Koskinen, W. C. 2014. Influence of biochar amendments on the sorption – desorption of aminocyclopyrachlor, bentazone and pyraclostrobin pesticides to an agricultural soil. *Science of the Total Environment*, 470–471:438–443.

Callieri, C., Eckert, E. M., Di Cesare, A., and Bertoni, F. 2019. Microbial Communities. *Encyclopedia of Ecology*, 1:126–134.

Cha, J. S., Park, S. H., Jung, S. C., Ryu, C., Jeon, J. K., Shin, M. C., and Park, Y. K. 2016. Production and utilization of biochar: A review. *Journal of Industrial and Engineering Chemistry*, 40:1–15.

Chen, J., Liu, X., Li, L., Zheng, J., Qu, J., Zheng, J., Zhang, X., and Pan, G. 2015. Consistent increase in abundance and diversity but variable change in community composition of bacteria in topsoil of rice paddy under short term biochar treatment across three sites from South China. *Applied Soil Ecology*, 91:68–79.

Chen, Y., Lan, T., Li, J., Yang, G., Zhang, K., and Hu, D. 2021. Effects of biochar produced from cornstalk, rice husk and bamboo on degradation of flumioxazin in soil. *Soil and Sediment Contamination: An International Journal*, 1–15.

Cheng, H., Jones, D. L., Hill, P., and Bastami, M. S. 2017b. Biochar concomitantly increases simazine sorption in sandy loam soil and lowers its dissipation. *Archives of Agronomy and Soil Science*, 63(8):1082–1092.

Cheng, J., Lee, X., Gao, W., Chen, Y., Pan, W., and Tang, Y. 2017a. Effect of biochar on the bioavailability of difenoconazole and microbial community composition in a pesticide-contaminated soil. *Applied Soil Ecology*, 121:185–192.

Chin-Pampillo, J. S., Perez-Villanueva, M., Masis-Mora, M., Mora-Dittel, T., Carazo-Rojas, E., Alcañiz, J. M., Chinchilla-Soto, C., and Domene, X. 2021. Amendments with pyrolyzed agrowastes change bromacil and diuron's sorption and persistence in a tropical soil without modifying their environmental risk. *Science of the Total Environment*, 772:145515.

Chintala, R., Mollinedo, J., Schumacher, T. E., Malo, D. D., and Julson, J. L. 2014. Effect of biochar on chemical properties of acidic soil. *Archives of Agronomy and Soil Science*, 60:393–404.

Cotta, S. R. 2016. O solo como ambiente para a vida microbiana. In: *Microbiologia do solo*, ed. Cardoso, E. J. B. N., and Andreote, F. D. 2ª Edição. Piracicaba, Brazil: ESALQ.

Cox, L., Cecchi, A., Celis, R., Hermos´ın, M., Koskinen, W., and Cornejo, J. 2001. Effect of exogenous carbon on movement of simazine and 2,4-D in soils. *Soil Science Society of America Journal*, 65:1688–1695.

Cui, H., Zhou, J., Zhao, Q., Si, Y., Mao, J., Fang, G., and Liang, J. 2013. Fractions of Cu, Cd, and enzyme activities in a contaminated soil as affected by applications of micro and nanohydroxyapatite. *Journal of Soils and Sediments*, 13(4):742–752.

Dai, Z., Wang, Y., Muhammad, N., Yu, X., Xiao, K., Meng, J., Liu, X., Xu, J., and Brookes, P. C. 2014. The effects and mechanisms of soil acidity changes, following incorporation of biochars in three soils differing in initial pH. *Soil Science Society of America Journal*, 78:1606–1614.

Dan, Y., Ji, M., Tao, S., Luo, G., Shen, Z., Zhang, Y., and Sang, W. 2021. Impact of rice straw biochar addition on the sorption and leaching of phenylurea herbicides in saturated sand column. *Science of the Total Environment*, 769:144536.

Dechene, A., Rosendahl, I., Laabs, V., and Amelung, W. 2014. Sorption of polar herbicides and herbicide metabolites by biochar-amended soil. *Chemosphere*, 109:180–186.

Demisie, W., Liu, Z., and Zhang, M. 2014. Effect of biochar on carbon fractions and enzyme activity of red soil. *Catena*, 121:214–221.

Diao, Z. H., Zhang, W. X., Liang, J. Y., Huang, S. T., Dong, F. X., Yan, L., Quian, W., and Chu, W. 2021. Removal of herbicide atrazine by a novel biochar based iron composite coupling with peroxymonosulfate process from soil: Synergistic effect and mechanism. *Chemical Engineering Journal*, 409:127684.

Eibisch, N., Schroll, R., and Fuß, R. 2015. Effect of pyrochar and hydrochar amendments on the mineralization of the herbicide isoproturon in an agricultural soil. *Chemosphere*, 134:528–535.

Ekenler, M., and Tabatabai, M. A. 2004. β-Glucosaminidase activity as an index of nitrogen mineralization in soils. *Communications in Soil Science and Plant Analysis*, 35(7–8):1081–1094.

Elzobair, K. A., Stromberger, M. E., Ippolito, J. A., and Lentz, R. D. 2016. Contrasting effects of biochar versus manure on soil microbial communities and enzyme activities in an Aridisol. *Chemosphere*, 142:145–152.

Fernandes, B. C. C., Mendes, K. F., Tornisielo, V. L., Teófilo, T. M. S., Takeshita, V., das Chagas, P. S. F., Lins, H. A., Souza, M. F., and Silva, D. V. 2021. Effect of pyrolysis temperature on eucalyptus wood residues biochar on availability and transport of hexazinone in soil. *International Journal of Environmental Science and Technology*, 1–16.

Ferreira, E. P. B., Stone, L. F., and Martin-Didonet, C. G. 2017. Population and microbial activity of the soil under an agro-ecological production system. *Revista Ciência Agronômica*, 1:22–31.

Francisco, J. G., Mendes, K. F., Pimpinato, R. F., Tornisielo, V. L., and Guimarães, A. C. D. 2018. Soil factors effects on the mineralization, extractable residue, and bound residue formation of aminocyclopyrachlor in three tropical soils. *Agronomy*, 8(1):1–10.

Gámiz, B., Hall, K., Spokas, K. A., and Cox, L. 2019. Understanding activation effects on low-temperature biochar for optimization of herbicide sorption. *Agronomy*, 9(10):588.

Gámiz, B., Velarde, P., Spokas, K. A., Hermosín, M. C., and Cox, L. 2017. Biochar soil additions affect herbicide fate: Importance of application timing and feedstock species. *Journal of Agricultural and Food Chemistry*, 65:3109–3117.

Gao, S., and DeLuca, T. H. 2018. Wood biochar impacts soil phosphorus dynamics and microbial communities in organically-managed croplands. *Soil Biology and Biochemistry*, 126:144–150.

García-Jaramillo, M., Cox, L., Cornejo, J., and Hermosín, M. C. 2014. Effect of soil organic amendments on the behavior of bentazone and tricyclazole. *Science of the Total Environment*, 466–467:906–913.

García-Jaramillo, M., Trippe, K. M., Helmus, R., Knicker, H. E., Cox, L., Hermosín, M. C., Parsons, J. R., and Kalbitz, K. 2020. An examination of the role of biochar and biochar water-extractable substances on the sorption of ionizable herbicides in rice paddy soils. *Science of the Total Environment*, 706:135682.

Gaur, A., and Adholeya, A. 2000. Effects of the particle size of soil-less substrates upon AM fungus inoculum production. *Mycorrhiza*, 10:43–48.

Ghodake, G. S., Shinde, S. K., Kadam, A. A., Saratale, R. G., Saratale, G. D., Kumar, M., Palem, R. R., Al-Schwaiman, H. A., Elgorban, A. M., Syed, A., and Kim, D. Y. 2021. Review on biomass feedstocks, pyrolysis mechanism and physicochemical properties of biochar: State-of-the-art framework to speed up vision of circular bioeconomy. *Journal of Cleaner Production*, 126645.

Gorovtsov, A. V., Minkina, T. M., Mandzhieva, S. S., Perelomov, L. V., Soja, G., Zamulina, I. V., Rajput, V. D., Sushkova, S. N., Mohan, D., and Yao, J. 2020. The mechanisms of biochar interactions with microorganisms in soil. *Environmental Geochemistry and Health*, 42(8):2495–2518.

Graber, E. R., Tsechansky, L., Gerstl, Z., and Lew, B. 2012. High surface area biochar negatively impacts herbicide efficacy. *Plant and Soil*, 353(1):95–106.

Gul, S., Whalen, J. K., Thomas, B. W., Sachdeva, V., and Deng, H. 2015. Physico-chemical properties and microbial responses in biochar-amended soils: Mechanisms and future directions. *Agriculture, Ecosystems & Environment*, 206:46–59.

Guo, M., Song, W., and Tian, J. 2020. Biochar-facilitated soil remediation: Mechanisms and efficacy variations. *Frontiers in Environmental Science*, 8:183.

Hammer, E. C., Balogh-Brunstad, Z., Jakobsen, I., Olsson, P. A., Stipp, S. L. S., and Rillig, M. C. 2014. A mycorrhizal fungus grows on biochar and captures phosphorus from its surfaces. *Soil Biology and Biochemistry*, 77:252–260.

Haskis, P., Mantzos, N., Hela, D., Patakioutas, G., and Konstantinou, I. 2019. Effect of biochar on the mobility and photodegradation of metribuzin and metabolites in soil–biochar thin-layer chromatography plates. *International Journal of Environmental Analytical Chemistry*, 99:310–327.

Hassan, M., Liu, Y., Naidu, R., Parikh, S. J., Du, J., Qi, F., and Willett, I. R. 2020. Influences of feedstock sources and pyrolysis temperature on the properties of biochar and functionality as adsorbents: A meta-analysis. *Science of the Total Environment*, 140714.

He, L., Fan, S., Müller, K., Hu, G., Huang, H., Zhang, X., Lin, X., Che, L., and Wang, H. 2016. Biochar reduces the bioavailability of di-(2-ethylhexyl) phthalate in soil. *Chemosphere*, 142:24–27.

Hussain, S., Arshad, M., Springael, D., SøRensen, S. R., Bending, G. D., Devers-Lamrani, M., Maqbool, Z., and Martin-Laurent, F. 2015. Abiotic and biotic processes governing the fate of phenylurea herbicides in soils: A review. *Critical Reviews in Environmental Science and Technology*, 45(18):1947–1998.

Ippolito, J. A., Cui, L., Kammann, C., Wrage-Mönnig, N., Estavillo, J. M., Fuertes-Mendizabal, T., Cayuela, M. L., Sigua, G., Novak, J., Spokas, K., and Borchard, N. 2020. Feedstock choice, pyrolysis temperature and type influence biochar characteristics: A comprehensive meta-data analysis review. *Biochar*, 2(4):421–438.

Jacoby, R., Peukert, M., Succurro, A., Koprivova, A., and Kopriva, S. 2017. The role of soil microorganisms in plant mineral nutrition – current knowledge and future directions. *Frontiers in Plant Science*, 8:1617.

Jiang, J., Yuan, M., Xu, R., and Bish, D. L. 2015. Mobilization of phosphate in variable-charge soils amended with biochars derived from crop straws. *Soil and Tillage Research*, 146:139–147.

Jing, X., Wang, T., Yang, J., Wang, Y., and Xu, H. 2018. Effects of biochar on the fate and toxicity of herbicide fenoxaprop-ethyl in soil. *Royal Society Open Science*, 5(5):171875.

Johnson, N. C. 1993. Can fertilization of soil select less mutualistic mycorrhizae? *Ecological Applications*, 3(4):749–757.

Jones, D. L., Edwards-Jones, G., and Murphy, D. V. 2011. Biochar mediated alterations in herbicide breakdown and leaching in soil. *Soil Biology and Biochemistry*, 43:804–813.

Joseph, S., Pow, D., Dawson, K., Rust, J., Munroe, P., Taherymoosavi, S., Mitchell, D. R. G., Robb, S., and Solaiman, Z. M. 2020. Biochar increases soil organic carbon, avocado yields and economic return over 4 years of cultivation. *Science of the Total Environment*, 724:138153.

Junqueira, L. V., Mendes, K. F., Sousa, R. N. D., Almeida, C. D. S., Alonso, F. G., and Tornisielo, V. L. 2020. Sorption-desorption isotherms and biodegradation of glyphosate in two tropical soils aged with eucalyptus biochar. *Archives of Agronomy and Soil Science*, 66:1651–1667.

Kang, S., Jian-chun, J., and Dan-dan, C. 2011. Preparation of activated carbon with highly developed mesoporous structure from Camellia oleifera shell through water vapor gasification and phosphoric acid modification. *Biomass and Bioenergy*, 35(8):3643–3647.

Kanissery, R. G., and Sims, G. K. 2011. Biostimulation for the enhanced degradation of herbicides in soil. *Applied and Environmental Soil Science*, 2011:843450.

Khalid, S., Shahid, M., Murtaza, B., Bibi, I., Naeem, M. A., and Niazi, N. K. 2020. A critical review of different factors governing the fate of pesticides in soil under biochar application. *Science of the Total Environment*, 711:134645.

Khodadad, C. L. M., Zimmerman, A. R., Green, S. J., Uthandi, S., and Foster, J. S. 2011. Taxa-specific changes in soil microbial community composition induced by pyrogenic carbon amendments. *Soil Biology and Biochemistry*, 43(2):385–392.

Khorram, M., Zhang, Q., Lin, D., Zheng, Y., Fang, H., and Yu, Y. 2016. Biochar: A review of its impact on pesticide behavior in soil environments and its potential applications. *Journal of Environmental Sciences*, 44:269–279.

Kolton, M., Harel, Y. M., Pasternak, Z., Graber, E. R., Elad, Y., and Cytryn, E. 2011. Impact of biochar application to soil on the root-associated bacterial community structure of fully developed greenhouse pepper plants. *Applied and Environmental Microbiology*, 77(14):4924–4930.

Kookana, R. S. 2010. The role of biochar in modifying the environmental fate, bioavailability, and efficacy of pesticides in soils: A review. *Soil Research*, 48(7):627–637.

Koufopanos, C. A., Lucchesi, A., and Maschio, G. 1989. Kinetic modelling of the pyrolysis of biomass and biomass components. *The Canadian Journal of Chemical Engineering*, 67(1):75–84.

Kumar, A., Kumar, J., and Bhaskar, T. 2020b. High surface area biochar from *Sargassum tenerrimum* as potential catalyst support for selective phenol hydrogenation. *Environmental Research*, 186:109533.

Kumar, A., Saini, K., and Bhaskar, T. 2020a. Hydochar and biochar: Production, physicochemical properties and techno-economic analysis. *Bioresource Technology*, 310:123442.

Kupryianchyk, D., Hale, S., Zimmerman, A. R., Harvey, O., Rutherford, D., Abiven, S., Knicker, H. Schmidt, H., Rumpel, C., and Cornelissen, G. 2016. Sorption of hydrophobic organic compounds to a diverse suite of carbonaceous materials with emphasis on biochar. *Chemosphere*, 144:879–887.

Lattao, C., Cao, X., Mao, J., Schmidt-Rohr, K., and Pignatello, J. J. 2014. Influence of molecular structure and adsorbent properties on sorption of organic compounds to a temperature series of wood chars. *Environmental Science & Technology*, 48(9):4790–4798.

Lawrinenko, M., Laird, D. A., Johnson, R. L., and Jing, D. 2016. Accelerated aging of biochars: Impact on anion exchange capacity. *Carbon*, 103:217–227.

Lee, Y. G., Shin, J., Kwak, J., Kim, S., Son, C., Cho, K. H., and Chon, K. 2021. Effects of NaOH activation on adsorptive removal of herbicides by biochars prepared from ground coffee residues. *Energies*, 14:1297.

Lehmann, J., and Joseph, S. 2009. Biochar for environmental management: An introduction. In: *Biochar for environmental management: Science and Technology*, ed. Lehmann, J. and Joseph, S., 33–46. 1st ed. London: Earthscan.

Leng, L., and Huang, H. 2018. An overview of the effect of pyrolysis process parameters on biochar stability. *Bioresource Technology*, 270:627–642.

Li, D. C., and Jiang, H. 2017. The thermochemical conversion of non-lignocellulosic biomass to form biochar: A review on characterizations and mechanism elucidation. *Bioresource Technology*, 246:57–68.

Li, J., Li, Y., Wu, M., Zhang, Z., and Lü, J. 2013. Effectiveness of low-temperature biochar in controlling the release and leaching of herbicides in soil. *Plant and Soil*, 370(1):333–344.

Li, M., Zhang, J., Yang, X., Zhou, Y., Zhang, L., Yang, Y., Luo, L., and Yan, Q. 2021. Responses of ammonia-oxidizing microorganisms to biochar and compost amendments of heavy metals-polluted soil. *Journal of Environmental Sciences*, 102:263–272.

Li, W., Shan, R., Fan, Y., and Sun, X. 2020. Effects of tall fescue biochar on the adsorption and desorption of atrazine in different types of soil. *Environmental Science and Pollution Research* 28(4):4503–4514.

Lian, F., Cui, G., Liu, Z., Duo, L., Zhang, G., and Xing, B. 2016. One-step synthesis of a novel N-doped microporous biochar derived from crop straws with high dye adsorption capacity. *Journal of Environmental Management*, 176:61–68.

Lian, F., and Xing, B. 2017. Black carbon (biochar) in water/soil environments: Molecular structure, sorption, stability, and potential risk. *Environmental Science & Technology*, 51(23):13517–13532.

Liu, Y., Lonappan, L., Brar, S. K., and Yang, S. 2018. Impact of biochar amendment in agricultural soils on the sorption, desorption, and degradation of pesticides: A review. *Science of the Total Environment*, 645:60–70.

Liu, Z., Wu, X., Liu, W., Bian, R., Ge, T., Zhang, W., Zheng, J., Drosos, M., Liu, X., Zhang, X., Cheng, K., Li, L., and Pan, G. 2020. Greater microbial carbon use efficiency and carbon sequestration in soils: Amendment of biochar versus crop straws. *GCB Bioenergy*, 12(12):1092–1103.

Lu, W., Ding, W., Zhang, J., Li, Y., Luo, J., Bolan, N., and Xie, Z. 2014. Biochar suppressed the decomposition of organic carbon in a cultivated sandy loam soil: A negative priming effect. *Soil Biology and Biochemistry*, 76:12–21.

Mahdi, Z., El Hanandeh, A., and Yu, Q. J. 2019. Preparation, characterization and application of surface modified biochar from date seed for improved lead, copper, and nickel removal from aqueous solutions. *Journal of Environmental Chemical Engineering*, 7(5):103379.

Manna, S., Singh, N., Purakayastha, T. J., and Berns, A. E. 2020. Effect of deashing on physico-chemical properties of wheat and rice straw biochars and potential sorption of pyrazosulfuron-ethyl. *Arabian Journal of Chemistry*, 13(1):1247–1258.

Mao, J. D., Johnson, R. L., Lehmann, J., Olk, D. C., Neves, E. G., Thompson, M. L., and Schmidt-Rohr, K. 2012. Abundant and stable char residues in soils: Implications for soil fertility and carbon sequestration. *Environmental Science & Technology*, 46(17):9571–9576.

Mendes, K. F., Almeida, C. de S., Inoue, M. H., Mertens, T. B., and Tornisielo, V. L. 2018. Impacto do biochar no comportamento de herbicidas em solos: Um enfoque no Brasil. *Revista Brasileira de Herbicidas*, 17(1):106–117.

Mendes, K. F., Furtado, I. F., Sousa, R. N., Lima, A. C., Mielke, K. C., and Brochado, M. G. S. 2021. Cow bonechar decreases indaziflam pre-emergence herbicidal activity in tropical soil. *Journal of Environmental Science and Health, Part* B, 56:1–8.

Mendes, K. F., Hall, K. E., Spokas, K. A., Koskinen, W. C., and Tornisielo, V. L. 2017. Evaluating agricultural management effects on alachlor availability: Tillage, green manure, and biochar. *Agronomy*, 7(4):64.

Mendes, K. F., Soares, M. B., Sousa, R. N. de, Mielke, K. C., Brochado, M. G. da S., and Tornisielo, V. L. 2021. Indaziflam sorption – desorption and its three metabolites from biochars- and their raw feedstock-amended agricultural soils using radiometric technique. *Journal of Environmental Science and Health, Part B*, 1–10.

Mendes, K. F., Sousa, R. N. de, Goulart, M. O., and Tornisielo, V. L. 2020. Role of raw feedstock and biochar amendments on sorption-desorption and leaching potential of three ^{3}H- and ^{14}C-labelled pesticides in soils. *Journal of Radioanalytical and Nuclear Chemistry*, 324(3):1373–1386.

Meng, L., Sun, T., Li, M., Saleem, M., Zhang, Q., and Wang, C. 2019. Soil-applied biochar increases microbial diversity and wheat plant performance under herbicide fomesafen stress. *Ecotoxicology and Environmental Safety*, 171:75–83.

Meyer, S., Glaser, B., and Quicker, P. 2011. Technical, economical, and climate-related aspects of biochar production technologies: A literature review. *Environmental Science & Technology*, 45(22):9473–9483.

Mohan, D., Sarswat, A., Ok, Y. S., and Pittman Jr, C. U. 2014. Organic and inorganic contaminants removal from water with biochar, a renewable, low cost and sustainable adsorbent – a critical review. *Bioresource Technology*, 160:191–202.

Mojiri, A., Zhou, J. L., Robinson, B., Ohashi, A., Ozaki, N., Kindaichi, T., Ferraji, H., and Vakili, M. 2020. Pesticides in aquatic environments and their removal by adsorption methods. *Chemosphere*, 253:126646.

Moorman, T., Cowan, J., Arthur, E., and Coats, J. 2001. Organic amendments to enhance herbicide biodegradation in contaminated soils, *Biology and Fertility of Soils*, 33:541–545.

Muter, O., Berzins, A., Strikauska, S., Pugajeva, I., Bartkevics, V., Dobele, G., Truu, J., Truu, M., and Steiner, C. 2014. The effects of woodchip- and straw-derived biochars on the persistence of the herbicide 4-chloro-2-methylphenoxyacetic acid (MCPA) in soils. *Ecotoxicology and Environmental Safety*, 109:93–100.

Nag, S. K., Kookana, R., Smith, L., Krull, E., Macdonald, L. M., and Gill, G. 2011. Poor efficacy of herbicides in biochar-amended soils as affected by their chemistry and mode of action. *Chemosphere*, 84(11):1572–1577.

Nannipieri, P., Giagnoni, L., Renella, G., Puglisi, E., Ceccanti, B., Masciandaro, G., Fornasier, F., Moscatelli, M. C., and Marinari, S. 2012. Soil enzymology: Classical and molecular approaches. *Biology and Fertility of Soils*, 48(7):743–762.

Obregón-Alvarez, D., Mendes, K. F., Tosi, M., Fonseca de Souza, L., Campos Cedano, J. C., de Souza Falcão, N. P., Dunfield, K., Tsai, S. M., and Tornisielo, V. L. (2021). Sorption-desorption and biodegradation of sulfometuron-methyl and its effects on the bacterial communities in Amazonian soils amended with aged biochar. *Ecotoxicology and Environmental Safety*, 207, 111222.

Oliveira Jr, R. S., Koskinen, W. C., and Ferreira, F. A. 2001. Sorption and leaching potential of herbicides on Brazilian soils. *Weed Research*, 41:97–110.

Ortiz-Hernández, M. L., Sánchez-Salinas, E., Dantán-González, E., and Castrejón-Godínez, M. L. (2013). Pesticide biodegradation: Mechanisms, genetics and strategies to enhance the process. Biodegradation-life of science. In: *Biodegradation – life of science*, ed. Chamy, R., and Rosenkranz, 251–287, London: IntechOpen.

Pareja, L., Pérez-Parada, A., Agüera, A., Cesio, V., Heinzen, H., and Fernández-Alba, A. R. 2012. Photolytic and photocatalytic degradation of quinclorac in ultrapure and paddy field water: Identification of transformation products and pathways. *Chemosphere*, 87(8):838–844.

Pokharel, P., Ma, Z., and Chang, S. X. 2020. Biochar increases soil microbial biomass with changes in extra- and intracellular enzyme activities: A global meta-analysis. *Biochar*, 2(1):65–79.

Qiu, Y., Cheng, H., Xu, C., and Sheng, G. D. 2008. Surface characteristics of crop-residue-derived black carbon and lead (II) adsorption. *Water Research*, 42(3):567–574.

Quilliam, R. S., Glanville, H. C., Wade, S. C., and Jones, D. L. 2013. Life in the 'charosphere' – Does biochar in agricultural soil provide a significant habitat for microorganisms? *Soil Biology and Biochemistry*, 65:287–293.

Sánchez, M., Estrada, I., Martinez, O., Martin-Villacorta, J., Aller, A., and Moran, A. 2004. Influence of the application of sewage sludge on the degradation of pesticides in the soil. *Chemosphere*, 57, 673–679.

Shaaban, M., Van Zwieten, L., Bashir, S., Younas, A., Núñez-Delgado, A., Chhajro, M. A., Kubar, K. A., Ali, U., Rana, M. S., and Hu, R. 2018. A concise review of biochar application to agricultural soils to improve soil conditions and fight pollution. *Journal of Environmental Management*, 228:429–440.

Si, Y., Wang, M., Tian, C., Zhou, J., and Zhou, D. 2011. Effect of charcoal amendment on adsorption, leaching and degradation of isoproturon in soils. *Journal of Contaminant Hydrology*, 123:75–81.

Sims, G. K., and Cupples, A. M. 1999. Factors controlling degradation of pesticides in soil. *Pesticide Science*, 55:598–601.

Sims, G. K., Radosevich, M., He, X. T., and Traina, S. J. 1991. The effects of sorption on the bioavailability of pesticides. In: *Biodegradation of natural and synthetic materials*, ed. Betts, W. B. London: Springer Verlag.

Sopeña, F., Semple, K., Sohi, S., and Bending, G. 2012. Assessing the chemical and biological accessibility of the herbicide isoproturon in soil amended with biochar. *Chemosphere*, 88:77–83.

Spokas, K. A., Koskinen, W. C., Baker, J. M., and Reicosky, D. C. 2009. Impacts of woodchip biochar additions on greenhouse gas production and sorption/degradation of two herbicides in a Minnesota soil. *Chemosphere*, 77(4):574–581.

Sun, J., He, F., Pan, Y., and Zhang, Z. 2017. Effects of pyrolysis temperature and residence time on physicochemical properties of different biochar types. *Acta Agriculturae Scandinavica, Section B—Soil & Plant Science*, 67(1):12–22.

Sun, D., Meng, J., Liang, H., Yang, E., Huang, Y., Chen, W., Jiang, L., Lan, Y., and Gao, J. 2015. Effect of volatile organic compounds absorbed to fresh biochar on survival of *Bacillus mucilaginosus* and structure of soil microbial communities. *Journal of Soils and Sediments*, 15(2):271–281.

Sun, H., Xu, J., Yang, S., Liu, G., and Dai, S. 2004. Plant uptake of aldicarb from contaminated soil and its enhanced degradation in the rhizosphere. *Chemosphere*, 54(4):569–574.

Sun, Z., Bruun, E. W., Arthur, E., de Jonge, L. W., Moldrup, P., Hauggaard-Nielsen, H., and Elsgaard, L. 2014. Effect of biochar on aerobic processes, enzyme activity, and crop yields in two sandy loam soils. *Biology and Fertility of Soils*, 50(7):1087–1097.

Suo, F., Xie, G., Zhang, J., Li, J., Li, C., Liu, X., Zhang, Y., Ma, Y., and Ji, M. 2018. A carbonised sieve-like corn straw cellulose – graphene oxide composite for organophosphorus pesticide removal. *RSC Advances*, 8(14):7735–7743.

Takaya, C. A., Fletcher, L. A., Singh, S., Okwuosa, U. C. and Ross, A. B., 2016. Recovery of phosphate with chemically modified biochars. *Journal of Environmental Chemical Engineering*, 4(1):1156–1165.

Takeshita, V., Mendes, K. F., Junqueira, L. V., Pimpinato, R. F., and Tornisielo, V. L. 2020. Quantification of the fate of aminocyclopyrachlor in soil amended with organic residues from a sugarcane system. *Sugar Tech*, 22(3):428–436.

Tan, Z., Lin, C. S., Ji, X., and Rainey, T. J. 2017. Returning biochar to fields: A review. *Applied Soil Ecology*, 116:1–11.

Tan, X., Liu, Y., Zeng, G., Wang, X., Hu, X., Gu, Y., and Yang, Z. 2015. Application of biochar for the removal of pollutants from aqueous solutions. *Chemosphere*, 125:70–85.

Tao, Y., Han, S., Zhang, Q., Yang, Y., Shi, H., Akindolie, M. S., Jiao, Y., Qu, J., Jiang, Z., Han, W., and Zhang, Y. 2020. Application of biochar with functional microorganisms for enhanced atrazine removal and phosphorus utilization. *Journal of Cleaner Production*, 257:120535.

Tao, Y., Hu, S., Han, S., Shi, H., Yang, Y., Li, H., Jiao, Y., Zhang, Q., Akindolie, M. S., Ji, M., Chen, Z., and Zhang, Y. 2019. Efficient removal of atrazine by iron-modified biochar loaded *Acinetobacter lwoffii* DNS32. *Science of the Total Environment*, 682:59–69.

Tatarková, V., Hiller, E., and Vaculík, M. 2013. Impact of wheat straw biochar addition to soil on the sorption, leaching, dissipation of the herbicide (4-chloro-2-methylphenoxy)acetic acid and the growth of sunflower (*Helianthus annuus* L.). *Ecotoxicology and Environmental Safety*, 92:215–221.

Terzaghi, E., Zanardini, E., Morosini, C., Raspa, G., Borin, S., Mapelli, F., Vergani, L., and Di Guardo, A. 2018. Rhizoremediation half-lives of PCBs: Role of congener composition, organic carbon forms, bioavailability, microbial activity, plant species and soil conditions, on the prediction of fate and persistence in soil. *Science of the Total Environment*, 612:544–560.

Tong, Y., Mayer, B. K., and McNamara, P. J. 2019. Adsorption of organic micropollutants to biosolids-derived biochar: Estimation of thermodynamic parameters. *Environmental Science: Water Research & Technology*, 5(6):1132–1144.

Verheijen, F., Jeffery, S., Bastos, A., Van Der Velde M., and Diafas, I. 2010. *Biochar application to soils – A critical scientific review of effects on soil properties, processes and functions. EUR 24099 EN*. Luxembourg: European Commission, JRC55799.

Viti, M. L., Mendes, K. F., dos Reis, F. C., Guimarães, A. C. D., Soria, M. T. M., and Tornisielo, V. L. 2021. Characterization and metabolism of bound residues of three herbicides in soils amended with sugarcane waste. *Sugar Tech*, 23(1):23–37.

Voorhees, J. P., Phillips, B. M., Anderson, B. S., Tjeerdema, R. S., and Page, B. 2020. Comparison of the relative efficacies of granulated activated carbon and biochar to reduce chlorpyrifos and imidacloprid loading and toxicity using laboratory bench scale experiments. *Bulletin of Environmental Contamination and Toxicology*, 104(3):327–332.

Wahla, A. Q., Anwar, S., Mueller, J. A., Arslan, M., and Iqbal, S. 2020. Immobilization of metribuzin degrading bacterial consortium MB_3R on biochar enhances bioremediation of potato vegetated soil and restores bacterial community structure. *Journal of Hazardous Materials*, 390:121493.

Wallstedt, A., Coughlan, A., Munson, A. D., Nilsson, M.-C., and Margolis, H. A. 2002. Mechanisms of interaction between Kalmia angustifolia cover and *Picea mariana* seedlings. *Canadian Journal of Forest Research*, 32(11):2022–2031.

Wang, J., Li, X., Li, X., Wang, H., Su, Z., Wang, X., and Zhang, H. 2018. Dynamic changes in microbial communities during the bioremediation of herbicide (chlorimuron-ethyl and atrazine) contaminated soils by combined degrading bacteria. *PLoS One*, 13(4):e0194753.

Wang, J., and Wang, S. 2019. Preparation, modification and environmental application of biochar: A review. *Journal of Cleaner Production*, 227:1002–1022.

Wang, P., Liu, X., Yu, B., Wu, X., Xu, J., Dong, F., and Zheng, Y. 2020. Characterization of peanut-shell biochar and the mechanisms underlying its sorption for atrazine and nicosulfuron in aqueous solution. *Science of the Total Environment*, 702:134767.

Wang, S., Dai, G., Yang, H., and Luo, Z. 2017. Lignocellulosic biomass pyrolysis mechanism: A state-of-the-art review. *Progress in Energy and Combustion Science*, 62:33–86.

Wang, X., Song, D., Liang, G., Zhang, Q., Ai, C., and Zhou, W. 2015. Maize biochar addition rate influences soil enzyme activity and microbial community composition in a fluvo-aquic soil. *Applied Soil Ecology*, 96:265–272.

Wei, L., Huang, Y., Huang, L., Li, Y., Huang, Q., Xu, G., Müller, K., Wang, H., Ok, Y. S., and Liu, Z. 2020. The ratio of H/C is a useful parameter to predict adsorption of the herbicide metolachlor to biochars. *Environmental Research*, 184:109324.

Wu, C., Liu, X., Wu, X., Dong, F., Xu, J., and Zheng, Y. 2019. Sorption, degradation and bioavailability of oxyfluorfen in biochar-amended soils. *Science of the Total Environment*, 658:87–94.

Wu, F., Jia, Z., Wang, S., Chang, S. X., and Startsev, A. 2013. Contrasting effects of wheat straw and its biochar on greenhouse gas emissions and enzyme activities in a Chernozemic soil. *Biology and Fertility of Soils*, 49(5):555–565.

Xiao, F., and Pignatello, J. J. 2015a. Interactions of triazine herbicides with biochar: Steric and electronic effects. *Water Research*, 80:179–188.

Xiao, F., and Pignatello, J. J. 2015b. π+ – π Interactions between (Hetero) aromatic Amine cations and the graphitic surfaces of pyrogenic carbonaceous materials. *Environmental Science & Technology*, 49(2):906–914.

Xie, Z., Xu, Y., Liu, G., Liu, Q., Zhu, J., Tu, C., Amonette, J. E., Cadisch, G., Yong, J. W. H., and Hu, S. 2013. Impact of biochar application on nitrogen nutrition of rice, greenhouse-gas emissions and soil organic carbon dynamics in two paddy soils of China. *Plant and Soil*, 370(1):527–540.

Xu, L., Fang, H., Deng, X., Ying, J., Lv, W., Shi, Y., Zhou, G., and Zhou, Y. 2020. Biochar application increased ecosystem carbon sequestration capacity in a Moso bamboo forest. *Forest Ecology and Management*, 475:118447.

Xu, R. K., Zhao, A. Z., Yuan, J. H., and Jiang, J. 2012. pH buffering capacity of acid soils from tropical and subtropical regions of China as influenced by incorporation of crop straw biochars. *Journal of Soils and Sediments*, 12(4):494–502.

Yang, R., Liu, G., Xu, X., Li, M., Zhang, J., and Hao, X. 2011. Surface texture, chemistry and adsorption properties of acid blue 9 of hemp (*Cannabis sativa* L.) bast-based activated carbon fibers prepared by phosphoric acid activation. *Biomass and Bioenergy*, 35(1):437–445.

Yang, X., Liu, J., McGrouther, K., Huang, H., Lu, K., Guo, X., He, L., Lin, X., Che, L., Ye, Z., and Wang, H. 2015. Effect of biochar on the extractability of heavy metals (Cd, Cu, Pb, and Zn) and enzyme activity in soil. *Environmental Science and Pollution Research*, 23(2):974–984.

Yavari, S., Sapari, N. B., Malakahmad, A., and Yavari, S. 2019. Degradation of imazapic and imazapyr herbicides in the presence of optimized oil palm empty fruit bunch and rice husk biochars in soil. *Journal of Hazardous Materials*, 366:636–642.

Yu, X. Y., Mu, C. L., Gu, C., Liu, C., and Liu, X. J. 2011. Impact of woodchip biochar amendment on the sorption and dissipation of pesticide acetamiprid in agricultural soils. *Chemosphere*, 85(8):1284–1289.

Yuan, J. H., and Xu, R. K. 2011. The amelioration effects of low temperature biochar generated from nine crop residues on an acidic Ultisol. *Soil Use Manage*, 27:110–115.

Zhang, C., Liu, L., Zhao, M., Rong, H., and Xu, Y. 2018b. The environmental characteristics and applications of biochar. *Environmental Science and Pollution Research*, 25(22):21525–21534.

Zhang, M., Riaz, M., Liu, B., Xia, H., El-desouki, Z., and Jiang, C. 2020. Two-year study of biochar: Achieving excellent capability of potassium supply via alter clay mineral composition and potassium-dissolving bacteria activity. *Science of the Total Environment*, 717:137286.

Zhang, P., Sheng, G., Feng, Y., and Miller, D. M. 2005. Role of wheat-residue-derived char in the biodegradation of benzonitrile in soil: Nutritional stimulation versus adsorptive inhibition. *Environmental Science & Technology*, 39(14):5442–5448.

Zhang, P., Sun, H., Min, L., and Ren, C. 2018a. Biochars change the sorption and degradation of thiacloprid in soil: Insights into chemical and biological mechanisms. *Environmental Pollution*, 236:158–167.

Zhang, P., Sun, H., Yu, L., and Sun, T. 2013. Adsorption and catalytic hydrolysis of carbaryl and atrazine on pig manure-derived biochars: Impact of structural properties of biochars. *Journal of Hazardous Materials*, 244:217–224.

Zhao, L., Cao, X., Mašek, O., and Zimmerman, A. 2013. Heterogeneity of biochar properties as a function of feedstock sources and production temperatures. *Journal of Hazardous Materials*, 256:1–9.

Zhelezova, A., Cederlund, H., and Stenström, J. 2017. Effect of biochar amendment and ageing on adsorption and degradation of two herbicides. *Water, Air, & Soil Pollution*, 228:111–222.

Zheng, H., Zhang, Q., Liu, G., Luo, X., Li, F., Zhang, Y., and Wang, Z. 2019. Characteristics and mechanisms of chlorpyrifos and chlorpyrifos-methyl adsorption onto biochars: Influence of deashing and low molecular weight organic acid (LMWOA) aging and co-existence. *Science of the Total Environment*, 657:953–962.

Zhou, G., Xu, X., Qiu, X., and Zhang, J. 2019. Biochar influences the succession of microbial communities and the metabolic functions during rice straw composting with pig manure. *Bioresource Technology*, 272:10–18.

8 Impacts of Biochar Addition on Herbicides' Efficacy for Weed Control in Agriculture

*Gabriel da Silva Amaral[1], Manuel Alejandro Ix-Balam[2,3], Kassio Ferreira Mendes[4], Maria Fátima das Graças Fernandes da Silva[1], and Ricardo Alcántara-de la Cruz[5]**

[1]Departamento de Química, Universidade Federal de São Carlos, 13565–905 São Carlos, SP, Brazil.
[2]Departamento de Entomologia, Universidade Federal de Viçosa, 36570–900, Viçosa, MG, Brazil.
[3]Instituto de Investigación para el Desarrollo Sustentable de Ceja de Selva (INDES-CES), Universidad Nacional Toribio Rodríguez de Mendoza de Amazonas, 01001, Amazonas, Peru.
[4]Departamento de Agronomia, Universidade Federal de Viçosa, 36570–900, Viçosa, MG, Brazil.
[5]Centro de Ciências da Natureza, Universidade Federal de São Carlos – Campus Lagoa do Sino, 18290–000 Buri, SP, Brazil.
* Corresponding author: ricardo.cruz@ufscar.br

CONTENTS

DOI: 10.1201/9781003202073-8

8.1 INTRODUCTION

The mechanization of agricultural tasks and the use of pesticides for pest control began at the end of the nineteenth century (Pingali 2007; Tudi et al. 2021); however, these practices only became popular worldwide on a large scale in the following decades of the Second World War (Tauger 2010). The major agricultural transition occurred with the Green Revolution (1960–1980) (Tauger 2010), which was characterized by the implementation of technological packages (high-yielding variety seeds, chemical fertilizers and pesticides, large-scale irrigation, and mechanization) to increase the agricultural productivity to supply the growing demand for food due to the postwar population explosion and the fight against world hunger (Patel 2013). Those technological packages promoted a great change in the use of land and agricultural inputs in the world, shaping the modern agri-food system of today (Meiyappan et al. 2014; Tudi et al. 2021).

After the Second World War, pesticide consumption rose from 0.2 million tons in the 1950s (Tudi et al. 2021) to 3.08 million tons in 2000 (FAOSTAT 2021), although estimates by some researchers exceed 5 million tons for the same year (Carvalho 2017). According to the FAOSTAT (2021), the consumption of pesticides was 4.2 million tons in 2019, of which 2.2 million were only herbicides. In general terms, it is estimated that only 1% of all pesticides applied in agricultural fields act in controlling pests (Bernardes et al. 2015). In the case of herbicides, the amounts that reach the target weeds vary from 1% to 10% (medium from 2% to 5%), depending on whether they are PRE or POS emergence (Monquero and Silva 2021), but even so the greater part of herbicides are deposited in environments and nontarget organisms (Tudi et al. 2021), with the soil being the main receptor and accumulator of these compounds (Monquero and Silva 2021).

The soil is an open, extensive, complex, and dynamic system made up of three phases (solid – minerals and organic matter (OM); liquid – water; and gaseous – atmospheric air) in which various vital processes occur such as the storage and cycling of nutrients, retention and movement of water, maintaining habitat of the microfauna and its activity, remediation and immobilization of pollutants, in addition to being the physical support of plants (Silva et al. 2007). Once herbicide molecules come into contact with the soil, they are subject to processes of retention, transport, and transformation, which result in their rapid dissipation or long persistence in the soil (Curran 2016). The behavior of the herbicides in the soil due to the interactions between environmental conditions with the physicochemical characteristics of the soil and herbicides, as well as their influence on weed control, was described in detail in Chapter 3 of this book. While that chapter mentions the main environmental impacts of herbicides, it only presented agricultural actions related to the use of herbicides themselves to mitigate them. Over the past two decades, health and environmental authorities around the world have encouraged joint efforts to explore additional alternatives to solve these impacts (Panahi et al. 2020; Joseph et al. 2021). In this regard, the use of biochars has gained relevance to mitigate the environmental issues caused by the use of herbicides (Gámiz et al. 2017; Kochanek et al. 2022).

Biochars are porous solids rich in carbon (C) produced from biomass by pyrolysis in the absence of oxygen (Weber and Quicker 2018). Biochars have generally been used as amendments to improve soil quality, increase crop yields, reduce irrigation and fertilizer requirements, and even mitigate greenhouse gas emissions (Kookana et al. 2011; Xu et al. 2012; Brtnicky et al. 2021; Joseph et al. 2021). On the other hand, these compounds have shown the potential to immobilize herbicide molecules in contaminated soils (Clay and Malo 2012; Martin et al. 2012; Gámiz et al. 2017; Mendes et al. 2018). The porosity, specific surface area (SSA), pH, abundance of functional groups, and aromatic structure of the biochars influence the sorption and desorption of herbicides in the soil (Gámiz et al. 2017; Mendes et al. 2018). These characteristics depend on the feedstock of the biochar but mainly the pyrolysis temperature (Subedi et al. 2016; Ippolito et al. 2020). Both high and low pyrolysis temperatures have advantages and disadvantages over the adsorption of herbicides in the soil: the lower one due to the formation of a larger SSA and the other due to a large number of functional groups (Graber et al. 2012; Chen et al. 2015). In both cases, the bioavailability of herbicides in the soil is affected, which can affect weed management (Gámiz et al. 2017; Mendes et al. 2018); therefore, a clear understanding of these effects and mechanisms is necessary to produce biochars with desirable properties (Li et al. 2019), i.e., a satisfactory level of weed control is maintained, but the environmental impact of herbicides is reduced.

This chapter will describe the main properties of biochars that participate directly or indirectly in the sorption and biodegradation of herbicides in order to understand the potential impacts of biochars on the dynamics and fates of herbicides in the environment, but mainly in the control weed.

8.2 AGRICULTURAL AND ENVIRONMENTAL BENEFITS OF BIOCHARS

The most common use of biochars is as a soil amendment to increase fertility and improve water retention (Kookana et al. 2011; Xu et al. 2012). The improvement in water retention is due to the modification of the soil texture by biochars and fertility due to the C/P and C/N relationships, depending on the feedstock of the biochar (Kookana et al. 2011). Therefore, soils modified with biochars generally have a higher nutrient and water content than unamended soils (Yang et al. 2018; Liao and Thomas 2019; Razzaghi et al. 2020). In addition, biochars can act as conditioners by modifying the physical–chemical and biological properties of the soil, mainly the soil pH, the cation exchange capacity (CEC), and buffering of the soil (Weber and Quicker 2018; Panahi et al. 2020). The improvement in water retention and nutrient content results in higher crop yields. For example, corn yield increased from 70% to 150% in biochar-amended soils (Uzoma et al. 2011). Improved plant growth and increased yield are possibly related to improved soil properties by biochars (Kookana et al. 2011; Xu et al. 2012). On the other hand, the improvement of the soil's water retention capacity can reduce drought stress in arid, semi-arid regions, or in general without irrigation (Razzaghi et al. 2020). For example, wood biochars (3–6 ton ha^{-1}) improved water retention increasing wheat yield in sandy clay loam soils (Blackwell et al. 2010; Solaiman et al. 2010), i.e., biochars have the same potential to improve crop productivity in highly degraded/eroded soils.

In addition to improving water retention and nutrient content, biochars also are capable of increasing the resistance of crops to diseases (Alaylar et al. 2021). The incidence of fungal foliar diseases caused by *Botrytis cinerea* and *Oidiopsis sicula* in tomatoes and peppers was lower in biochar-amended soils (Elad et al. 2010). In strawberry cultivation, yields of up to 35% were obtained in biochar-amended soils due to the lower incidence of *B. cinerea*, *Colletotrichum acutatum*, and *Podosphaera aphanis* (Harel et al. 2012). The principle is simple: better nourished individuals are healthier and, therefore, more tolerant to diseases than those who suffer some stress or are not well nourished (Dordas 2008). In the case of biochars, molecular studies suggest these amendments stimulate various general defense pathways (Liu et al. 2018).

On the other hand, biochars also have the potential to reduce soil nutrient losses through leaching, sequester C from the atmosphere, reduce greenhouse gas emissions, but above all to reduce soil pollution by sorption and sequestering herbicides or its metabolites (Ippolito et al. 2012; Khorram et al. 2016b; Liu et al. 2018). Biochar reduces the bioavailability of herbicides in the soil by sorption of their molecules onto their particles. The sequestration of herbicides by biochars and their consequent decrease in bioavailability in the soil are one of the most appropriate methods to remediate soil and water contaminated with herbicides because it reduces secondary effects and accumulation in nontarget environmental organisms (Gámiz et al. 2017; Mendes et al. 2018), since it reduces the transport, leaching, and runoff processes of these pollutants (Jones et al. 2011). Biochar from bovine manure adsorbed 77% of the simazine from the aqueous solution of the soil (Cao and Harris 2010), which reduced the leaching of this herbicide (Tatarková et al. 2013). In addition, obtaining biochars is an efficient way to manage animal or plant waste, reducing the burden of pollution associated with the environment (Liu et al. 2018). The most common sources of feedstock for biochars production are crop residues, forest residues, animal manure, food-processing residues, paper mill residues, urban solid residues, and sewage sludge (Kwak et al. 2019; Ippolito et al. 2020). However, the environmental functions of biochars as well as their ability to sip herbicides are influenced by the type of feedstock and the pyrolysis conditions (temperature and burning time) (Subedi et al. 2016; Kwak et al. 2019; Ippolito et al. 2020). Therefore, before using any type of biochar to reduce the mobility/bioavailability of herbicides in contaminated soils or other environments, it is necessary to know the main characteristics of biochar.

8.3 BIOCHARS AND THEIR CHARACTERISTICS

Biochars are fine-grained, porous coals produced from plant and animal waste pyrolyzed at high temperatures (>350 °C) (Khalid et al. 2020; Joseph et al. 2021). Pyrolysis is a process of thermal degradation of a substance in the absence of oxygen, i.e., substances are decomposed by heat without combustion reactions taking place (Peters et al. 2015). Therefore, biochars are by-products of feedstock pyrolysis, in some cases used as fuel in some industrial processes (green energy), rich in C (Khalid et al. 2020). The herbicide sorption and immobilization capacity of a biochar are determined by its specific physicochemical properties, mainly porosity, types of functional groups (hydroxyl, phenolic, carboxyl), C content, SSA, pH, aromatic structure, and mineralogical composition (Gámiz et al. 2017; Liu et al. 2018; Mendes et al. 2018). These properties vary widely depending on the type and conditions of the feedstock, temperature, and time of pyrolysis, as well as the possible treatments before and/or after pyrolysis (Ippolito et al. 2020).

The resulting biochar, as well as its behavior, function, and fate in the soil, depends, first of all, on the chemical and structural composition of the feedstock, elemental composition, density, grain size, mineral content, composition of the three main ones (Kwak et al. 2019; Liao et al. 2019; Li et al. 2019), polysaccharide polymers (cellulose, hemicelluloses, and lignin), calorific value, and mechanical resistance (Mendes et al. 2018); and second, depends on the scope of the physical and chemical alterations suffered during pyrolysis (final temperature, pressure, residence time in the heating zone, heating rate, thermal fluxes, and heat transfer coefficients resulting from the heating rate, chemical or thermal pretreatments in feedstock) such as wear, cracking, microstructural rearrangements of feedstock (Kookana et al. 2011), since the parameters of the pyrolytic process, such as temperature, heating rate, and pressure, can modify the amount recovered from the final product (Yaman 2004; Somerville and Jahanshahi 2015). In relation to the feedstock, wood-based biochars generally have a larger surface area with a high C content, those made of straw have a higher CEC, and those made of manure have a higher content of N and P (Ippolito et al. 2020; Werdin et al. 2020). In relation to the pyrolysis conditions, feedstock pyrolyzed at more than 500°C produces biochars more persistent in the soil with higher ash content, high pH, increasing aromaticity and C/O and C/H ratios (Keiluweit et al. 2010; Ippolito et al. 2020). In this section, we will address how the main characteristics of biochars participate in the sorption of herbicides in the soil.

8.3.1 Specific Surface Area and Porosity

Specific surface (SSA) and porosity are some of the main properties of biochars that regulate most soil–biochar interactions (Kookana et al. 2011; Weber and Quicker 2018). The SSA and porosity depend on the type of feedstock and the conditions under which the biochar is produced (Ippolito et al. 2020). The chemical composition of the feedstock is highly variable depending on the botanical species and the part of the feedstock (leaves, branches, wood, bagasse, residues from the extraction of vegetable oil, among others) (Heitkötter and Marschner 2015; Kwak et al. 2019; Ippolito et al. 2020), i.e., the same pyrolysis conditions produce biochars with different characteristics of SS, porosity, grain size, among others. On the other hand, the SSA and the porosity of a biochar are strongly related, i.e., both the SSA and the porosity increase with temperature; therefore, biochars produced with the same feedstock may have different SSA and porosity depending on the pyrolysis conditions (Kookana et al. 2011; Chen et al. 2015).

Generally, temperatures higher than 500°C produce biochars with higher SSA and less abundance of surface functional groups, while temperatures lower than 400°C produce biochars with low SSA (Mendes et al. 2018), which implies partitioning and specific interactions with functional groups on the surface of biochar (Li, Sheng et al. 2013b; Gámiz et al. 2017). Each feedstock can reach a maximum SSA at a certain temperature, after which the SSA decreases to levels similar to those obtained at low temperatures (Kookana et al. 2011). For example, in biochars from wheat straw pyrolyzed for 6 h at temperatures between 300 and 700°C, it was observed that the SSA increased with the increase when the temperatures increased from 300 to 600°C, but the SSA of the biochar produced at 700°C was lower than biochar produced at 600°C, possibly because the microporous structures were destroyed at 700°C (Chun et al. 2004). In another study, the SSA of biochar produced at 450°C was less than 10 m^2 g^{-1}, while at 600–750°C they had an SSA of 400 m^2 g^{-1} (Brown et al. 2006). This occurs because as the temperature increases, a greater number of micropores with smaller diameter are formed, and consequently the SSA increases. The number of micropores less than 2 nm in diameter increases the SSA of the biochars with increasing temperatures (Kookana et al. 2011). In relation to feedstock, wood generally provides SSA up to twice as high as annual plant residues (Graber et al. 2012), possibly due to the increase in lignin content (Mendes et al. 2018). Biochars with high SSA and porosity have more active sites capable of sorbing herbicide molecules, which, depending on the situation, can be beneficial by immobilizing large amounts of herbicides or can be counterproductive, diminishing weed control (Graber et al. 2012).

The increase in SSA is directly influenced by the porosity of the biochar. As previously commented, high pyrolysis temperatures produce more porous biochars than at low temperatures (Gámiz et al. 2017), i.e., with the increase in temperature, there is the effect of ramping speed on the pore structure (Brown et al. 2006). The porosity of a biochar is the measurement of void space, which directly influences the hygroscopicity, reactivity, combustion efficiency, and sorption capacity of herbicides, and which can influence other properties, such as mechanical resistance (Mendes et al. 2018). The microporosity of wood biochar produced at 450°C was low with large pores (~1.1 nm diameter), while that of biochar produced at 850°C was high, and most of the pores were less than 1 nm in diameter (~ 0.49 nm) (Bornemann et al. 2007). This is because the increase in porosity is the result of an increase in the proportion of micropores, contributing to the sorption capacity of these carbonaceous materials.

Sorption is related to the high porosity of biochar, caused by the difference between its real specific mass and its apparent specific mass (Mendes et al. 2018), depending on the temperature and the duration of the pyrolysis causing the restructuring and ordering of the material, determining the degree of completion of these reactions (Keiluweit et al. 2010; Kwak et al. 2019; Li et al. 2019). Some authors have observed a strong and almost linear correlation between SSA and the volume of micropores of biochar (Graber et al. 2012; Weber and Quicker 2018). Therefore, both SSA and micropores play an important role in the dissipation of herbicides in the environment and ecotoxicological impacts on soil organisms, as discussed later. However, SSA and porosity decrease as biochar ages due to pore blockage and binding site saturation, mainly, with high molecular weight OM (Yavari et al. 2015), which can affect the environmental functionality of biochars in the medium and long term.

8.3.2 Apparent and Specific Densities

Density is a characteristic that determines the mobility of biochar in the environment, as well as its water retention capacity (Brewer et al. 2014). On the other hand, it is important to differentiate between apparent density and real density. The apparent density considers the specific weight of the volume of a material and includes both the pores in the solid structure and the gaps between the different particles of the mass (Weber and Quicker 2018), while the real density considers the specific mass discounting the volume of internal porosity (Mendes et al. 2018). Because the true density is related to the apparent density, it is necessary to measure the porosity, a parameter discussed in the previous subsection (Somerville and Jahanshahi 2015).

The density of the biochars is determined by the density of the feedstock and the pyrolysis (Mendes et al. 2018), i.e., the densities contain some information on the structural changes of the feedstock during carbonization (Weber and Quicker 2018). The increase in the density of the biochars is related to the lignin content of the feedstock, which is rich in C and favors the conversion, which implies a higher conversion yield in biochar (Werdin et al. 2020). The higher the pyrolysis temperature, the higher the true density and therefore the higher the porosity of the biochar produced (Somerville and Jahanshahi 2015). In this regard, wood feedstock generally produces biochars with higher density and better conversion performance, because, chemically, wood is mainly composed of hemicellulose, cellulose, and lignin, which decompose during pyrolysis, releasing liquid volatile fractions and soft drinks and solid aromatic C compounds (Werdin et al. 2020). Wood biochars produced at temperatures higher than 500°C generally have more than 80% (w/w) C with abundance of aromatic C, which contributes to long-term biochar stability by improving water retention (Weber and Quicker 2018). On the other hand, the density of the wood is related to the thickness of the cell wall of the fiber and the diameter of the fiber lumen. Lower-density eucalyptus wood biochar retains 35% more water than higher-density eucalyptus wood biochar (Werdin et al. 2020). However, as previously mentioned, in addition to wood, biochars can also be produced from other plant and animal residues, which, due to low or no lignin content, produce biochars with lower specific and apparent density (Brewer et al. 2014).

8.3.3 pH

Biochar pH is an important property for agricultural applications as soil amendment (Weber and Quicker 2018). The pH of biochars is generally alkaline, but it can also be acidic, neutral, or vary from acidic to alkaline with pH values ranging from less than 4 to 12 (most range from 5 to 9) (Fidel et al. 2017). pH biochar depends on the degree of oxidation of the feedstock and the pyrolysis temperature (Cybulak et al. 2019; Li et al. 2019). Temperatures higher than 400°C produce biochars with a neutral to acidic pH (Li et al. 2019), while those produced at 600–750°C have an alkaline pH that can reach up to 12.0 (Fidel et al. 2017; Joseph et al. 2021). For example, the biochar pH of corn and *Panicum virgatum* straws produced at 650°C was highly alkaline (pH>9), regardless of the processing time, while at temperatures less than 550°C, it was less than 5 (Clay and Malo 2012).

The increase in the pH of the biochar as the pyrolysis temperature increases is due to rapid (first 5–10 min) release of carboxyl, hydroxyl, or formyl functional groups, predominantly acidic in nature, during carbonization increasing the ash content (Li, Shen et al. 2013; Weber and Quicker 2018). For example, in *Conocarpus* waste biochars produced at temperatures ranging from 200 to 800°C, the basic functional groups increased from 0.15 to 3.55 mmol g^{-1} with increasing temperature, while the functional groups acids decreased from 4.17 to 0.22 mmol g^{-1} (Al-Wabel et al. 2013). The decrease in the pH of the biochars is due to the formation of acidic carboxyl groups from the oxidation of C (Li, Sheng et al. 2013), while the increase in the pH of the biochar is influenced by the content of cations and total alkaline minerals (Yuan et al. 2011; Fidel et al. 2017). Furthermore, the ashes are basic in nature; therefore, biochars with a high content of mineral ash show high pH values and vice versa (Liao and Thomas 2019).

The pH of biochars can increase or decrease depending on the feedstocks and the properties of the soil (Li, Sheng et al. 2013; Weber and Quicker 2018). This is because the properties of the different feedstocks vary from materials poor in minerals, generally woody, to crop residues rich in minerals, and consequently the ability to modify the pH of the soil when incorporated immediately (Somerville and Jahanshahi 2015; Werdin et al. 2020). For example, a mineral-poor oak wood biochar lowered the soil pH from 4.9 to 4.7, while a mineral-rich corn stubble biochar increased the pH from 6.7 to 8.1 (Nguyen and Lehmann 2009). The ability of biochar to change its pH is determined by factors such as soil pH and the composition of alkaline substances in the biochar itself. For example, the addition of 1% biochar to the soil did not affect the soil pH (6.4), while the addition of 10% biochar decreased or increased the soil pH by up to 12% as a function of alkalinity or acidity of the applied biochar (Clay and Malo 2012). Therefore, biochars obtained from high pyrolysis temperatures have a greater ability to raise the pH of soils.

On the other hand, the pH of the biochar also determines the ability to sorb herbicide molecules in contaminated soils (Khorram et al. 2016b). High pH values can increase the hydrolysis of herbicides (carbamates) in the soil through the dissolution of alkaline minerals and alkaline catalysis mechanisms (Zhang et al. 2012). The terbutylazine sorption was higher in *Pinus radiata* biochar produced at 700°C than in those obtained at 350°C (Wang et al. 2010). Leaching of 2,4-dichlorophenoxyacetic acid (2,4-D) and acetochlor was lower in biochars produced at 350°C, which amplifies the efficacy of herbicides (Li, Li et al. 2013). Glyphosate sorption in rice husk biochar ranged from 75% to 85% at pH 3–5 and decreased to 75–65% at pH 6–8 and to 55% at pH 9 (Herath et al. 2019). This suggests that, like SSA and porosity, the pH of the biochar influences the functionality of the biochar in the soil (Li, Shen et al. 2013; Li et al. 2019), whether it is used only for agricultural purposes or for the biodegradation of herbicides. However, the pH of the biochar diminishes during the aging process by the CO_2 sorption and acidic functional groups (Xiang et al. 2021), i.e., the agricultural and environmental functionality of the biochars is finite.

8.3.4 Chemical Elements, Molar Ratios, and Ashes of Biochar

Chemical elements of biochars and their relationships determine their stability and interaction with the soil (Li, Sheng et al. 2013; Leng and Huang 2018). Biochars have between 45% and 90% of C when they come from plant feedstock. Generally, the C content is higher in biochars produced at temperatures higher than 600°C than in those produced at 400°C. For example, the C content in pine chip biochar at 600 and 400°C was 92% and 79%, respectively (Heitkötter and Marschner 2015). In general, the C content of the biochar depends on the content of lignocellulosic materials in the feedstock (Ippolito et al. 2020; Werdin et al. 2020). Biochars from corn presented 65% and 64% of C when produced at 600 and 400°C, respectively (Heitkötter and Marschner 2015), i.e., the final C content of the biochars is determined mainly by the feedstock, and not both by the temperature of the pyrolysis.

The C content and aromatic structures regulate the sorption capacity of biochars (Jesus et al. 2019; Wu, Zhang et al. 2019), because these characteristics participate in the formation of porous networks in soils (Brewer et al. 2014; Khalid et al. 2020). Biochars can be made up of relatively leachable C, recalcitrant C, aromatic C, and a high content of mineral ash (Jiang et al. 2013). The fused aromatic structures of biochars can exist as amorphous C and turbo-stratic C. Amorphous C is common in biochars produced at low temperatures, while turbo-stratic C is dominant in biochars produced at high temperatures (Keiluweit et al. 2010; Jiang et al. 2013).

Chemical variations of C, mainly aromatic C, influence the content of other elements in biochar (Chen et al. 2015; Li, Tang et al. 2018). When the C concentration is less than 40%, as in biochars from organic waste from sewage sludge (Figure 8.1a), the H and O content decreases from 40 to just 7% as the pyrolysis temperature increases (Figure 8.1b and c). The content of N and H is variable depending on the feedstock, but in general it does not exceed 6% (Figure 8.1c and d), i.e., in general terms, biofuels have high content of C and O and low H and N content (Wang et al. 2018).

The chemical elements and molar and mineral relationships of biochar influence the stability and resistance to oxidation of biochars and their stability and interaction with soils (Mendes et al. 2018). H/C molar ratios are an indicator of carbonization, and those of O/C and (O+N)/C are polarity indices (Yang et al. 2018), which decrease with increasing pyrolysis temperature. The H/C ratio ranges from 0.1 to 1.5, but this ratio is less than 1 for most biochars (Figure 8.2a) and can be a useful parameter to predict adsorption of the metolachlor (Wei et al. 2020). The O/C and (O+N)/C ratios generally do not exceed 0.3 in biofuels obtained between 300 and 800°C (Figure 8.2b and d). The N/C ratio seems to be more associated with the biochar feedstock than with the pyrolysis temperature. The N/C ratio is less than 0.04 in biochars of plant origin but increases between 0.08 and 0.14 in biochars from sewage sludge (Figure 8.2c).

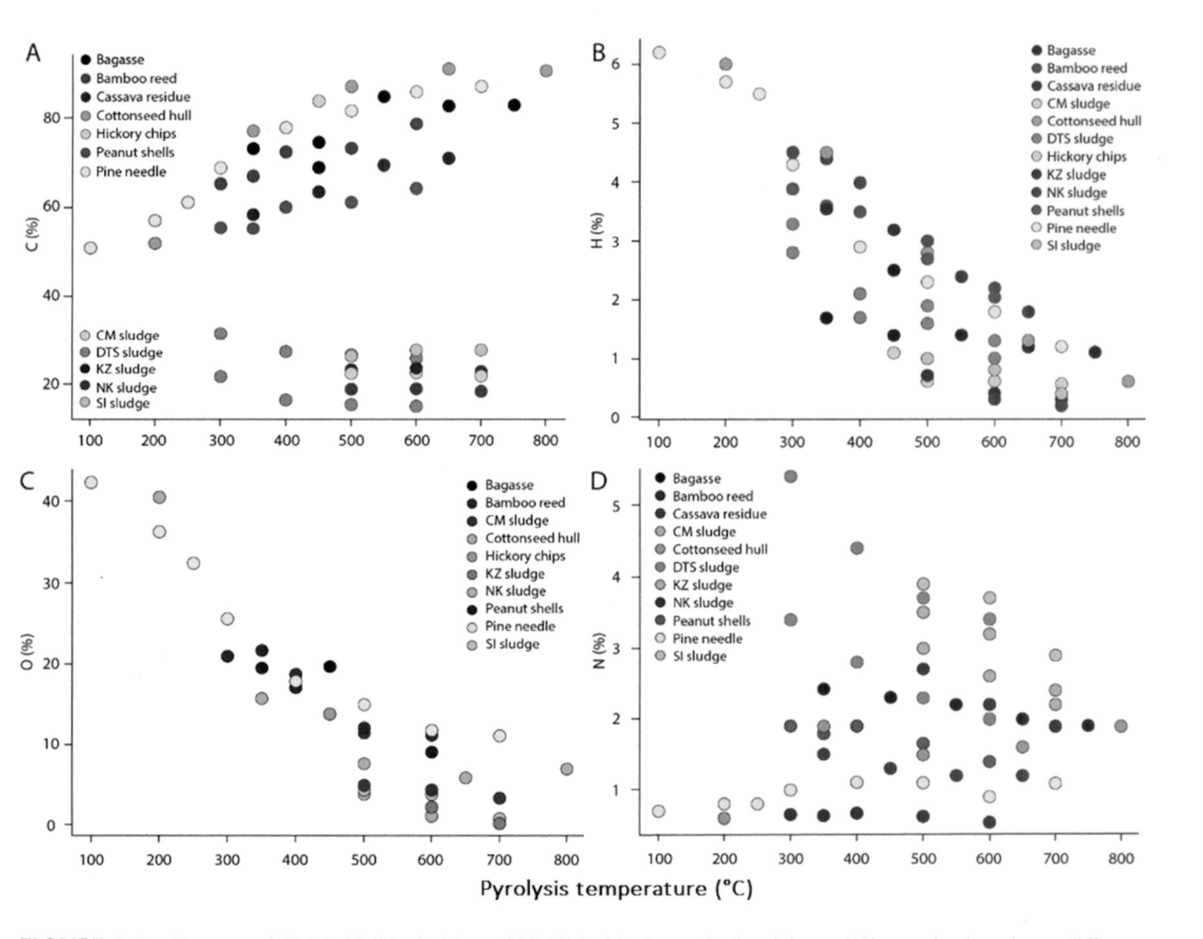

FIGURE 8.1 Content of C (a), H (b), O (c), and N (d) in biochars obtained from different feedstocks at different pyrolysis temperatures (adapted from Wang et al. 2018).

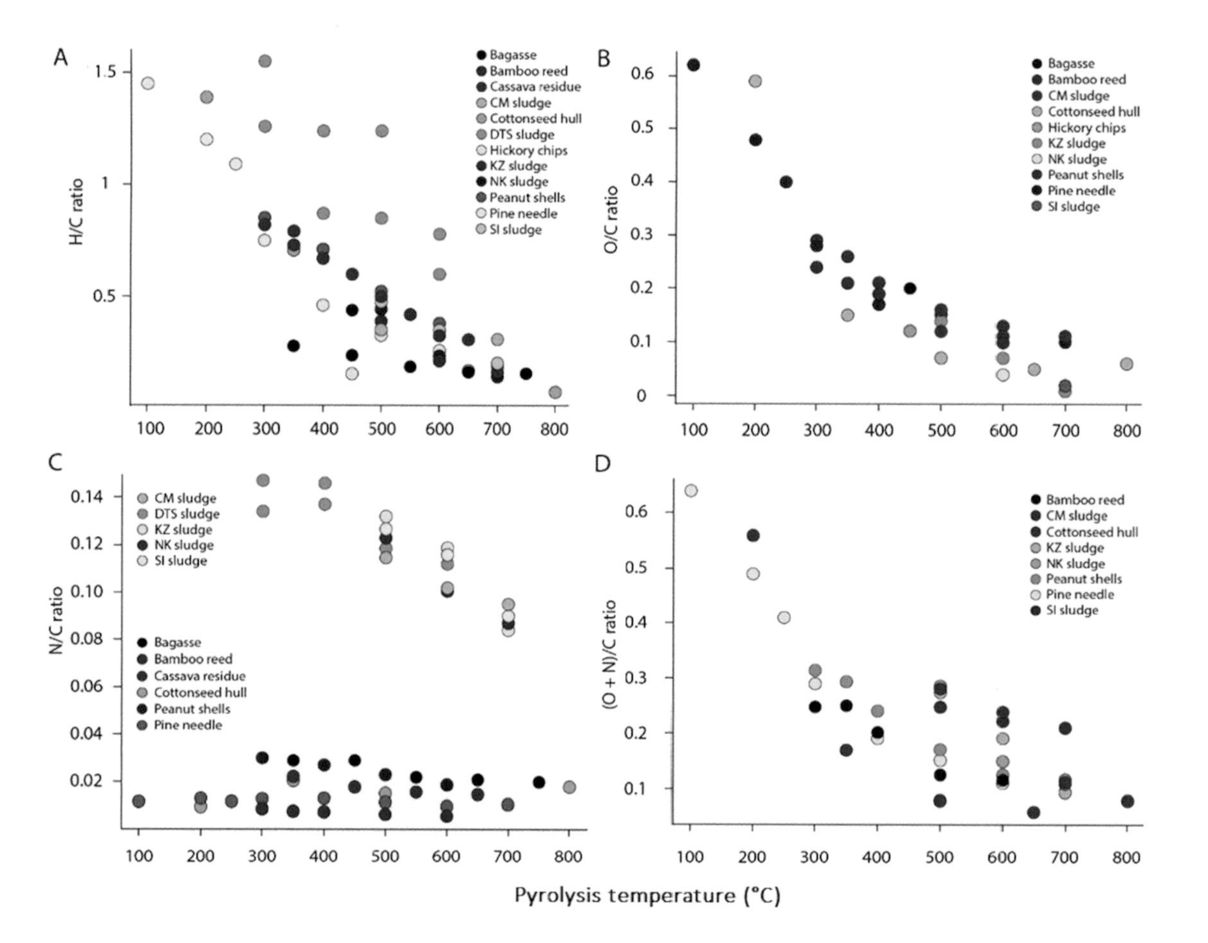

FIGURE 8.2 Molar ratios of H/C (a), O/C (b), N/C (c), and (O+N)/C (d) of biochars derived from different feedstocks at different pyrolysis temperatures (adapted from Wang et al. 2018).

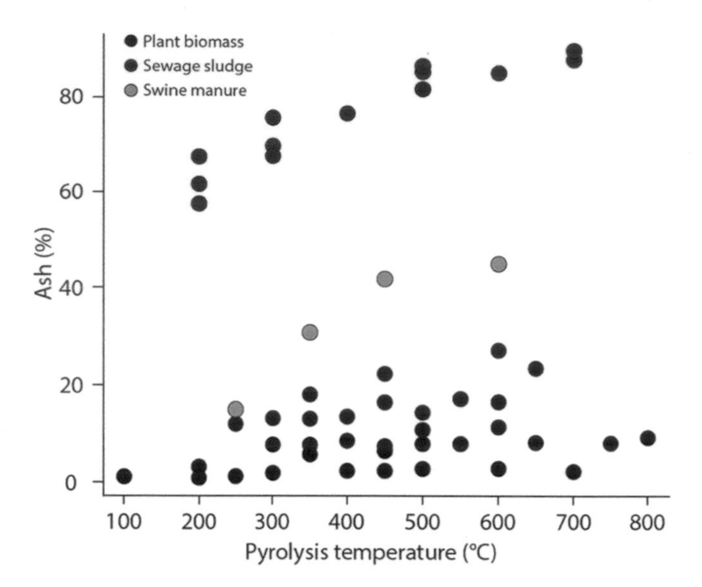

FIGURE 8.3 Ash content in biochars derived from different feedstocks at different pyrolysis temperatures. The graphs were constructed using the plant feedstock data available in Wang et al. (2018) and Yang et al. (2018). Pig manure data were obtained from Yang et al. (2018). The wastewater data were obtained from Li, Tang et al. (2018) and Fan et al. (2020).

The ashes of the biochar are sourced from the mineral elements and inorganic matter of the feedstock, with a predominance of potassium, calcium, phosphorus, and sodium (Mendes et al. 2018). The ash content of biochar ranges from 1% to 20% for most biochars derived from plant feedstocks. In biochar based on pig manure, the ash content ranges between 15% and 45%, increasing with the pyrolysis temperature (Yang et al. 2018). Biochars from sewage sludge have a high ash content ranging between 57% and 90% (Figure 8.3). In all cases, the mineral ash content increases with increasing temperature.

On the other hand, biochars with a high content of C have a greater capacity to sorb organic pollutants (Wu et al. 2013), i.e., the production of biochar with desirable characteristics for the sorption of herbicides in the soil requires high pyrolysis temperatures (Bornemann et al. 2007). In this sense, biochars obtained at high temperatures present high aromatization, carbonization, more stable structures, and very low degradation (Leng and Huang 2018), which allow to immobilize pollutants in the soil for a longer time. Furthermore, stable biochar limits access to C, N, and other nutrients present in the C structure of biochar to microorganisms (Lehmann et al. 2011).

8.3.4 Immediate Chemical Composition

One of the main objectives of biochar production is the change in chemical composition of the feedstock, especially the increase in C content (Weber and Quicker 2018). The fixed content of C, ash, and volatile matter make up the immediate chemical composition of biochar (Mendes et al. 2018). The fixed C content, a quality index, increases with temperature and the elimination of volatile matter, favoring the C stability in the feedstock. However, the formation of large amounts of ash at high temperatures and the rate of heating can reduce the fixed C content of biochar (Yang et al. 2020).

The volatile matter content ranges from 20% to 60% for most biochars, which decreases as the pyrolysis temperature increases (Spokas et al. 2011), i.e., the volatile matter content has an inverse relationship to temperature. Volatile matter content varies more in biochars from wood, but this variation is reduced in biochars from animal waste (Ghidotti et al. 2017). The volatile matter content causes changes in the structure and chemical composition of biochar with effects on the biological and biochemical properties of soils (Maaz et al. 2021). Higher content of fixed C and lower content of volatile matter and ash confer better properties and stability to biochar (Mendes et al. 2018); however, the content of ash and volatile matter are among the most important factors of biochar to improve the physicochemical properties of the soil (Li et al. 2019).

The ash content is related to the pyrolysis temperature, but it also varies depending on the inorganic minerals obtained after decomposition of the C, O, and H of the feedstock (Li et al. 2019; Panahi et al. 2020). The high ash content in biochars contributes to increasing the soil pH (Liao and Thomas 2019; Fan et al. 2020), reducing the sorption capacity of pollutants by blocking of internal pores by inorganic mineral particles (Li et al. 2019) and the content of fixed C (Yang et al. 2020), as well as incorporating heavy metals in the functional groups (Panahi et al. 2020). The physical and mechanical structure of the biochars is also associated with the ash content (Keiluweit et al. 2010).

The mineralogical composition that is responsible for the high CEC of biochar is also considered very important for the absorption of herbicides from the soil (Li, Tang et al. 2018; Yang et al. 2018), as they are capable of reducing the availability of herbicides through chelation mechanisms and/or surface acidity (Gámiz et al. 2017). Both the pyrolysis temperature and the feedstock influence the concentrations of mineral components in the biochar, because the temperature enriches the minerals in the biochar (Domingues et al. 2017; Ippolito et al. 2020). In general, plant residue biochars contain more P (1.82–3.60%) than wood biochars (0.03–0.06%) (Hossain et al. 2011; Li et al. 2019), while the K content in biochars from poultry and pig manure (1.6–5.9%) is higher than that of biochars from other feedstocks (Domingues et al. 2017). On the other hand, the variation in temperature can confer unwanted properties to the biochars, affecting the absorption of herbicides. For example, biochars obtained at high temperatures have more sites for herbicide uptake due to the larger SSA, but that uptake could be affected by functional groups on the surface of the biochar (Graber et al. 2012; Chen et al. 2015). Therefore, it is necessary to understand how biochar affects the sorption–desorption processes of herbicides in the soil to obtain biochars with desirable properties for the elimination of chemical products (Martin et al. 2012; Liu et al. 2018; Mendes et al. 2018).

8.3.5 Surface Functional Groups

The carbonization process is an evolution of the atomic relationships, which implies changes in the chemical structure of the feedstock pyrolyzed in biochars (Keiluweit et al. 2010; Yang et al. 2020), due to the detachment and release of functional groups of H and O, thus decreasing the respective relationships of these elements with C (Weber and Quicker 2018). The functional groups present on the surface of biochar can be characterized by Fourier transform infrared spectroscopy (FTIR) or nuclear magnetic resonance (NMR) (Chen et al. 2015; Liu et al. 2018). The main functional groups on the surface of biochars include carboxylic (–COOH), phenolic, hydroxyl (–OH), lactonic, and amide and amine groups (Li, Shen et al. 2013). The pyrolysis temperature influences the formation/degradation of functional groups on the surface of the biochar (Chen et al. 2015). Generally, as pyrolysis temperatures increase, the O/C, N/C, and H/C atomic ratios and functional groups on the surface of the biochar decrease (Wei et al. 2020). For example, the abundance of functional groups' feedstock pyrolyzed in biochars produced at 100–200°C was similar; however, the increase in temperature decreased the number of functional groups (Chen et al. 2015). The *O*-alkyl C groups decreased from 56% to 20% and from 13% to 6.9% with an increase in the pyrolysis temperature (200 to 600°C), respectively (Li, Shen et al. 2013).

The feedstock determines the type and abundance of functional groups (Janu et al. 2021). Wood biochar produced at 300°C presented large amounts of functional groups, lignin, hemicellulose, and dehydrated cellulose, and at 400°C lignocellulosic products appeared due to the degradation of phenolic and carboxylic compounds in lignin (Keiluweit et al. 2010). During the pyrolysis process, most of the functional groups of lignocellulosic materials are lost as the temperature increases (Suliman et al. 2016). However, at high temperatures, the abundance of functional groups is low and does not present great differences regardless of the type of feedstock (Janu et al. 2021). The abundance and type of functional groups in soybean, corn, canola, and peanut straw biochars produced at 300°C differed depending on the feedstock, but, at 700°C, the amount and type of functional groups were similar for all biochars obtained (Yuan et al. 2011).

Abundance and type of functional groups determine the herbicide sorption capacity of biochar (Martin et al. 2012; Gámiz et al. 2017). Phytoaccumulation and poisoning of plants grown in sewage sludge decreased after the addition of biochar due to the sorption of herbicide molecules on the biochar particles (Zielińska and Oleszczuk 2015). Herbicide sorption decreased on the biochar surface as the *O*-alkyl C functional groups decreased (Liu, Shen et al. 2013). The mechanisms that regulate the sorption of herbicides into biochars particles are not fully understood, but it has been suggested that herbicide molecules bind to the surface macro and micro pores of biochars, which are suitable sites for the sorption of contaminants due to the presence of functional groups (Oleszczuk et al. 2012). Therefore, the decrease in functional groups on the surface of biochars reduces the sorption of herbicides (Shi et al. 2015; Liu, Tang et al. 2018).

The functional groups on the SSA and the chemistry of biochar can be altered due to aging, oxidation, or microbial degradation (Martin et al. 2012), resulting in a higher CEC, allowing biochars to be able to retain nutrients (for example, Ca^{2+}, K^{+}, and Mg^{2+}) (Hailegnaw et al. 2019). The CEC is an important parameter to understand the sorption process of herbicides in soil colloids and/or organic material (Silva et al. 2007); however, the CEC of biochar is more difficult to measure than the CEC of soils due to its variable charge properties dependent on pH and the presence of soluble salts (Joseph et al. 2021). Therefore, it is necessary to understand the sorption–desorption processes of herbicides in the soil to obtain biochar with desirable properties for the sorption of herbicides.

8.4 DIRECT EFFECTS OF BIOCHAR ON THE BEHAVIOR AND FATE OF HERBICIDES IN THE ENVIRONMENT

Biochars have a high capacity to sorb herbicides, reducing their concentration in the soil solution (Mendes et al. 2018; Jones et al. 2011), property for which biochars are being used to immobilize these compounds in agricultural soils and other environments, with the purpose to minimize the risk of human exposure to herbicides (Martin et al. 2012; Gámiz et al. 2017). However, the addition of biochars to the soil can directly or indirectly affect the behavior of herbicides in the soil by altering the sorption, desorption, degradation, leaching, and bioavailability of herbicides in the soil, affecting weed control (Mendes et al. 2018).

8.4.1 Herbicide Sorption

Sorption is a physicochemical process that refers to the herbicide's ability to bind to soil colloids (Gerstl 1990), which is regulated by hydrophobic interactions, hydrogen bonds, ionic and covalent bonds, and van der Waals forces (Monquero and Silva 2021). Sorption coefficients (K_d) and its normalization in relation to the organic carbon (OC) content of the soil (K_{oc}) are the basic parameters to estimate the behavior and fate of herbicides in the soil (Gerstl 1990) – i.e., these indices allow predicting the mobility of the herbicides in the soil as a function of the OM content. Therefore, sorption plays an important role in the transport, degradation, and retention of herbicides in the soil and the magnitude of its impacts on the environment (Curran 2016).

The addition of biochars can enhance the retention of herbicides in the soil (Mendes et al. 2018), making them a potential repository of herbicides for the recovery of contaminated soils (Khalid et al. 2020). The addition of biochars from crop residues increased the sorption of simazine and atrazine (Zheng et al. 2010) and diuron (Liu et al. 2018). The diuron sorption capacity of rice straw and wheat biochars was up to 2500 times greater than in soil without the addition of biochar (Yang and Sheng 2003). The addition of 1% wheat biochar increased the sorption of ametryn, bromacil, and diuron by 70%, 86%, and 80%, respectively (Sheng et al. 2005). Biochars of soybean residues, sugarcane bagasse, and wood chips increased the sorption of indaziflam in the soil by 7%, 55% and 69%, respectively (Mendes et al. 2021c). However, although the retention and possibly the persistence time of herbicides in the soil increase with the addition of biochars due to sorption, the bioavailability of the molecules decreases directly affecting the agricultural efficiency of these products in weed control (Kookana et al. 2011; Khorram et al. 2016b).

The factors involved in the sorption of herbicides by biochar in the soil are related to the physicochemical characteristics of both the biochar itself (aromaticity, porosity, SSA, polar C content), as well as the herbicide (hydrophobicity) and soil (content mineral, soil pH) in question and environmental conditions (Mendes et al. 2018;

Khalid et al. 2020). Furthermore, the sorption capacity of biochars changes over time since the physicochemical characteristics of biochar in the soil can change due to mineralization processes, interaction with OM, and oxidation of minerals and the presence of soil microorganisms (Yang et al. 2018; Joseph et al. 2021).

Herbicide sorption on biochars particles can occur in different ways. At this point, the presence of micropores and mesopores is essential, because due to the hydrophobic nature of most herbicides, the hydrophilic groups of biochar favor the sorption of herbicide molecules in the biochar particles (Semple et al. 2013). For example, the high adsorption capacity of biochar from herbicide aromatic molecules such as paraquat and diquat (Shi et al. 2015) depends on pore-filling and electron interaction mechanisms (Brewer et al. 2014). On the other hand, the oxygen residues present in the biochar can favor the sorption of herbicides through hydrogen bonds (Wang et al. 2010). In this way, herbicides from the group of imidazolinones, α-triazines, and amine/heterocyclic nitrogen atoms of pyridines that behave as electron donors (Senesi 1992) can be more easily sorbed into biochar particles. Therefore, the aromatic structures with the electron donor–acceptor interactions of the functional groups also play an important role in the sorption of herbicides (Keiluweit et al. 2010). On the other hand, the sorption of herbicides such as atrazine depends on the aryl C content in the biochar (Sun et al. 2010).

As mentioned before, the sorption capacity of an herbicide in the soil is regulated by the C content in the OM (Gherstl 1990); therefore, the percentage of OM in the soil will determine the amount of herbicide absorbed in the biochar. This is due, in part, to the fact that physicochemical characteristics of biochar in the soil differ in comparison with the OM with neutral charge (Lehmann et al. 2011), i.e., the biochar sorption capacity increases or decreases depending on of polar C content (Khalid et al. 2020). Therefore, the herbicide sorption capacity of a biochar is defined by the non-carbonized and polar carbonized fractions (Sun, Gao et al. 2012). This means that biochars with low C content, such as sewage sludge, will have less sorption capacity than those with high C content, such as biochars obtained from plant feedstocks (Zielińska and Oleszczuk 2015). In summary, the sorption capacity of a biochar is regulated by pore density and volume, SSA, functional groups, and particle size. Other mechanisms involved in biochar–herbicide interactions that favor sorption are the polymeric matrix, surface coverage, multilayer sorption, and condensation in capillary pores (Khalid et al. 2020). Furthermore, the biochar sorption capacity could be reduced with aging or weathering in soils (Martin et al. 2012).

8.4.2 Herbicide Desorption

Contrary to sorption, desorption is the detachment of herbicide molecules that were retained in solid soil particles, releasing them back into the soil solution (Khorram et al. 2016b), which increases the bioavailability, efficiency, and transport of herbicides by leaching, as well as the severity of environmental impacts (Kookana et al. 2011). The amount of herbicide sorbed is always greater than the desorbed (Curran 2016); therefore, it is important to know the equilibrium of sorption and desorption to determine the bioavailability of herbicides only in the solution.

The addition of biochar in the soil results in high sorption and sequestration of herbicide molecules, severely affecting the herbicide desorption process (Mendes et al. 2018). Furthermore, the sorption and desorption processes of herbicides from a biochar depend on the type of feedstock and the temperature and time of pyrolysis (Gámiz et al. 2017). Desorption of sulfometuron-methyl was approximately 5% lower in soil treated with 40 or 80 t/ha of biochar of industrial waste compared to unamended soil (desorption = 13%) (Alvarez et al. 2021). Desorption of imazapic and imazapyr in soils amended with biochars produced from agricultural wastes was lower than in unamended soils (Yavari et al. 2019). The addition of biochars of soybean residues, sugarcane bagasse, or wood chips increased the sorption of indaziflam, depending on the type of soil, and decreased desorption between 5% and 10% (Mendes et al. 2021c). However, although the addition of 1% natural biochar, with a low and irregular SSA (1.6–10.3 m^2 g^{-1}), pores of different sizes (3.1–30.4 nm), and low volumes (0.001–0.008 cm^3 g^{-1}) increased the sorption of diclosulam by 16% and did not reduce the desorption (16.5%) of this herbicide; for pendimethalin, no differences were observed in sorption (>95%) and desorption (<2.5%) both in biochar-amended and biochar -unamended soils (Mendes et al. 2019). As can be seen, some herbicides, depending on their physical–chemical characteristics, are more strongly sorbed by biochar particles than others with or without effects on their desorption process. The sorption–desorption processes of metabolites of imazamox and metazachlor were not affected by biochars, possibly as a result of their net negative charge as they are polar compounds (Dechene et al. 2014). From an agronomic point of view, biochars do not affect the bioavailability of imazamox, pendimethalin, and diclosulam to control weeds.

8.4.3 Herbicide Biodegradation

Herbicide biodegradation, after they have exerted their biological activity in controlling weeds satisfactorily, is essential to reduce carryover problems to susceptible crops that come from rotation and environmental pollution (Mendes et al. 2021b). The addition of biochar to the soil increases the sorption of herbicides, inducing the progressive accumulation of these substances in the soil (Jones et al. 2011). However, the high sorption of herbicide molecules into soil particles reduces the physical, chemical, and biological degradation and mineralization of herbicides (Khalid et al. 2020). In addition, biochars also reduce the bioavailability of herbicides, reducing their effectiveness in controlling weeds (Yang et al. 2006).

Several studies have shown that the addition of biochars reduces the biodegradation of herbicides in the soil, since the strong binding of herbicide molecules to the biochars particles reduces their bioavailability (Khalid et al. 2020). For example, diuron biodegradation was 15% lower in soil modified with 1% biochar compared to unmodified soil (Yang et al. 2006). In the case of sulfometuron-methyl, the addition of biochar from industrial residues reduced bacterial richness and therefore biodegrading of the herbicide, increasing its half-life time from 36 to more than 50 days (Alvarez et al. 2021). Isoproturon biodegradation decreased by up to 50% in soil amended with 1% biochar due to low bioavailability for autochthonous soil microbiota due to the high sorption and low desorption processes of the herbicide (Sopeña et al. 2012).

On the other hand, the impact on the biodegradation of herbicides depends on the feedstock of the biochar. For example, rice husk biochar had a greater impact on imidazolinone biodegradation than on oil palm empty bunch biochars (Yavari et al. 2019). Hardwood biochars of different tree species completely suppressed the biodegradation of simazine (Jones et al. 2011). The biodegradation rate influenced by the addition of biochar can also differ depending on the herbicide and/or the type of soil. For example, the biodegradation of atrazine was very high, but that of trifluralin was nil (Jones et al. 2011). The diuron biodegradation in clayey, sandy, and sandy-aged soils amended with 10%, 20%, or 30% biochar, respectively, did not differ (40–60 days); however, glyphosate biodegradation ranged from 49 to 83 days depending on the amount of biochar and the type of soil (Zhelezova et al. 2017).

As commented in the previous section, the desorption of herbicides is generally lower in biochar-amended soil; therefore, the addition of biochar may reduce the concentration of extractable herbicides in the soil (Khalid et al. 2020). This condition may be suitable for very persistent active ingredients with few degradation pathways such as indaziflam. However, the sorption of non-persistent herbicides (half-life time ≤30 days) in the biochars particles and that are rapidly degraded by microorganisms such as glyphosate and 2,4-D is not desirable, since their persistence could be increased by reducing its biodegradation rate. For example, the increased sorption of 4-chloro-2-methylphenoxyacetic acid (MCPA) increased its half-life time from 5.2 days in unamended soil to 21.5 days in amended soil with biochar of wheat straw at 1% (Tatarková et al. 2013). In the case of fomesafen, its half-life time went from 34.6 days in unamended soil to 50.8, 82.7, and 160.3 days in amended soils with 0.5%, 1%, and 2% of shell biochar of rice, respectively (Khorram et al. 2016a).

8.4.4 Herbicide Leaching

Leaching and runoff are two types of transport processes of herbicides, generally the non-volatile, soluble, and soil mobile ones (Jones et al. 2011). In leaching, herbicide molecules take advantage of water infiltration (irrigation or rain) through soil profiles, being able to reach underground water bodies, while runoff, the transport of herbicides, occurs on the soil surface after heavy rains (Monquero and Silva 2021). Both herbicide transport processes can account for losses of up to 15%, which have caused great environmental and public health concerns due to the possible contamination of groundwater and surface waters (Curran 2016).

The addition of biochar contributes to reducing the leaching and runoff of herbicides due to increased sorption (Li, Li et al. 2013), owing to the strong binding of herbicide molecules on the biochars particles (low-rate desorption), and there is very little chance that the herbicide will return to the soil solution (Mendes et al. 2018). Atrazine leaching was 54% lower in amended soil with biochar of pine shavings homogenized in the soil, while in heterogeneous soils the leaching levels were variable and uncertain in the long term (Delwiche et al. 2014). The reduction in leaching is due to the strong binding and sorption of herbicides in biochar (Khalid et al. 2020). On the other hand, the size of the biochar particle and the deformation of the pores influence the leaching of herbicides. Particles of sizes less than 2 mm reduced simazine leaching better than particles of sizes more than 2 mm (Jones et al. 2011).

On the other hand, heavy rains can favor the movement of colloidal herbicidal soil particles to other areas (Kookana et al. 2011), and, because biochars have a very low bulk density (Brewer et al. 2014), they can also be leached or surface transported (Xiang et al. 2021), i.e., biochar particles can act as a means of herbicide transport for nonagricultural environments (Khalid et al. 2020). Therefore, the potential of biochars to reduce herbicide leaching or runoff depends on the frequency and intensity of rainfall.

8.4.5 Herbicide Efficacy to Weed Control

For herbicides to control weeds, they must be bioavailable. However, the application of biochar, in addition to reducing the microbial degradation of herbicides, also affects its transfer to plants due to high sorption (Xiang et al. 2021). The decrease in the effectiveness of herbicides due to the effect of biochar is not a desired effect (Gámiz et al. 2017; Mendes et al. 2018), since it is generally necessary to increase the herbicide dose to obtain control levels similar to those obtained in biochar-unamended soils, raising production costs and increasing potential environmental risks (Mendes et al. 2021a).

The negative effects of biochars in weed control are more perceptible when it comes to PRE herbicides because they are strongly sorbed in the biochar particles. In an experiment evaluating the addition of cow bonechar (cow bone charcoal) ranging from 1, 2, 5, 10, and 20 t ha^{-1} in a tropical soil, only 1.4 t ha^{-1} of bonechar reduced the control of eight weed species by 50% (Figure 8.4) (Mendes et al. 2021a). Biochar

FIGURE 8.4 Control of weeds in a tropical soil amended with cow bonechar (0, 1, 2, 5, 10, and 20 t/ha) at 21 days after application of indaziflam (75 g ha^{-1}) compared to the unamended soil (adapted from Mendes et al. 2021a).

of pine wood chips applied at 0.5 kg m^2 reduced weed control by 75% and 60% with atrazine and pendimethalin, respectively (Soni et al. 2015). The addition of wheat straw biochar reduced the control of *Lolium rigidum* with trifluralin and atrazine, making it necessary to increase the herbicide doses by three to four times (Nag et al. 2011). A similar phenomenon was observed for diuron controlling *Echinochloa crus-galli*, requiring twice the dose to be applied (Yang et al. 2006).

On the other hand, the improvement of soil quality with biochars is to favor crops, but, indirectly, it also may favor weeds. For example, the addition of biochars derived from *Zea mays, Panicum vigatum*, and *Pinus ponderosa* improved the germination of lettuce, radish, and wheat due to the high sorption of atrazine and 2,4-D, which suggested that weeds of the Asteraceae, Brasicaceae, and Poaceae could also germinate better if soil are amended with biochars (Clay et al. 2016). Higher weed infestations have been observed in biochar-amended soils than unamended soils (Arif et al. 2013). The application of biochar at 0.5 and 2.5 kg m^{-2} did not affect the germination of *Senna obtusifolia* and *Digitaria ciliaris* but increased the germination of *Amaranthus palmeri* (Soni et al. 2014). Another similar case was observed for *Andropogon gerardii* and *Lespedeza cuneata*, where the growth of the first one improved as the percentage of biochar increased (0%, 1%, 2%, and 4%), while for the second species, biochar did not influence its growth (Adams et al. 2013). Wheat straw biochar at 100 Mg/ha reduced the germination of *Trifolium subterraneum* and *Vigna radiata* by 46% and 10%, respectively; however, both the shoot and root biomass increased compared to the unamended soil when biochar was applied at 10 Mg ha^{-1} (Solaiman et al. 2011). The favorable or unfavorable response of weeds depends on the species and the amount of biochar applied.

In the event that weeds are favored by biochars, higher doses of herbicides may be required to control them and, contrary to expectations, may favor evolution for herbicide resistance (Gressel 2017), i.e., biochars may act as an additional agent to select resistance. Weed scientists classify those weed plants as resistant that survive the recommended doses of herbicide on the label under field conditions (Gaines et al. 2020). However, the high herbicide sorption capacity of biochars may prevent increasing herbicide doses from improving overall weed control (Khalid et al. 2020), exposing weeds to sublethal concentrations, which also contribute to selection of resistance (Manalil et al. 2011), mainly to PRE herbicides, which are already more prone to select resistance.

High herbicide doses generally select for monogenic resistance (governed by a single gene) of the specific site of action of the applied herbicide, known as target-site (TS) resistance, while low doses select for nontarget site (NTS) resistance, which is governed by multiple genes (Kudsk and Moss 2017). TS resistance is technically *easy* to manage by changing the herbicide action mode; however, the polygenic nature of NTS resistance makes it more complex to understand and manage, since all the genes involved in resistance are difficult to identify (Ghanizadeh and Harrington 2017). In addition, each gene provides a certain level of resistance, which together could confer resistance to multiple herbicides with different mode of action (Ghanizadeh and Harrington 2017). In addition, low doses of herbicides can cause stimulating effects on many physiological and growth processes (hormesis) (Duke 2017). In this sense, it is necessary to maintain a balance between the sorption capacity of biochar and the efficacy of the herbicide, maintaining satisfactory weed control levels with low environmental impacts and reduced risk of resistance selection.

8.5 INDIRECT EFFECTS OF BIOCHAR ON THE BEHAVIOR OF HERBICIDES AND THEIR FATE IN THE SOIL

Biochar-amended soils, in addition to causing indirect effects on the efficacy of herbicides, as well as their transport, degradation, and retention, can also affect these processes indirectly (Khalid et al. 2020).

8.5.1 Soil Microbial Community

The microbiological degradation of herbicides is an important route of dissipation of these compounds (Singh and Singh 2014). The activity of microorganisms, such as bacteria, fungi, and algae, is influenced by the OM content, pH, fertility, temperature, and humidity of the soil, the last two factors being the most important (Mendes et al. 2021b). Therefore, any alteration of these parameters can affect the microbial activity of the soil and consequently the retention and persistence of herbicides in the soil (Lehmann et al. 2011). Amended soils with biochars can improve or reduce the abundance and microbial activity of the soil (Khalid et al. 2020), which can alter the sorption or desorption of herbicides in the soil (Lehmann et al. 2011).

When global microbial activity is measured, biochars are generally reported to enhance the diversity and abundance of microorganisms (Palansooriya et al. 2019); however, sometimes, this improvement of soil properties by biochar favors some species and disadvantages others. For example, biochar from sewage sludge, soybean straw, rice straw, and peanut shells improved the activity and abundance of *Actinobacteria, Gemmatimonadetes,* and *Proteobacteria firmicutes* but decreased that of *Chloroflexi* and *Acidobacteria* (Ali et al. 2019). Yeast-derived biochar promoted fungi in the soil, while glucose-derived biochar was utilized by Gram-negative bacteria (Steinbeiss et al. 2009). The application of biochar to the soil increased the relative abundance of bacteria, while the relative abundance of arbuscular mycorrhizal fungi decreased (Song et al. 2019). Some classes of microorganisms are generalists and capable of degrading a wide range of herbicide active ingredients; however, there are also very specific species (Singh and Singh 2014). This implies that, if the microorganisms favored by biochar are generalists, the biodegradation of herbicides may increase. Otherwise, the addition of biochar can reduce the biodegradation of herbicides in relation to unamended soils, increasing its persistence in the environment, as observed for acetochlor (Li, Liu et al. 2018) and other herbicides (Tatarková et al. 2013; Khorram et al. 2016a; Alvarez et al. 2021).

In general, herbicide biodegradation is lower in amended soils due to the low bioavailability of strongly sorbed molecules in the biochar particles, as discussed before; however, there are also studies reporting that herbicide degradation may be high in amended soils. The biodegradation of atrazine increased as a function of the increase in wheat biochar in the soil from 0.1% to 1% (Qiu et al. 2009). The addition of biochar and compost of oil mill (olive) waste decreased the half-life time and leaching rate of MCPA due to high microbial activity (López-Pineiro et al. 2013). The increase in atrazine degradation and mineralization in hardwood biochar-amended soils were also attributed to the biodegradation (Jablonowski et al. 2013). The half-life time of oxyfluorfen decreased from 108 days in unamended soil to only 35 days, depending on the soil type and amount of biochar due to degradation (Wu, Liu et al. 2019). As can be seen, soil biochar amendments can affect weed control by favoring the biodegradation of herbicides in some cases.

Reports of decreased abundance and microbial activity in biochar-amended soils are rare (Khalid et al. 2020), but, when this occurs, it is speculated that it is due to the improvement in the availability of water and nutrients for plants, the strong mycorrhizal symbiosis not being necessary (Lehmann et al. 2011). It is important to note that microbial activity is also determined by soil properties, climatic conditions, and herbicide management schemes (Muter et al. 2014); therefore, it cannot be generalized that the effects described here were a direct consequence of the use of biochars – i.e., if any of the edaphic and climatic variables is altered, it is possible that different biological activities of the soil are observed on the same type of biochar and herbicide.

8.5.2 Soil Organic Carbon

Soil C is found in two main forms: inorganic carbon (IC) and OC (Singh and Cowie 2014). OC has better adsorption capacity for herbicides (Curran 2016); therefore, any variation in the amount or composition of OC can affect the sorption and desorption of herbicides in the soil (Sun, Gao et al. 2012). Because biochar presents high levels of OC (Jiang et al. 2013), its addition to the soil increases OC reserves (Cheng et al. 2017). OC has high affinities for herbicides due to the presence of functional groups (Liu et al. 2018); therefore, OC can form various types of complexes with herbicides indirectly affecting the sorption and desorption of herbicides in the soil (Khalid et al. 2020). Biochar-amended soils, which have an aromatic nature, improves the absorption of herbicides in the soil (Shi et al. 2015; Jesus et al. 2019). However, it is important to remember that the abundance of functional groups decreases as a function of increasing the pyrolysis temperature (Chen et al. 2015).

Application of biochars with low OC content in soils rich in OM may have resulted in the lower sorption of herbicides. Aminocyclopyrachlor and bentazone were better sorbed in unamended silty loam soil than in wood pellet biochar with low content of OC (Cabrera et al. 2014). This occurs because OC (from OM) dissolved in the soil competes with herbicide molecules for the binding sorption sites of biochars (Cox et al. 2004) – i.e., dissolved OC from the soil confers selective herbicide sorption to biochars by blocking micropores (Lou et al. 2017). In addition, herbicide sorption capacity of biochars in the soil varies over time, which is generally greater in fresh biochars than in aged ones (Martin et al. 2012). Picloram dissipated better in aged biochar-amended soil than in soil unamended of fresh biochar-amended soil (Gámiz et al. 2019a). Contrary, an aged biochar had a higher sorption capacity of mesotrione than fresh biochar, increasing the half-life time of the herbicide slightly (Gámiz et al. 2019b). In addition, dissolved OC increases the functional groups rich in O on the surfaces of aged biochar, which decreases the sorption of herbicides (Lou et al. 2017).

It is important to consider the aforementioned information, because it will be difficult to modify the soil with biochars at the same time as the herbicides are applied – i.e., it is most likely that the molecules would interact with biochar particles that have undergone a period of aging in the soil, which have been able to undergo alterations in their physical–chemical properties. The herbicide sorption–desorption capacity of biochars changes over time due to soil exposure, altering the efficacy and bioavailability of herbicides applied to the soil as well as the time of application of the herbicide (Gámiz et al. 2019a, 2019b).

8.5.3 Soil pH

Most biochars have an alkaline pH, although there are some with a slightly acidic pH (Fidel et al. 2017); therefore, most biochars have the ability to increase soil pH. Sometimes biochars are even used as liming agents to increase the pH of acidic soils (Gámiz et al. 2019a). Soil pH is an important factor influencing chemical solubility and bioavailability of essential plant nutrients, CEC and anion exchange capacity (AEC), OM decomposition and microbial activity, regulating the herbicide performance and leaching (Hailegnaw et al. 2019; Brtnicky et al. 2021). In addition, when biochars are applied in alkaline soils, the pH is generally not altered, but when it does occur, the pH may decrease depending on the type of biochar feedstock, which can cause effects contrary to those expected. For example, the alkaline pH of sandy desert soils (7.1–8.5) decreased between by 0.92 and 0.95 units with a biochar of pine sawdust (Laghari et al. 2015).

The increase/decrease in soil pH depends on the biochar pH and the amount applied (Liao et al. 2019), which is determined by the nature of the feedstock and the pyrolysis temperature (Kwak et al. 2019), as previously explained. Modifying the soil pH with biochar alters a number of chemical processes such as oxidation/reduction that may affect the sorption–desorption processes of herbicides (Khalid et al. 2020), depending on the ability of the herbicide to form ions in the soil solution, since ionic forms (polar and more hydrophilic) of herbicides behave differently from non-ionized (neutral or nonpolar and more lipophilic) forms (Mendes et al. 2021b). The impact of modifying the soil pH with biochars will depend on the chemical nature of the herbicides (basic, acidic, or nonionizable) (see Chapter 3); but in general, weak-acid herbicides are less sorbed into the soil, while weak-base and nonionic herbicides tend to be more sorbed (Silva et al. 2007). Acidic herbicides dissociate into anions while basic herbicides dissociate into cations, acidifying or alkalizing the medium, respectively (Mehdizadeh et al. 2021). Generally acidic soils better sorb herbicides with acidic properties and high ionization capacity, and vice versa (Mehdizadeh et al. 2021).

When soil pH is increased, herbicides with a weak acid strength hardly become anionic, do not dissociate, and remain little available in the soil (Mendes et al. 2021b). However, the electrolytic charges of the biochar may limit the sorption of weak acid herbicides. For example, dissociation of picloram and imazamox occurred at pH 8.6–8.8 due to repulsion between negatively charged herbicides and negatively charged biochar surfaces (Gámiz et al. 2017 and 2019a). In this sense, the ability of biochar to reduce desorption and mitigate leaching of polar herbicides such as imidazolinones may be limited (Dechene et al. 2014), unless the increase in soil pH is significant. In addition, increasing the soil pH also with biochars can increase the hydrolysis of certain herbicides such as carbamates (barban, carbetamide, and chlorpropham) (Zhang et al. 2012).

Although generally increasing the pH of the soil with biochars improves the sorption of herbicides, sometimes the sorption may not be affected or may even decrease. For example, increasing the soil pH with wood and grass biochars did not affect the sorption of norflurazon but did reduce the sorption of fluridone (Sun et al. 2011). Diuron sorption decreased with increasing pH with wheat straw biochar from 2.42 to 5.30, possibly due to the deprotonation of functional groups of the biochar (Yang et al. 2004). Another similar behavior was observed for bromoxynil, where the sorption of the herbicide was lower in the soil amended with wheat biochar (pH = 7.0) than in with the unamended soil (pH = 3.0) (Sheng et al. 2005). Therefore, determining the pH of biochar and soil, as well as characterizing/knowing the physicochemical characteristics of biochars, soil and herbicides, and the environmental conditions that participate in the sorption process, is important to predict the possible impact of biochars on the behavior of herbicides to obtain both agricultural and environmental benefits (Wang et al. 2015).

8.6 CONCLUDING REMARKS

In general, the scientific literature widely highlights the agricultural and environmental "benefits" of biochars due to their ability to improve soil quality, increase water retention, and sorb organic pollutants such as herbicides. However, as has been discussed in this chapter, the "benefits" of biochars are relative and highly variable depending on the nature of feedstock, pyrolysis temperature, environmental conditions, soil biological activity, as well as physicochemical characteristics of the soil and of the herbicide to be immobilized. Therefore, the combinations between all these variables can be infinite, observing divergent and heterogeneous responses by the use of biochars.

Although environmentally biochars are a sustainable alternative for remediation of soils contaminated with herbicides, for which many environmental and health authorities bet to solve this problem, they can reduce weed control. The addition of biochars to the soil can affect the efficacy of herbicides directly by altering the sorption, desorption, biodegradation, and leaching of these chemical products, or indirectly by modifying the microbial community and activity, the OC content, and the pH of the soil. The modification of these parameters often results in the high sorption of herbicide molecules in the biochars particles, decreasing the bioavailability of herbicide diminishing the weed control. In addition, the improvement of the soil characteristics, which favor cultivated plants, can also favor the germination and growth of weeds.

Both when biochar reduces the bioavailability of herbicides and when they favor weed infestation, it is necessary to increase the doses to obtain control levels similar to those observed in unamended soils. Increasing herbicide doses, in addition to increasing production costs and giving rise to new possible environmental pollution problems, can favor the evolution of weed biotypes resistant to herbicides – i.e., biochars can be a selection agent for herbicide resistance, a phytosanitary problem that already makes weed management difficult and causes large agricultural, environmental, and economic impacts worldwide. Therefore, it is necessary to know the physicochemical characteristics of the biochar, the soil, and the herbicides, as well as the possible interactions between these factors, in order to achieve satisfactory levels of weed control with low environmental impacts by exploiting the environmental functionality of the biochars to sorb the herbicide molecules without representing a risk of resistance selection. Therefore, it is prudent to carry out a complete analysis of the physicochemical characteristics of the biochars before their addition to the soil and maintain a continuous monitoring of their behavior in the field to obtain the best environmental and agronomic benefits.

REFERENCES

Adams MM, Benjamin TJ, Emery NC, Brouder SJ, Gibson KD. 2013. The effect of biochar on native and invasive Prairie Plant Species. *Invasive Plant Sci Manag*. 6: 197–207.

Alaylar B, Güllüce M, Egamberdieva D, Wirth S, Bellingrath-Kimura SD. 2021. Biochar mediated control of soil-borne phytopathogens. *Environ Sustain*. 4: 329–334.

Ali N, Khan S, Li Y, Zheng N, Yao H. 2019. Influence of biochars on the accessibility of organochlorine pesticides and microbial community in contaminated soils. *Sci Total Environ*. 647: 551–560.

Alvarez DO, Mendes KF, Tosi M, et al. 2021. Sorption-desorption and biodegradation of sulfometuron-methyl and its effects on the bacterial communities in Amazonian soils amended with aged biochar. *Ecotoxicol Environ Saf*. 207: 111222.

Al-Wabel MI, Al-Omran A, El-Naggar AH, Nadeem M, Usman AR. 2013. Pyrolysis temperature induced changes in characteristics and chemical composition of biochar produced from conocarpus wastes. *Bioresour Technol*. 131: 374–379.

Arif M, Ali K, Haq MS, Khan Z. 2013. Biochar, FYM and nitrogen increases weed infestation in wheat. *Pak J Weed Sci Res*. 19: 411–418.

Bernardes MFF, Pazin M, Pereira LC, Dorta DJ. (2015). Impact of pesticides on environmental and human health. In *Toxicology studies-cells, drugs and environment*, eds. Andreazza AC, Scola G, 195–233 (1st ed). London: IntcehOpen.

Blackwell P, Krull E, Butler G, Herbert A, Solaiman Z. 2010. Effect of banded biochar on dryland wheat production and fertilizer use in south-western Australia: An agronomic and economic perspective. *Aust J Soil Res*. 48: 531–545.

Bornemann LC, Kookana RS, Welp G. 2007. Differential sorption behaviour of aromatic hydrocarbons on charcoals prepared at different temperatures from grass and wood. *Chemosphere* 67: 1033–1042.

Brewer CE, Chuang VJ, Masiello CA, et al. 2014. New approaches to measuring biochar density and porosity. *Biomass Bioenergy* 66: 176–185.

Brown RA, Kercher AK, Nguyen TH, Nagle DC, Ball WP. 2006. Production and characterization of synthetic wood chars for use as surrogates for natural sorbents. *Org Geochem*. 37: 321–333.

Brtnicky M, Datta R, Holatko J, et al. 2021. A critical review of the possible adverse effects of biochar in the soil environment. *Sci Total Environ*. 796: 148756.

Cabrera A, Cox L, Spokas K, et al. 2014. Influence of biochar amendments on the sorption – desorption of aminocyclopyrachlor, bentazone and pyraclostrobin pesticides to an agricultural soil. *Sci Total Environ*. 470–471: 438–443.

Cao X, Harris W. 2010. Properties of dairy-manure-derived biochar pertinent to its potential use in remediation. *Bioresour Technol*. 101: 5222–5228.

Carvalho FP. 2017. Pesticides, environment, and food safety. *Food Energy Secur.* 6: 48–60.

Chen Z, Xiao X, Chen B, Zhu L. 2015. Quantification of chemical states, dissociation constants and contents of oxygen-containing groups on the surface of biochars produced at different temperatures. *Environ Sci Technol.* 49: 309–317.

Cheng H, Hill PW, Bastami MS, Jones DL. 2017. Biochar stimulates the decomposition of simple organic matter and suppresses the decomposition of complex organic matter in a sandy loam soil. *GCB Bioenergy* 9: 1110–1121.

Chun Y, Sheng GY, Chiou CT, Xing BS. 2004. Compositions and sorptive properties of crop residue-derived chars. *Environ Sci Technol.* 38: 4649–4655.

Clay SA, Krack KK, Bruggeman SA, Papiernik S, Schumacher TE. 2016. Maize, switchgrass, and ponderosa pine biochar added to soil increased herbicide sorption and decreased herbicide efficacy. *J Environ Sci Health B* 51: 497–507.

Clay SA, Malo DD. 2012. The influence of biochar production on herbicide sorption characteristics. In *Herbicides-properties, synthesis and control of weeds*, ed. Hasaneen MN, 57–74 (1st ed). Rijeka: InTech Europe.

Cox L, Fernandes MC, Zsolnay A, Hermosin MC, Cornejo J. 2004. Changes in dissolved organic carbon of soil amendments with aging: Effect on pesticide adsorption behaviour. *J Agric Food Chem.* 52: 5635–5642.

Curran WS. 2016. Persistence of herbicides in soil. *Crops Soils* 49: 16–21.

Cybulak M, Sokołowska Z, Boguta P, Tomczyk A. 2019. Influence of pH and grain size on physicochemical properties of biochar and released humic substances. *Fuel* 240: 334–338.

Dechene A, Rosendahl I, Laabs V, Amelung W. 2014. Sorption of polar herbicides and herbicide metabolites by biochar-amended soil. *Chemosphere* 109: 180–186.

Delwiche KB, Lehmann J, Walter MT. 2014. Atrazine leaching from biochar-amended soils. *Chemosphere* 95: 346–352.

Domingues RR, Trugilho PF, Silva CA, et al. 2017. Properties of biochar derived from wood and high-nutrient biomasses with the aim of agronomic and environmental benefits. *PLoS One* 12: e0176884.

Dordas C. 2008. Role of nutrients in controlling plant diseases in sustainable agriculture. A review. *Agron Sustain Dev.* 28: 33–46.

Duke SO. 2017. Pesticide dose – A parameter with many implications. In *Pesticide dose: Effects on the environment and target and non-target organisms*, eds. Duke SO, Kudsk P, Solomon K, 1–13 (1st ed). Washington, DC: ACS Publications.

Elad Y, David DR, Harel YM, et al. 2010. Induction of systemic resistance in plants by biochar, a soil-applied carbon sequestering agent. *Phytopathol.* 100: 913–921.

Fan J, Li Y, Yu H, et al. 2020. Using sewage sludge with high ash content for biochar production and Cu (II) sorption. *Sci Total Environ.* 713: 136663.

FAOSTAT. 2021. Pesticide use. Available from: www.fao.org/faostat/en/#data/RP (accessed November 5, 2021).

Fidel RB, Laird DA, Thompson ML, Lawrinenko M. 2017. Characterization and quantification of biochar alkalinity. *Chemosphere* 167: 367–373.

Gaines TA, Duke SO, Morran S, et al. 2020. Mechanisms of evolved herbicide resistance. *J Biol Chem.* 295: 10307–10330.

Gámiz B, Velarde P, Spokas KA, Celis R, Cox L. 2019a. Changes in sorption and bioavailability of herbicides in soil amended with fresh and aged biochar. *Geoderma.* 337: 341–349.

Gámiz B, Velarde P, Spokas KA, Cox L. 2019b. Dynamic effect of fresh and aged biochar on the behavior of the herbicide mesotrione in soils. *J Agric Food Chem.* 67: 9450–9459.

Gámiz B, Velarde P, Spokas KA, Hermosín MC, Cox L. 2017. Biochar soil additions affect herbicide fate: Importance of application timing and feedstock species. *J Agric Food Chem.* 65: 3109–3117.

Gerstl Z. 1990. Estimation of organic chemical sorption by soils. *J Contam Hydrol.* 6: 357–375.

Ghanizadeh H, Harrington KC. 2017. Non-target site mechanisms of resistance to herbicides. *Crit Rev Plant Sci*. 36: 24–34.

Ghidotti M, Fabbri D, Hornung A. 2017. Profiles of volatile organic compounds in biochar: Insights into process conditions and quality assessment. *ACS Sustain Chem Eng*. 5: 510–517.

Graber ER, Tsechansky L, Gerstl Z, Lew B. 2012. High surface area biochar negatively impacts herbicide efficacy. *Plant Soil* 353: 95–106.

Gressel J. 2017. All doses select for resistance; when will this happen and how to slow evolution. *Am Chem Soc Symp Ser*. 1249: 61–72.

Hailegnaw NS, Mercl F, Pračke K, Száková J, Tlustoš P. 2019. Mutual relationships of biochar and soil pH, CEC, and exchangeable base cations in a model laboratory experiment. *J Soils Sediments* 19: 2405–2416.

Harel YM, Elad Y, Davide DR, et al. 2012. Biochar mediates systemic response of strawberry to foliar fungal pathogens. *Plant Soil* 357: 245–257.

Heitkötter J, Marschner B. 2015. Interactive effects of biochar ageing in soils related to feedstock, pyrolysis temperature, and historic charcoal production. *Geoderma* 245: 56–64.

Herath GAD, Poh LS, Ng WJ. 2019. Statistical optimization of glyphosate adsorption by biochar and activated carbon with response surface methodology. *Chemosphere* 227: 533–540.

Hossain MK, Strezov V, Chan KY, Ziolkowski A, Nelson PF. 2011. Influence of pyrolysis temperature on production and nutrient properties of wastewater sludge biochar. *J Environ Manage*. 92: 223–228.

Ippolito JA, Cui L, Kammann C. et al. 2020. Feedstock choice, pyrolysis temperature and type influence biochar characteristics: A comprehensive meta-data analysis review. *Biochar* 2: 421–438.

Ippolito JA, Laird DA, Busscher WJ. 2012. Environmental benefits of biochar. *J Environ Qual*. 41: 967–972.

Jablonowski ND, Borchard N, Zajkoska P, et al. 2013. Biochar-mediated ^{14}C-atrazine mineralization in atrazine adapted soils from Belgium and Brazil. *J Agric Food Chem*. 61: 512–516.

Janu R, Mrlik V, Ribitsch D, et al. 2021. Biochar surface functional groups as affected by biomass feedstock, biochar composition and pyrolysis temperature. *Carbon Resour Convers*. 4: 36–46.

Jesus JHF, Matos TDS, Cunha GDC, Mangrich AS, Romão LPC. 2019. Adsorption of aromatic compounds by biochar: Influence of the type of tropical biomass precursor. *Cellulose* 26: 4291–4299.

Jiang J, Zhang L, Wang X, et al. 2013. Highly ordered macroporous woody biochar with ultrahigh carbon content as supercapacitor electrodes. *Electrochim Acta* 113: 481–489.

Jones DL, Edwards-Jones G, Murphy DV. 2011. Biochar mediated alterations in herbicide breakdown and leaching in soil. *Soil Biol Biochem*. 43: 804–813.

Joseph S, Cowie AL, Van Zwieten L, et al. 2021. How biochar works, and when it doesn't: A review of mechanisms controlling soil and plant responses to biochar. *GCB Bioenergy* 13: 1731–1764.

Keiluweit M, Nico PS, Johnson MG, Kleber M. 2010. Dynamic molecular structure of plant biomass-derived black carbon (biochar). *Environ Sci Technol*. 44: 1247–1253.

Khalid S, Shahid M, Murtaza B, et al. 2020. A critical review of different factors governing the fate of pesticides in soil under biochar application. *Sci Total Environ*. 711: 134645.

Khorram MS, Zheng Y, Lin D, et al. 2016a. Dissipation of fomesafen in biochar-amended soil and its availability to corn (*Zea mays* L.) and earthworm (*Eisenia fetida*). *J Soils Sediments* 16: 2439–2448.

Khorram SM, Zhang Q, Lin D, et al. 2016b. Biochar: A review of its impact on pesticide behavior in soil environments and its potential applications. *J Environ Sci*. 44: 269–279.

Kochanek J, Soo RM, Martinez C, Dakuidreketi A, Mudge AM. 2022. Biochar for intensification of plant-related industries to meet productivity, sustainability and economic goals: A review. *Resour Conserv Recy*. 179: 106109.

Kookana RS, Sarmah AK, Van Zwieten L, Krull E, Singh B. 2011. Biochar application to soil: Agronomic and environmental benefits and unintended consequences. *Adv Agron*. 112: 103–143.

Kudsk P, Moss S. 2017. Herbicide dose: What is a low dose? In *Pesticide dose: Effects on the environment and target and non-target organisms*, eds. Duke SO, Kudsk P, Solomon K, 15–24 (1st ed). Washington, DC: ACS Publications.

Kwak J-H, Islam MS, Wang S, et al. 2019. Biochar properties and lead (II) adsorption capacity depend on feedstock type, pyrolysis temperature, and steam activation. *Chemosphere* 231: 393–404.

Laghari M, Mirjat MS, Hu Z, et al. 2015. Effects of biochar application rate on sandy desert soil properties and sorghum growth. *Catena* 135: 313–320.

Lehmann J, Rillig MC, Thies J, et al. 2011. Biochar effects on soil biota – A review. *Soil Biol Biochem*. 43: 1812–1836.

Leng L, Huang H. 2018. An overview of the effect of pyrolysis process parameters on biochar stability. *Bioresour Technol*. 270: 627–642.

Li J, Li Y, Wu M, Zhang Z, Lü J. 2013. Effectiveness of low-temperature biochar in controlling the release and leaching of herbicides in soil. *Plant Soil* 370: 333–344.

Li M, Tang Y, Ren N, Zhang Z, Cao Y. 2018. Effect of mineral constituents on temperature-dependent structural characterization of carbon fractions in sewage sludge-derived biochar. *J Clean Prod*. 172: 3342–3350.

Li S, Harris S, Anandhi A. and Chen G. 2019. Predicting biochar properties and functions based on feedstock and pyrolysis temperature: A review and data syntheses. *J Clean Prod*. 215: 890–902.

Li XM, Shen QR, Zhang DQ, et al. 2013. Functional groups determine biochar properties (pH and EC) as studied by two dimensional ^{13}C NMR correlation spectroscopy. *PLoS One* 8: e65949.

Li Y, Liu X, Wu X, et al. 2018. Effects of biochars on the fate of acetochlor in soil and on its uptake in maize seedling. *Environ Pollut*. 241: 710–719.

Liao W, Thomas SC. 2019. Biochar particle size and post-pyrolysis mechanical processing affect soil pH, water retention capacity, and plant performance. *Soil Syst*. 3: 14.

Liu Y, Lonappan L, Brar SK, Yang S. 2018. Impact of biochar amendment in agricultural soils on the sorption, desorption, and degradation of pesticides: A review. *Sci Total Environ*. 645: 60–70.

López-Pineiro A, Peña D, Albarran A, Sánchez-Llerena J, Becerra D. 2013. Behavior of MCPA in four intensive cropping soils amended with fresh, composted, and aged olive mill waste. *J Contam Hydrol*. 152: 137–146.

Luo L, Lv J, Chen Z, Huang R, Zhang S. 2017. Insights into the attenuated sorption of organic compounds on black carbon aged in soil. *Environ Pollut*. 231: 1469–1476.

Maaz TM, Hockaday WC, Deenik JL. 2021. Biochar volatile matter and feedstock effects on soil nitrogen mineralization and soil fungal colonization. *Sustainability* 13: 2018.

Manalil S, Busi R, Renton M, Powles SB. 2011. Rapid evolution of herbicide resistance by low herbicide dosages. *Weed Sci*. 59: 210–217.

Martin SM, Kookana RS, Van Zwieten L, Krull E. 2012. Marked changes in herbicide sorption – desorption upon ageing of biochars in soil. *J Hazard Mater*. 231–232: 70–78.

Mehdizadeh M, Mushtaq W, Siddiqui SA, et al. 2021. Herbicide residues in agroecosystems: Fate, detection, and effect on non-target plants. *Rev Agric Sci*. 9: 157–167.

Meiyappan P, Dalton M, O'Neill BC, Jain AK. 2014. Spatial modeling of agricultural land use change at global scale. *Ecol Model*. 291: 152–174.

Mendes KF, Dias Jr AF, Takeshita V, Rêgo APJ, Tornisielo VL. 2018. Effect of biochar amendments on the sorption and desorption herbicides in agricultural soil. In *Advanced sorption process applications*, ed. Edebali S, 87–103 (1st ed). London: IntechOpen.

Mendes KF, Furtado IF, Sousa RN, et al. 2021a. Cow bonechar decreases indaziflam pre-emergence herbicidal activity in tropical soil. *J Environ Sci Health B* 56: 532–539.

Mendes KF, Mielke KC, Barcellos LHJ, Alcántara-de la Cruz R, Sousa RN. 2021b. Anaerobic and aerobic degradation studies of herbicides and radiorespirometry of microbial activity in soil. In *Radioisotopes in Weed Research*, ed. Mendes KF, 95–126 (1st ed). Boca Raton: CRC Press Taylor & Francis Group.

Mendes KF, Olivatto GP, Sousa RN, Junqueira LV, Tornisielo VL. 2019. Natural biochar effect on sorption – desorption and mobility of diclosulam and pendimethalin in soil. *Geoderma* 347: 118–125.

Mendes KF, Soares MB, Sousa RN, et al. 2021c. Indaziflam sorption – desorption and its three metabolites from biochars-and their raw feedstock-amended agricultural soils using radiometric technique. *J Environ Sci Health B*, 56: 731–740.

Monquero PA, Silva PV. 2021. Comportamento de herbicidas no ambiente. In *Matologia: Estudos sobre Plantas Daninhas*, eds. Barroso AAM, Murata AT, 253–294 (1st ed). Jaboticabal, SP, Brazil: Fábrica da Palavra.

Muter O, Berzins A, Strikauska S, et al. 2014. The effects of woodchip- and straw-derived biochars on the persistence of the herbicide 4-chloro-2-methylphenoxyacetic acid (MCPA) in soils. *Ecotoxicol Environ Saf* 109: 93–100.

Nag SK, Kookana R, Smith L. et al. 2011. Poor efficacy of herbicides in biochar-amended soils as affected by their chemistry and mode of action. *Chemosphere* 84: 1572–1577.

Nguyen BT, Lehmann J. 2009. Black carbon decomposition under varying water regimes. *Org Geochem*. 40: 846–853.

Oleszczuk P, Rycaj M, Lehmann J, Cornelissen G. 2012. Influence of activated carbon and biochar on phytotoxicity of air-dried sewage sludges to *Lepidium sativum*. *Ecotoxicol Environ Saf*. 80: 321–326.

Palansooriya, K.N, Wong, J.T.F, Hashimoto, Y. et al. 2019. Response of microbial communities to biochar-amended soils: A critical review. *Biochar* 1: 3–22.

Panahi H.K.S, Dehhaghi M, Ok Y.S, et al. 2020. A comprehensive review of engineered biochar: Production, characteristics, and environmental applications. *J Clean Prod*. 270: 122–462.

Patel R. 2013. The long Green Revolution. *J Peasant Stud*. 40: 1–63.

Peters JF, Iribarren D, Dufour J. 2015. Biomass pyrolysis for biochar or energy applications? A life cycle assessment. *Environ Sci Technol*. 49: 5195–5202.

Pingali, P. 2007. Agricultural mechanization: Adoption patterns and economic impact. *Handb Agric Econ*. 3: 2779–2805.

Qiu Y, Pang H, Zhou Z, et al. 2009. Competitive biodegradation of dichlobenil and atrazine coexisting in soil amended with a char and citrate. *Environ Pollut*. 157: 2964–2969.

Razzaghi F, Obour PB, Arthur E. 2020. Does biochar improve soil water retention? A systematic review and meta-analysis. *Geoderma*. 361: 114055.

Semple KT, Riding MJ, McAllister LE, Sopena-Vazquez F, Bending GD. 2013. Impact of black carbon on the bioaccessibility of organic contaminants in soil. *J Hazard Mater*. 261: 808–816.

Senesi N. 1992. Binding mechanisms of pesticides to soil humic substances. *Sci Total Environ*. 123: 63–76.

Sheng G, Yang Y, Huang M, Yang K. 2005. Influence of pH on pesticide sorption by soil containing wheat residue-derived char. *Environ Pollut*. 134: 457–463.

Shi K, Xie Y, Qiu Y. 2015. Natural oxidation of a temperature series of biochars: Opposite effect on the sorption of aromatic cationic herbicides. *Ecotoxicol Environ Saf*. 114: 102–108.

Silva AA. Viviam R, Oliveira Jr, RS. 2007. Herbicidas: Comportamento no solo. In *Tópicos em manejo de plantas daninhas*, eds. Silva AA, Silva JF, 155–209 (1st ed). Viçosa: Editora UFV.

Singh B, Singh K. 2014. Microbial degradation of herbicides. *Crit Rev Microbiol*. 42: 245–261.

Singh BP, Cowie AL. 2014. Long-term influence of biochar on native organic carbon mineralisation in a low-carbon clayey soil. *Sci Rep*. 4: 3687.

Solaiman ZM, Blackwell P, Abbott LK, Storer P. 2010. Direct and residual effect of biochar application on mycorrhizal root colonization, growth and nutrition of wheat. *Aust J Soil Res*. 48: 546–554.

Solaiman ZM, Murphy DV, Abbott LK. 2011 Biochars influence seed germination and early growth seedlings. *Plant Soil* 353: 273–287.

Somerville M, Jahanshahi S. 2015. The effect of temperature and compression during pyrolysis on the density of charcoal made from Australian eucalypt wood. *Renew Energy* 80: 471–478.

Song D, Xi X, Zheng Q, et al. 2019. Soil nutrient and microbial activity responses to two years after maize straw biochar application in a calcareous soil. *Ecotoxicol Environ Saf*. 180: 348–356.

Soni N, Leon RG, Erickson JE, Ferrell JA, Silveira ML. 2015. Biochar decreases atrazine and pendimethalin preemergence herbicidal activity. *Weed Technol*. 29: 359–366.

Soni N, Leon RG, Erickson JE, Ferrell JA, Silveira ML, Giurcanu M. 2014. Vinasse and biochar effects on germination and growth of palmer amaranth (*Amaranthus palmeri*), sicklepod (*Senna obtusifolia*), and southern crabgrass (*Digitaria ciliaris*). *Weed Technol*. 28: 694–702.

Sopeña F, Semple K, Sohi S, Bending G. 2012. Assessing the chemical and biological accessibility of the herbicide isoproturon in soil amended with biochar. *Chemosphere* 88: 77–83.

Spokas KA, Novak JM, Stewart CE, et al. 2011. Qualitative analysis of volatile organic compounds on biochar. *Chemosphere* 85: 869–882.

Steinbeiss S, Gleixner G, Antonietti M. 2009. Effect of biochar amendment on soil carbon balance and soil microbial activity. *Soil Biol Biochem*. 41: 1301–1310.

Subedi, R, Taupe, N, Pelissetti, S, et al. 2016. Greenhouse gas emissions and soil properties following amendment with manure derived biochars: Influence of pyrolysis temperature and feedstock type. *J Environ Manag*. 166: 73–83.

Suliman, W, Harsh, JB, Abu-Lail, NI, et al. 2016. Modification of biochar surface by air oxidation: Role of pyrolysis temperature. *Biomass Bioenerg*. 85: 1–11.

Sun K, Gao B, Zhang Z, et al. 2010. Sorption of atrazine and phenanthrene by organic matter fractions in soil and sediment. *Environ Pollut*. 158: 3520–3526.

Sun K, Keiluweit M, Kleber M, Pan Z, Xing B. 2011. Sorption of fluorinated herbicides to plant biomass-derived biochars as a function of molecular structure. *Bioresour Technol*. 102: 9897–9903.

Sun K, Gao B, Ro KS, et al. 2012. Assessment of herbicide sorption by biochars and organic matter associated with soil and sediment. *Environ Pollut*. 163: 167–173.

Sun K, Zhang Z, Gao B, et al. 2012. Adsorption of diuron, fluridone and norflurazon on single-walled and multi-walled carbon nanotubes. *Sci Total Environ*. 439: 1–7.

Tatarková V, Hiller E, Vaculík M. 2013. Impact of wheat straw biochar addition to soil on the sorption, leaching, dissipation of the herbicide (4-chloro-2-methylphenoxy) acetic acid and the growth of sunflower (*Helianthus annuus* L.). *Ecotoxicol Environ Saf*. 92: 215–221.

Tauger MB. 2010. Boom and crisis: Agriculture from the Second World War to the twenty-first century. In *Agriculture in world history,* 148–189. New York: Routledge.

Tudi M, Ruan HD, Wang L, et al. 2021. Agriculture development, pesticide application and its impact on the environment. *Int J Environ Res Public Health* 18: 1112.

Uzoma KC, Inoue M, Andry H, et al. 2011. Effect of cow manure biochar on maize productivity under sandy soil condition. *Soil Use Manag*. 27: 205–212.

Wang D, Mukome FN, Yan D, et al. 2015. Phenylurea herbicide sorption to biochars and agricultural soil. *J Environ Sci Health B*. 50: 544–551.

Wang HL, Lin KD, Hou ZN, Richardson B, Gan J. 2010. Sorption of the herbicide terbuthylazine in two New Zealand forest soils amended with biosolids and biochars. *J Soils Sediments* 10: 283–289.

Wang M, Zhu Y, Cheng L, et al. 2018. Review on utilization of biochar for metal-contaminated soil and sediment remediation. *J Environ Sci*. 63: 156–173.

Weber K, Quicker P. 2018. Properties of biochar. *Fuel* 217: 240–261.

Wei L, Huang Y, Huang L, et al. 2020. The ratio of H/C is a useful parameter to predict adsorption of the herbicide metolachlor to biochars. *Environ Res*. 184: 109324.

Werdin J, Fletcher TD, Rayner JP, Williams NS, Farrell C. 2020. Biochar made from low density wood has greater plant available water than biochar made from high density wood. *Sci Total Environ*. 705: 135856.

Wu C, Liu X, Wu X. et al. 2019. Sorption, degradation and bioavailability of oxyfluorfen in biochar-amended soils. *Sci Total Environ*. 658: 87–94.

Wu M, Pan B, Zhang D, et al. 2013. The sorption of organic contaminants on biochars derived from sediments with high organic carbon content. *Chemosphere* 90: 782–788.

Wu Z, Zhang X, Dong Y, Li B, Xiong Z. 2019. Biochar amendment reduced greenhouse gas intensities in the rice-wheat rotation system: Six-year field observation and meta-analysis. *Agric For Meteorol*. 278: 107625.

Xiang L, Liu S, Ye S, et al. 2021. Potential hazards of biochar: The negative environmental impacts of biochar applications. *J Hazard Mater*. 420: 126611.

Xu G, Lv Y, Sun J, Shao H, Wei L. 2012. Recent advances in biochar applications in agricultural soils: Benefits and environmental implications. *Clean Soil Air Water* 40: 1093–1098.

Yaman S. 2004. Pyrolysis of biomass to produce fuels and chemical feedstocks. *Energ Conver Manag*. 45: 651–671.

Yang X, Kang K, Qiu L, Zhao L, Sun R. 2020. Effects of carbonization conditions on the yield and fixed carbon content of biochar from pruned apple tree branches. *Renew Energy* 146: 1691–1699.

Yang Y, Chun Y, Sheng G, Huang M. 2004. pH-dependence of pesticide adsorption by wheat-residue-derived black carbon. *Langmuir* 20: 6736–6741.

Yang Y, Sheng G. 2003. Enhanced pesticide sorption by soils containing particulate matter from crop residue burns. *Environ Sci Technol*. 37: 3635–3639.

Yang Y, Sheng G, Huang M. 2006. Bioavailability of diuron in soil containing wheat-straw-derived char. *Sci Total Environ*. 354: 170–178.

Yang Y, Sun K, Han L, et al. 2018. Effect of minerals on the stability of biochar. *Chemosphere* 204: 310–317.

Yavari S, Malakahmad A, Sapari NB. 2015. Biochar efficiency in pesticides sorption as a function of production variables—a review. *Environ Sci Pollut Res*. 22: 13824–13841.

Yavari S, Sapari NB, Malakahmad A, Yavari S. 2019. Degradation of imazapic and imazapyr herbicides in the presence of optimized oil palm empty fruit bunch and rice husk biochars in soil. *J Hazard Mater*. 366: 636–642.

Yuan J-H, Xu R-K, Zhang H. 2011. The forms of alkalis in the biochar produced from crop residues at different temperatures. *Bioresour Technol*. 102: 3488–3497.

Zhang P, Wu J, Li L, et al. 2012. Sorption and catalytic hydrolysis of carbaryl on pig-manure-derived biochars. *J Agro-Environ Sci*. 31: 416–421.

Zhelezova A, Cederlund H, Stenström J. 2017. Effect of biochar amendment and ageing on adsorption and degradation of two herbicides. *Water Air Soil Pollut*. 228: 216.

Zheng W, Guo M, Chow T, Bennett DN, Rajagopalan N. 2010. Sorption properties of greenwaste biochar for two triazine pesticides. *J Hazard Mater*. 181: 121–126.

Zielińska A, Oleszczuk P. 2015. The conversion of sewage sludge into biochar reduces polycyclic aromatic hydrocarbon content and ecotoxicity but increases trace metal content. *Biomass Bioenergy* 75: 235–244.

9 Removal of Herbicides from Water Using Biochar

Widad El Bouaidi[1], Ghizlane Enaime[2], Mohammed Loudiki[3], Abdelrani Yaacoubi[4], Mountasser Douma[5], and Manfred Lübken[6]*

[1]Laboratory of Water, Biodiversity and Climate Change; Phycology, Biotechnology and Environmental Toxicology research unit, Faculty of Sciences Semlalia, Cadi Ayyad University, Department of Biology, Marrakesh, Morocco

[2]Institute of Urban Water Management and Environmental Engineering, Ruhr-Universität Bochum, Bochum, Germany.

[3]Laboratory of Water, Biodiversity and Climate Change; Phycology, Biotechnology and Environmental Toxicology Research Unit, Faculty of Sciences Semlalia, Cadi Ayyad University, Department of Biology, Marrakesh, Morocco.

[4]Laboratory of Applied Organic Chemistry, Faculty of Sciences Semlalia, Cadi Ayyad University, Department of Chemistry, Marrakesh, Morocco.

[5]Polydisciplinary Faculty of Khouribga, Sultan Moulay Slimane University, Khouribga, Morocco.

[6]Institute of Urban Water Management and Environmental Engineering, Ruhr-Universität Bochum, Bochum, Germany.

* Corresponding author: manfred.luebken@rub.de

CONTENTS

DOI: 10.1201/9781003202073-9

9.1 INTRODUCTION

Recently, the global provision of drinking water has become increasingly important due to the increase in the world's population. This increase associated with the increase in anthropogenic activities leads to a rapid contamination of water sources. Several organic, inorganic, and microbial contaminants are discharged into water sources in which the degree of contamination is associated with the type and origin of discharged pollutants, as well as the sensitivity of water sources to degradation (Lenart-Boroń et al. 2017; Palansooriya et al. 2020). Exposure to these toxic substances can cause many environmental and health problems.

Herbicides, among many other organic contaminants, are widely released into the environment. The application of herbicides has been intensified to achieve higher agricultural yields and respond to the increased food demand (Li et al. 2014). Herbicides are mainly used to control broadleaf weeds and grasses in plantation crops and orchards, and are also used for general weed control on non-crop lands (Barchanska et al. 2017; Bromilow 2004; de Albuquerque et al. 2020; Kniss 2017; Vonberg et al. 2014). Most herbicides are introduced into the field or environment as a dilute solution. Nevertheless, only 10–30% of these herbicides may be absorbed by target plants or soil particles. In addition, the amount of herbicide residues is sometimes too high to be completely degraded by natural means such as metabolic, microbial, or photodegradation in sunlight (de Souza et al. 2020; John and Shaike 2015). These contaminants migrate downstream from agricultural fields and are more detected outside the application areas, which could consequently contaminate surface and ground waters (Cerejeira et al. 2003; Clark and Goolsby 2000; Götz et al. 1998; Lacorte et al. 2001; Papadakis et al. 2015). In water bodies, herbicides are not present as individual substances, but in mixtures (Zheng et al. 2016). The concentration of a particular herbicide may be below the so-called no-observed-effect concentration (NOEC), but the combined toxicity with other herbicides may still be present (Wang et al. 2014). This issue can be evaluated with the "concentration addition" model (Backhaus et al. 2004; Drescher and Boedeker 1995). In fact, there are over 300 active ingredients in herbicides and 1500 products on the market (Heap 2014), which explains the diversity of herbicides currently produced.

As a result, these herbicides constitute a major concern in water sources and therefore in water treatment plants. For this reason, several conventional methods, including precipitation, coagulation–flocculation, electrochemical decomposition, biological remediation, membrane technologies, sand filtration, adsorption and hybrid technologies, have been proposed to remove herbicides from contaminated water/wastewater (Abilarasu et al. 2021; Cosgrove et al. 2019; Ghanbarlou et al. 2020; Leovac et al. 2015; Morin-Crini et al. 2019; Vacca et al. 2020; Vieira et al. 2020). Even though some processes such as membranes and advanced oxidation processes can eliminate herbicides from contaminated water, the use of these techniques is not feasible at all treatment plants due to their high operating costs, energy consumption, and technical complexity (Hnatukova et al. 2011; Vanraes et al. 2018). Adsorption as a simple, easy-to-use, and versatile method could be considered as a preferred alternative for removing herbicides from aquatic sources (Foo and Hameed 2010; Huggins et al. 2016; Morin-Crini et al. 2019). Commercially activated carbon is one of the most used and studied adsorbent in water and wastewater decontamination due to its numerous advantages, including its large and controllable porous surface, its thermostability, and its low acid/base reactivity (Chingombe et al. 2005). Nevertheless, the use of activated carbon in wastewater remediation is mainly limited by its high cost, which makes the research of other alternatives a necessity. Biochar has recently attracted considerable attention as a novel material that can be produced from available and low-cost biomass with a low carbon footprint and can exhibit higher efficiency in different environmental applications (Hu et al. 2020; Shaheen et al. 2019; Enaime et al. 2020). Some investigations studied the benefits of biochar application in removing herbicides from contaminated wastewater. However, due to limited surface area, porosity, and surface functional groups of biochar, which make its efficiency in removing contaminants from wastewater lower as that of activated carbon, a further modification is then required. Many laboratory experiments have been performed for the modification and engineering of biochars with enhanced properties and improved adsorption capacities toward a wide range of contaminants including herbicides (Panahi et al. 2020; Yu et al. 2020).

According to the available literature data, the application of biochar in water treatment is not yet fully investigated. According to Tan et al. (2015), the majority of studies focus on a few groups of substances, such as the elimination of heavy metals (46%), organic pollutants (39%), nanoparticles (13%), and other pollutants (2%). This chapter provides comprehensive information about the application of biochar in the remediation of herbicides. Mechanisms and main factors affecting the adsorption of herbicides on biochars are briefly introduced. Besides providing information on biochar properties and application, this chapter also focuses on the limitations of biochar applications to remove herbicides.

9.2 BIOCHAR CHARACTERISTICS

Biochar is a solid material rich in carbon. It is produced by pyrolysis of carbonaceous biomass under low-oxygen conditions. The temperatures applied in its production range from 300 to 900°C (Cha et al. 2016; Hadi et al. 2015; Initiative 2012; McNamara et al. 2016; Meyer et al. 2011). Biochar could be thermochemically produced from any form of carbon-rich organic matter, including crop and forest residues, wood chips, algae, sewage sludge, manures, straw, and organic municipal solid waste (Colantoni et al. 2016; Qiu et al. 2009; Xiong et al. 2019; Zhang et al. 2013a).

Pyrolysis causes a number of changes in biomass structure at physical, chemical, and molecular levels. In certain cases, volatilization in pyrolysis results in mass loss and thus volume reduction and shrinkage without significant structural changes to the primary feedstock (Laine et al. 1991). Generally, when pyrolysis occurs, changes in surface area; porosity and pore volume; cation exchange capacity; C/N, O/C, and H/C ratios are observed following dehydration, dehydrogenation, and decarboxylation reactions.

Biochar surface area, porosity, and pore size are strongly influenced by the pyrolysis temperature (Gai et al. 2014). At high pyrolysis temperatures, more volatile substances are released, which induces the formation of pores (Brewer et al. 2014).

Moreover, a broadening of the microporous structures following the breakdown of adjacent wall pores occurs at high temperatures, which is followed by an increase in porosity volume and overall surface area (Ghodake et al. 2021). The pore size of biochar also varies depending on the temperature of pyrolysis (Li et al. 2016; Liang et al. 2021). Biochar produced at low temperature showed partially blocked pores with volatile organic compounds, which decreased its adsorption capacity toward contaminants (Figueiredo et al. 2018; Mohan et al. 2014). The value of the surface area and pores size vary also depending on the pyrolysis method. Biochar produced by fast pyrolysis was reported exhibiting a much higher surface area and well-developed porosity than biochar produced by fast pyrolysis (Tomczyk et al. 2020).

Although raw biomass is slightly acidic, pyrolysis at high temperatures tends to produce biochars with neutral-to-alkaline properties (Mendez et al. 2012; Sarfaraz et al. 2020). The partial detachment of the functional groups leads to the formation of unpaired negative charges such as carboxyl (COO−) and hydroxyl groups (OH−) that have the capability to attract positive charges (Singh et al. 2010). The increase in temperature during pyrolysis of corn straw, sugarcane straw, pine, poultry litter, and sewage sludge resulted in an increased pH of the produced biochars (Jatav et al. 2021; Melo et al. 2013; Mukherjee et al. 2011).

Changes in H/C and O/C ratios are used as indicators of biochar properties and of the degree of carbonization and dominant reactions occurring during pyrolysis (Harvey et al. 2012; Lehmann and Joseph 2015; Schimmelpfennig and Glaser 2012). H/C and O/C ratios decrease during pyrolysis with increasing pyrolysis temperature, which reflects the loss of volatile matter due to the intensification of dehydration and decarboxylation reactions (Manyà et al. 2014; Pereira et al. 2011; Schimmelpfennig and Glaser 2012). Lower H/C ratio indicates higher aromaticity and higher stability of biochar (Liu et al. 2015; Spokas 2010), while lower O/C ratio is an index of the structural arrangement of the aromatic rings (Manyà et al. 2014; Spokas 2010; Wu et al. 2012). Changes in oxygen content play a great role in shaping the surface chemistry behavior of biochar because of its close relationship with number and composition of substituted functional groups (Lehmann and Joseph 2015; Pereira et al. 2011; Spokas 2010). During pyrolysis, an increase in some functional groups on biochar surface such as aromatic C=C and a decrease in other groups like O-H and CH_3 have been reported (Baldock and Smernik 2002; Chan and Xu 2009; Graetz and Skjemstad 2003; Lua et al. 2004; Zabaniotou et al. 2008).

9.3 ENGINEERING OF BIOCHARS

Biochars may exhibit finite capacity to remove pollutants from wastewater especially at higher concentrations of contaminants. Engineering of biochar has been reported to improve biochar surface area, porosity, and surface functional groups and then enhance the biochar affinity toward different contaminants (Shakoor et al. 2021, 2020; Xiao et al. 2020). Biochar modification increases also its selectivity toward pollutants and reduces the effect of other competing compounds present in wastewater (Shakoor et al. 2021). The properties and the performance of the modified biochar toward different contaminants may differ depending on the modification method and the used agent. For biochar to be effectively modified, the physicochemical properties of the biochar must be understood as well as how its changes affect the binding of specific target compounds. Hence, care should be taken when selecting an appropriate biochar modification method or agent to remove a particular contaminant from water sources. Biochar modification involves various physical and chemical activation methods.

During physical activation, biochar is subjected to gasification at higher temperatures in the presence of activating agents such as steam, carbon dioxide, oxygen, nitrogen, and ammonia (Enaime et al. 2020; Rangabhashiyam and Balasubramanian 2019). Steam as the most physical activating agent is reported for its ability to significantly enhance surface area, porous structure, and surface properties of the biochar. During steam activation, an exchange of oxygen from water to carbon surface occurs, which leads to the oxidation of carbon and to the release of hydrogen and carbon dioxide (Ippolito et al. 2012). Steam activation results in enhancing biochar hydrophilicity, incorporating oxygen-functional groups such as ether, carbonyl, carboxylic, and phenolic hydroxyl groups on biochar surface, creating new micropores and widening of the existing pores (Zhang et al. 2014). Similar to steam activation, biochar activation using gases such as carbon dioxide, oxygen, nitrogen, ammonia, or their mixtures has been also reported to improve biochar surface properties and thus enhance its adsorption capacity (Enaime et al. 2020). According to the gas used, changes in the biochar properties could be observed. For instance, carbon dioxide modification can promote the microporous structure of biochar, while the breakdown of ammonia gas results in the production of radicals like NH_2 and NH, which react with biochar active sites and introduces nitrogen-containing groups in the biochar carbon matrix (Xiong et al. 2013). Results of Jung et al. (2013) showed that oxygen-activated biochar was more carbonized than nitrogen-activated biochar resulting in higher carbon content and lower H/C ratio for oxygen-activated biochar. The mixtures of gases during the activation process were also reported to enhance the porous and surface area and the surface functionalities of the biochar (Rangabhashiyam and Balasubramanian 2019). The acid or alkali activation of biochar is widely used to change and improve the surface characteristics of biochar and enhance its adsorption capacity. As compared to physical activation, chemical activation using acid or alkali agents could be performed at lower temperatures and in shorter time (Qian et al. 2015). According to literature data, chemical activation shows higher efficiency in improving biochar microporosity, reducing mineral fraction, and increasing surface functional groups like C-OH and C-H functional groups, which play a great role during the adsorption process (Hu et al. 2018). Surface area also increases significantly following activation with acid and alkali agent. According to some investigations, the specific surface area increases by 10 to 14 times after chemical activation of biochar (Cheng et al. 2021). Other modification methods such as clay mineral modification and ball milling modifications have been also reported to increase surface functional groups and pore structure of biochar. Ball milling methods consist of grinding biochars to nanometer size, which improves their surface properties and enhances their chemical adsorption capacity (Qin et al. 2020; Wang et al. 2017). Although this technology is reported to be a green modification method, few experiments have been carried out to confirm its efficiency in improving biochar performance in removing different contaminants from water sources (Cheng et al. 2021).

9.4 ENVIRONMENTAL APPLICATION OF BIOCHAR IN HERBICIDES-CONTAMINATED WATER

As a result of the rapid increase in the agricultural activities, agricultural contamination is becoming an increasingly serious problem due to the extensive use of chemicals, which find their way into water sources (Wei et al. 2018). As an adsorbent, biochar has been proven to be efficient to remove target pollutants, including herbicides (Zhelezova et al. 2017). It has been reported that the main source of herbicide pollution could be controlled by adding biochar to soil. Mixing biochar and soil impacts both the mobility and activity of herbicides in soils by providing sorption and binding sites with which the herbicides' molecules may interact. This will reduce the risk of herbicide runoff and leaching into surface and groundwater, respectively (Das et al. 2017). Diao et al. (2020) studied a novel zero valent iron-supported biochar used as a heterogeneous peroxymonosulfate activating catalyst for the removal of the herbicide atrazine from soil. The results revealed that iron-supported biochar was very effective in removing atrazine within 240 min with a corresponding removal rate of nearly 96% compared to another biochar coupled with peroxymonosulfate. The same study has shown that iron-supported biochar was capable to eliminate other herbicides, such as terbuthylazine and ametryn, with a removal rate of 100% and 99%, respectively. Immobilization of herbicides in soil by biochar application will indirectly protect water sources from contamination with herbicides (Zheng et al. 2019). The majority of research studies performed so far on the removal of herbicides by biochar are related to soil remediation, while still there is very little knowledge on the subject of biochar application for the removal of herbicides from contaminated waters and wastewater under actual conditions. Only few studies have been found in the literature where researchers have applied biochars and modified-biochars to remove herbicides from contaminated water. Woody biochar was examined for the elimination of glyphosate from aqueous system and showed an adsorption capacity of about 44 mg/g (Mayakaduwa et al. 2016). In the study of Essandoh et al. (2017a), switchgrass biochar (SGB) and magnetic switchgrass biochar (MSGB) were employed to remove 2,4-dichlorophenoxyacetic acid (2,4-D) and 2-methyl-4-chlorophenoxyacetic acid (MCPA) herbicides from aqueous solutions. The results confirmed the efficiencies of SGB and MSGB in removing both herbicides with maximum adsorption capacities of 134 and 50 mg/g, respectively at pH 2. Other studies demonstrated the efficiency of biochar-supported iron composites coupled with a persulfate or peroxymonosulfate process to remove atrazine herbicide from aqueous solutions (Jiang et al. 2020; Liu et al. 2018). Similarly, Tao et al. (2019) have demonstrated the effectiveness of iron-modified biochar combined with *Acinetobacter lwoffii* DNS32 as adsorbent in removing atrazine. The effectiveness of chitosan-modified biochars compared to the raw biochar in removing both imazapic and imazapyr has been demonstrated by Yavari et al. (2020).

Table 9.1 summarizes some recent studies on the efficiency of biochar and engineered biochar in removing herbicides in water sources. According to Table 9.1, various biochar samples achieve maximum removal efficiency at various equilibrium contaminant concentrations. Therefore, in order to determine the amount of biochar for drinking water treatment, the identification of the type of contaminants and their concentrations becomes crucial.

TABLE 9.1
Overview of Recent Studies Demonstrating Biochar's Ability to Remove Herbicides from Water Sources.

Biochar feedstock	Preparation conditions	Herbicide	Operating conditions	Adsorption capacity	References
Biochar supported zero valent iron (BC-nZVI)	n.a.*	Atrazine	Initial [Atrazine] =23.6 mg/kg, Initial [BC-nZVI] =2.0 g/L, initial soil pH =6.93, and reaction time of 240 min.	96%	Diao et al. (2021)
Ground coffee residue biochars (GCRB)	Pyrolysis at 800 °C for 2 h with a linear rise of 10°C/min	Alachlor Diuron Simazine	Concentration of each herbicide = 10, 20, 30, 40, and 50 μmol/L, working volume = 20 mL, absorbents dosage = 50 mg/L, agitation speed = 150 rpm, contact time = 24 h, temperature = 25 °C, and pH = 7.0.	– Qe,exp** of Alachlor = 11.74 μmol/g. – Qe,exp of Diuron = 9.95 μmol/g. – Qe,exp of Simazine = 6.53 μmol/g	Lee et al. (2021)
Ground coffee residue biochars with NaOH activation (GCRB-N)				– Qe,exp of Alachlor = 122.71 μmol/g. – Qe,exp of Diuron = 166.42 μmol/g. – Qe,exp of Simazine = 99.16 μmol/g).	
Functional biochar (FB)	Hydrothermal carbonization	Sulfonylurea chemical group	Adsorbent dosage = 10 mg of functional biochar (16.7%) Volume of iodosulfuron-methyl-solution (5 mL) at various concentrations (ranging from 0.05 to 6 mg/L), and then the mixture was shaken by thermostatic oscillator (24 h), and the results of adsorption isotherms were measured under a range of temperature conditions (24.85, 34.85, and 44.85 °C).	Over 95%	Zhang et al. (2021)

(Continued)

TABLE 9.1 (Continued)

Biochar feedstock	Preparation conditions	Herbicide	Operating conditions	Adsorption capacity	References
Rice straw biochar (RSB)	Pyrolysis under N_2 gas for 2 h at 500 °C at a rate of approximately 5 °C/min	Monuron, Diuron, and Linuron	Sorption conditions: (pH value at 3, ionic strength at 0.1 M, and RSB dosage at 60 mg) Herbicide's concentration = 25 mg/L	Monuron = 41.9% Diuron = 25% Linuron = 56.8%	Dan et al. (2021)
Hydrochar from *Prunus serrulate*	Hydrothermal carbonization for 6 h at 200°C	Atrazine	– The influence of pH on the adsorption process was analyzed for the pH values 3, 4, 6, 8, and 10, where 1 g/L of adsorbent was stirred with 50 mL of atrazine (50 mg/L) for 180 min at 24.85 °C). – The adsorbent dosage was conducted as follows: 50 mL of atrazine (50 mg/ L) solution at pH 3 (value determined from the pH test) was added into the five Erlenmeyer flasks containing 0.4, 0.6, 0.8, 1.0, and 1.2 g/L of hydrochar and stirred for 180 min at 24.85 °C. – The adsorption kinetics were 0.8 g/L of hydrochar and initial atrazine concentration of 25, 30, and 50 mg/L (initial pH 3).	Above 70% (at pH = 3, dosage of 0.8 g/L	Netto et al. (2021)

Grape biochar with H_2O_2 activation	Pyrolysis at 500°C and 900°C for 2 h under constant N_2	Clomazone and cyhalofop	– Adsorbent dosage: 40 mg Herbicide solution 8 mL	Clomazone removal efficiency was greater than cyhalofop removal efficiency on both the G-350 (65.0% versus 6.3%) and G-350 H_2O_2 (70.3% versus 35.4%) biochars	Gámiz et al. (2019)
Wood biochar	n.a.	Glyphosate	Initial glyphosate concentration = 0.7 mg/L Adsorbent dose = 12.3 g/L Operating pH = 5 Operating temperature = 50°C Contact time = 1.9 h	Max. removal capacity = 0.0569 mg/g Max. removal efficiency = 100%	Dissanayake Herath et al. (2019)
Biochar from corn straw (CSWP)	Pyrolysis	Triazine chemical group	– Adsorbent dosage = 5–120 mg – Triazine solution = 10 mL, 2 mg/L pH: 1–11	Max. adsorption capacity = up to 79.6 mg/g at 25°C	Suo et al. (2019)
Normal (RSBC) and phosphoric-acid-treated (T-RSBC) rice straw biochars	Pyrolysis at 600°C, under N_2 using heating rate of 3°C per min and 1 h residence time	Atrazine	Biochar (10 mg, oven dry basis) and 10 mL aqueous solution atrazine in 30 mL oak-ridge Concentration of herbicide ranged between 1 and 10 mg/L	– 95% of atrazine removal (10 mg/L)	Mandal et al. (2017)

n.a.*: not available, Qe,exp**: equilibrium adsorption capacities

9.5 MECHANISMS OF HERBICIDES REMEDIATION THROUGH BIOCHAR APPLICATION

The interaction of biochar with different types of herbicides is associated with different kind of mechanisms, which are correlated with the properties of biochar and the characteristics of the herbicide molecule. The most involved mechanisms that are reported to be predominating are pore-filling, electrostatic interaction, π–π interactions, hydrogen bonding, and hydrophobic interactions (Wei et al. 2018). Pore-filling mechanism is widely reported in the literature to control the adsorption of herbicides on biochar (Sun et al. 2016; Tan et al. 2015). In the study of Zhao et al. (2013), corn straw biomass pretreated by ammonium di-hydrogen phosphate (ADP) and subsequently carbonized at 450°C exhibited better sorption capacity for atrazine than corn straw biomass carbonized at 450°C. Authors concluded that biochar adsorption capacity is closely related to its surface area and micro-pore volume; the ADP pretreatment of corn straw biomass improved the surface area and micro-pore volume of biochar, which enhanced its adsorption capacity toward atrazine. Similar results were also reported by Zhang et al. (2013b) for the adsorption of atrazine and other herbicides like diuron and pentachlorophenol (Al Bahri et al. 2012; Devi and Saroha 2014). Wei et al. (2020) reported that biochar adsorption capacity toward metolachlor herbicide could be controlled by different mechanisms including pore-filling and polar functional groups' interactions, depending on the pyrolysis temperature. When biochar is produced at low pyrolysis temperature, the organic matter content is rising and functional groups are increasing, especially quinonyl groups, which are reported to be responsible for the higher adsorption of biochar. For biochar produced at high pyrolysis temperature, surface area and total pore volume have been shown to control the adsorption process. Interestingly, authors proposed H/C ratio as an indicator to predict the adsorption mechanisms of herbicide on biochar. When the value of H/C ratio is less than 0.5, the adsorption is controlled by pore-filling, while a value of H/C ratio more than 0.5 indicates that the herbicide removal is governed by surface chemical bond adsorption (Wei et al. 2020). In another study, Binh and Nguyen (2020) showed that the great proficiency of corn cob biochar in removing 2,4-D is due to pore-filling and other mechanisms such as π–π EDA interactions (electron–donor–acceptor) and H-bonding, which are controlled by pH. In the study of Zhang et al. (2013b), the adsorption of atrazine on biochar derived from pig manure and characterized by its high ash content has been tested. The results showed that pore-filling and the combination of herbicides and ash by specific interactions contribute to the adsorption process. However, the increased concentration of ash in biochar reduces the adsorption sites, which decreases the biochar adsorption capacity.

9.6 PARAMETERS INFLUENCING BIOCHAR ADSORPTION CAPACITY

Biochar performance in removing contaminants from water sources is closely dependent on many factors that could be related to biochar adsorption capacity, contaminant characteristics, and nature of wastewater to be treated.

Biochar adsorption capacity is significantly dependent on its physicochemical properties, which varies considerably according to the pyrolysis temperature and the nature of feedstocks (Tang et al. 2015; Tripathi et al. 2016). The selective use of feedstocks and operating conditions during pyrolysis allow designing specific biochars for specific purposes (Ahmad et al. 2014; Mohan et al. 2014). For instance, plant biomass possesses a high content of carbon and oxygen elements that can generate various functional groups (-COOH, -C-O-R, -C-OH) on the surface of the biochar; these groups are capable of serving as active sites to improve contaminants' removal efficiency (Takaya et al. 2016). It has been proven that biochars derived from wood biomass usually possess a larger surface area than grass biochars (Kloss et al. 2012; Mukherjee et al. 2011) and that the pH of biochar derived from the pyrolysis of animal manure is higher than that of plant biomass (Shinogi and Kanri 2003).

Mandal et al. (2016) examined the adsorption capacity of biochars derived from different agricultural wastes (corn cob, bamboo chips, rice straw, eucalyptus bark, and rice husk treated by phosphoric acid) for the elimination of atrazine herbicide. The results showed that among the five normal biochars, the rice-straw-derived biochar showed the maximum atrazine removal efficiency and that phosphoric acid treatment of rice-straw-derived biochar enhanced its adsorption capacity from 37.5–70.7% to about 59.5–89.8%. The adsorption capacity of studied biochar toward atrazine was affected by biochar aromaticity, polarity, pore volume, pore diameter, pH, and surface acidity, which strongly depend on the nature of the starting feedstock (Mandal et al. 2017).

An enhancement in biochar surface area, porosity, and carbon content is also reported due to the increase in pyrolysis temperature (Liu et al. 2015; Varjani et al. 2019). In a study performed by Wei et al. (2020), authors studied the variation in the affinity of metolachlor to biochar produced from traditional Chinese medicinal herb residue at three different temperatures (300, 500, and 750°C). The adsorption capacity was exhibited to be the strongest for biochar produced at 750°C, because of both higher pore volume and surface area compared to the two other biochars produced at lower temperatures. This is in line with many other studies (Mayakaduwa et al. 2016; Wang et al. 2010; Yang et al. 2010).Authors observed, however, that the adsorption capacity of biochar produced at 300°C was higher than that produced at 500°C; and even its surface area was lower. The authors attributed that to the higher organic matter content and to more functional groups especially quinonyl groups detected on the biochar produced at the lowest temperature (Wei et al. 2020). Pyrolysis temperature affects also the biochar surface charge and then its interaction with ionized contaminants. Yuan et al. (2011) studied the properties of biochars produced from crop residues at different temperatures (300, 500, and 700°C) and showed that biochars produced at low temperature are more negatively charged than those produced at high temperatures and that the negative surface charge of biochars decreased at lower-solution pH values. The solution pH in addition to the biochar surface charge affects significantly the speciation of chemicals and the biochar surface charge and then the behavior of biochar toward contaminants (Premarathna et al. 2019). If the solution has a very low pH, this favors the formation of herbicide cations, which increases the interaction with the negatively charged surface of the biochar by electrostatic attraction. This could significantly improve the sorption affinity of the biochar for herbicides (Oliveira Jr et al. 2001). In a study performed by Zheng et al. (2010), the adsorption of atrazine and simazine on biochar produced from green waste at 450°C was studied. Authors showed that the adsorption was more favored at low solution pH and that biochar adsorption capacity toward the two pesticides increased when the solid/solution ratio was decreasing. In another study, Essandoh et al. (2017b) showed that the adsorption of metribuzin on switchgrass biochar and magnetic switchgrass biochar from aqueous solutions was much higher at solution pH value of 2 as compared to higher pH values. Authors attributed that to the decrease in the concentration of protonated metribuzin as the pH in the solution increases.

In the study of Sun et al. (2011), the adsorption of two herbicides with two different functionalities was examined: norflurazon and fluridone that can be ionized at low pH. The results showed that the adsorption of norflurazon is unaffected by pH variation, and this is due to the fact that norflurazon is a polar but nonionic pyridazinone that does not undergo changes in protonation state, while the adsorption of fluridone, which is a ionized molecule, increased with the decrease in the pH value. It is assumed that the decrease in the solution pH induces a progressive protonation of negative ionic functions, mainly carboxyl groups and their conversion into neutral adsorption sites that are likely to bind with fluridone molecules, e.g., via H-bonding and/or non-specific London forces (Sun et al. 2011).

9.7 LIMITATIONS OF BIOCHAR APPLICATION

Application of biochar for environmental remediation has expanded nowadays due to its low cost, high adsorption efficiency, low energy consumption, and ease of use. Nevertheless, the subsequent use of biochar in environmental management and the possible negative effects must be considered before extending its use in drinking water treatment plants. Further studies demonstrated that biochar application to soil remediation to prevent raw runoff of contaminants to water sources has led to increase dissolved organic carbon (DOC) leaching. Consequently, dissolved organic carbon affects aquatic ecosystems through various mechanisms, including metal toxicity, increase in turbidity, temperature stratification, and decrease in light penetration to lakes (Evans et al. 2005; Lipczynska-Kochany 2018). Additionally, dissolved biochar molecules can produce toxic reactive oxygen species under sunlight in aquatic ecosystems, which negatively impacts the ecosystem community and subsequently public health (Fu et al. 2016; Wagner et al. 2018). Furthermore, carbon from biochar can be lost to CO_2 outside the water treatment field as well. DOC can be released by fresh and aged biochar (Liu et al. 2019; Mia et al. 2017), which possibly enters freshwater systems through leaching and transportation (Liu et al. 2019). Hence, the small aliphatic DOCs are degradable and can reduce the oxygen balance in water bodies (Wagner et al. 2018).

There are researchers who have studied the levels of extractable toxic elements in biochar and have provided some recommendations for minimizing the risk of potential toxic elements. It has been found that the yield and composition of toxic elements associated with biochars are greatly related to feedstock material and pyrolytic temperature process. For instance, grass biochar was found to contain substantially higher levels of several hazardous polycyclic aromatic hydrocarbons than wood biochars; high-temperature biochars appear to be safer than low-temperature biochars

(Hale et al. 2012). Further factors related to pyrolysis technology and pyrolysis residence time may impact the level of toxins in the biochars (Hale et al. 2012; Keiluweit et al. 2012).

In this context, the application of biochar in water treatment cannot be readily implemented. More and more specific and persuasive research is still needed to reduce the potential and expected negative impacts of biochar application on the environment.

Moreover, most of the studies performed so far to evaluate the effectiveness of biochar in removing herbicides are carried out in laboratory-scale, without considering the feasibility of their large-scale application in water treatment plants. As a result, substantial limitations could be encountered including, among others, the difficulty to accurately simulate actual processing scenarios, which may lead to lower performances as those found in laboratory scale (Kearns et al. 2020). Unfortunately, studies that used biochar column for effluents' treatment have generally not employed mass transfer modeling approaches for scaling down mass transfer parameters for full-size systems. The availability of biochar with a desirable quality could also limit its application in large scale.

Moreover, biochar production cost is a critical factor that should be taken into consideration for the application of biochar. In instances where biochar is the primary product, it needs to cover operating expenses, including production, maintenance, feedstock, transportation, labor, distribution costs, and others to ensure long-term commercial viability. In other instances, biochar could be an auxiliary by-product of processes designed to create efficiencies in established agricultural or land management operations as well as the treatment of water sources (Yoder et al. 2011).

Based on the results and discussion here, several factors can be identified that need to be considered for the real use of biochar taking into consideration the limitations of its application to water purification from herbicides. For these reasons, both local and international biochar policies should be directed to enhance the production and use of biochar on a commercial scale. Most importantly, expanded government investment in the research and development of biochar-based water treatment systems is needed. Besides, researchers ought to introduce new biochar policies to promote large-scale and standard biochar production to remove herbicides in drinking water treatment plants.

9.8 CONCLUDING REMARKS

Biochars derived from biowastes is gaining increasing attention as a promising and sustainable alternative to the costly conventional methods used to remove pollutants from contaminated water sources. Depending on biochar origin, pyrolysis-operating conditions, and modification methods, several types of biochars could be produced. Therefore, the biochar efficiency in removing pollutants is strongly dependent on their properties and the characteristics of the target contaminants. Thus, care must be taken when selecting an appropriate biochar to remove a particular contaminant from water sources.

The present chapter gives a systematical overview of the broad application of biochar in water and wastewater treatment to remove herbicides. The research discussed herein is not an exhaustive analysis of the application of biochar in the remediation of herbicides from water sources. Biochar is produced from a wide range of biomasses and under different pyrolysis conditions, and more research studies are still needed to develop real models that will enable predicting how any specific biochar will behave in the presence of any specific contaminant. The mechanisms controlling the adsorption of herbicides on biochars are not yet fully understood. Moreover, the engineering of biochars for its specific application to target herbicides is a new and interesting area of research that should be further investigated in the aim of enhancing its performance. Scale-up experiments and economic analysis of the potential use of biochar for the remediation of herbicides in contaminated waters are also important to address aspects of application rate and frequency that may be necessary to achieve specific end goals.

REFERENCES

Abilarasu, A., P.S. Kumar, D.-V.N. Vo, D. Krithika, et al. 2021. Enhanced photocatalytic degradation of diclofenac by Sn0. 15Mn0. 85Fe2O4 catalyst under solar light. *J. Environ. Chem. Eng* 9: 104875.

Ahmad, M., A.U. Rajapaksha, J.E. Lim, et al. 2014. Biochar as a sorbent for contaminant management in soil and water: A review. *Chemosphere* 99: 19–33.

Al Bahri, M., L. Calvo, M.A. Gilarranz and J.J. Rodríguez. 2012. Activated carbon from grape seeds upon chemical activation with phosphoric acid: Application to the adsorption of diuron from *water. Chem. Eng. J.* 203: 348–356.

Backhaus, T., Å. Arrhenius and H. Blanck. 2004. Toxicity of a mixture of dissimilarly acting substances to natural algal communities: Predictive power and limitations of independent action and concentration addition. *Environ. Sci. Technol.* 38: 6363–6370.

Baldock, J.A. and R.J. Smernik. 2002. Chemical composition and bioavailability of thermally altered *Pinus resinosa* (Red pine) wood. *Org. Geochem.* 33: 1093–1109.

Barchanska, H., M. Sajdak, K. Szczypka, A. Swientek, M. Tworek and M. Kurek. 2017. Atrazine, triketone herbicides, and their degradation products in sediment, soil and surface water samples in Poland. *Environ. Sci. Pollut. Res.* 24: 644–658.

Binh, Q.A. and H.-H. Nguyen. 2020. Investigation the isotherm and kinetics of adsorption mechanism of herbicide 2, 4-dichlorophenoxyacetic acid (2, 4-D) on corn cob biochar. *Bioresour. Technol. Reports* 11: 100520.

Brewer, C.E., V.J. Chuang, C.A. Masiello, et al. 2014. New approaches to measuring biochar density and porosity. *Biomass and bioenergy* 66: 176–185.

Bromilow, R.H. 2004. Paraquat and sustainable agriculture. *Pest Manag. Sci. Former. Pestic. Sci.* 60: 340–349.

Cha, J.S., S.H. Park, S.-C. Jung, et al. 2016. Production and utilization of biochar: A review. J. *Ind. Eng. Chem.* 40: 1–15.

Cerejeira, M.J., P. Viana, S. Batista, et al. 2003. Pesticides in Portuguese surface and ground waters. *Water Res.* 37: 1055–1063.

Chan, K.Y. and Z. Xu. 2009. *Biochar for environmental management: Science and technology: Biochar: Nutrient properties and their enhancement.* London, UK: Earthscan.

Cheng, N., B. Wang, P. Wu, et al. 2021. Adsorption of emerging contaminants from water and wastewater by modified biochar: A review. *Environ. Pollut.* 273: 116448.

Chingombe, P., B. Saha and R.J. Wakeman. 2005. Surface modification and characterisation of a coal-based activated carbon. *Carbon N. Y.* 43: 3132–3143.

Clark, G.M. and D.A. Goolsby. 2000. Occurrence and load of selected herbicides and metabolites in the lower Mississippi River. *Sci. Total Environ.* 248: 101–113.

Colantoni, A., N. Evic, R. Lord, et al. 2016. Characterization of biochars produced from pyrolysis of pelletized agricultural residues. *Renew. Sustain. Energy Rev.* 64: 187–194.
Cosgrove, S., B. Jefferson and P. Jarvis. 2019. Pesticide removal from drinking water sources by adsorption: A review. *Environ. Technol. Rev.* 8: 1–24.
Dan, Y., M. Ji, S. Tao, et al. 2021. Science of the Total Environment Impact of rice straw biochar addition on the sorption and leaching of phenylurea herbicides in saturated sand column. *Sci. Total Environ.* 769: 144536.
Das, S.K., G.K. Ghosh and R.K. Avasthe. 2017. Biochar amendments on physico-chemical and biological properties of soils. *Agrica* 6: 79–87.
de Albuquerque, F.P., J.L. de Oliveira, V. Moschini-Carlos and L.F. Fraceto. 2020. An overview of the potential impacts of atrazine in aquatic environments: Perspectives for tailored solutions based on nanotechnology. *Sci. Total Environ.* 700: 134868.
de Souza, R.M., D. Seibert, H.B. Quesada, F. de Jesus Bassetti, M.R. Fagundes-Klen and R. Bergamasco. 2020. Occurrence, impacts and general aspects of pesticides in surface water: A review. *Process Saf. Environ. Prot.* 135: 22–37.
Devi, P. and A.K. Saroha. 2014. Synthesis of the magnetic biochar composites for use as an adsorbent for the removal of pentachlorophenol from the effluent. *Bioresour. Technol.* 169: 525–531.
Diao, Z.-H., W. Qian, Z.-W. Zhang, et al. 2020. Removals of Cr (VI) and Cd (II) by a novel nanoscale zero valent iron/peroxydisulfate process and its Fenton-like oxidation of pesticide atrazine: Coexisting effect, products and mechanism. *Chem. Eng. J.* 397: 125382.
Diao, Z.H., W.X. Zhang, J.Y. Liang, et al. 2021. Removal of herbicide atrazine by a novel biochar based iron composite coupling with peroxymonosulfate process from soil: Synergistic effect and mechanism. *Chem. Eng. J.* 409: 127684.
Dissanayake Herath, G.A., L.S. Poh and W.J. Ng. 2019. Statistical optimization of glyphosate adsorption by biochar and activated carbon with response surface methodology. *Chemosphere* 227: 533–540.
Drescher, K. and W. Boedeker. 1995. Assessment of the combined effects of substances: The relationship between concentration addition and independent action. *Biometrics* 716–730.
Enaime, G., A. Baçaoui, A. Yaacoubi and M. Lübken. 2020. Biochar for wastewater treatment – conversion technologies and applications. *Appl. Sci.* 10: 3492.
Essandoh, M., D. Wolgemuth, C.U. Pittman, D. Mohan and T. Mlsna. 2017a. Phenoxy herbicide removal from aqueous solutions using fast pyrolysis switchgrass biochar. *Chemosphere* 174: 49–57.
Essandoh, M., D. Wolgemuth, C.U. Pittman, D. Mohan and T. Mlsna. 2017b. Adsorption of metribuzin from aqueous solution using magnetic and nonmagnetic sustainable low-cost biochar adsorbents. *Environ. Sci. Pollut. Res.* 24: 4577–4590.
Evans, C.D., D.T. Monteith and D.M. Cooper. 2005. Long-term increases in surface water dissolved organic carbon: Observations, possible causes and environmental impacts. *Environ. Pollut.* 137: 55–71.
Figueiredo, C., H. Lopes, T. Coser, et al. 2018. Influence of pyrolysis temperature on chemical and physical properties of biochar from sewage sludge. *Arch. Agron. Soil Sci.* 64: 881–889.
Foo, K.Y. and B.H. Hameed. 2010. Detoxification of pesticide waste via activated carbon adsorption process. *J. Hazard. Mater.* 175: 1–11.
Fu, H., H. Liu, J. Mao, et al. 2016. Photochemistry of dissolved black carbon released from biochar: Reactive oxygen species generation and phototransformation. *Environ. Sci. Technol.* 50: 1218–1226.
Gai, X., H. Wang, J. Liu, et al. 2014. Effects of feedstock and pyrolysis temperature on biochar adsorption of ammonium and nitrate. PloS one, 9: 12.

Gámiz, B., K. Hall, K.A. Spokas and L. Cox. 2019. Understanding activation effects on low-temperature biochar for optimization of herbicide sorption. *Agronomy* 9: 588.

Ghanbarlou, H., B. Nasernejad, M.N. Fini, M.E. Simonsen and J. Muff. 2020. Synthesis of an iron-graphene based particle electrode for pesticide removal in three-dimensional heterogeneous electro-Fenton water treatment system. *Chem. Eng. J.* 395: 125025.

Ghodake, G.S., S.K. Shinde, A.A. Kadam, et al. 2021. Review on biomass feedstocks, pyrolysis mechanism and physicochemical properties of biochar: State-of-the-art framework to speed up vision of circular bioeconomy. *J. Clean. Prod.* 126645.

Götz, R., O.H. Bauer, P. Friesel and K. Roch. 1998. Organic trace compounds in the water of the River Elbe near Hamburg Part I. *Chemosphere* 36: 2085–2101.

Graetz, R.D. and J.O. Skjemstad. 2003. The charcoal sink of biomass burning on the Australian continent. *CSIRO Atmospheric Research*. 64: 61.

Hadi, P., M. Xu, C. Ning, C.S.K. Lin and G. McKay. 2015. A critical review on preparation, characterization and utilization of sludge-derived activated carbons for wastewater treatment. *Chem. Eng. J.* 260: 895–906.

Hale, S.E., J. Lehmann, D. Rutherford, et al. 2012. Quantifying the total and bioavailable polycyclic aromatic hydrocarbons and dioxins in biochars. *Environ. Sci. Technol.* 46: 2830–2838.

Harvey, O.R., L.-J. Kuo, A.R. Zimmerman, P. Louchouarn, J.E. Amonette and B.E. Herbert. 2012. An index-based approach to assessing recalcitrance and soil carbon sequestration potential of engineered black carbons (biochars). *Environ. Sci. Technol.* 46: 1415–1421.

Heap, I. 2014. Herbicide resistant weeds. *Integrated pest management*. Dordrecht: Springer, 281–301.

Hnatukova, P., I. Kopecka and M. Pivokonsky. 2011. Adsorption of cellular peptides of Microcystis aeruginosa and two herbicides onto activated carbon: Effect of surface charge and interactions. *Water Res.* 45: 3359–3368.

Hu, B., Y. Ai, J. Jin, et al. 2020. Efficient elimination of organic and inorganic pollutants by biochar and biochar-based materials. *Biochar* 2: 47–64.

Hu, H., L. Yang, Z. Lin, Y. Zhao, X. Jiang and L. Hou. 2018. A low-cost and environment friendly chitosan/aluminum hydroxide bead adsorbent for fluoride removal from aqueous solutions. *Iran. Polym. J.* 27: 253–261.

Huggins, T.M., A. Haeger, J.C. Biffinger and Z.J. Ren. 2016. Granular biochar compared with activated carbon for wastewater treatment and resource recovery. *Water Res.* 94: 225–232.

Initiative, I.B. 2012. Standardized product definition and product testing guidelines for biochar that is used in soil. *IBI Biochar Stand.* 48. Available online: https://www.biochar-international.org/wp-content/uploads/2018/04/IBI_Biochar_Standards_V1.1.pdf.

Ippolito, J.A., D.A. Laird and W.J. Busscher. 2012. Environmental benefits of biochar. *J. Environ. Qual.* 41: 967–972.

Jatav, H.S., V.D. Rajput, T. Minkina, et al. 2021. Sustainable approach and safe use of biochar and its possible consequences. *Sustain.* 13: 10362.

Jiang, C., J. Bo, X. Xiao, et al. 2020. Converting waste lignin into nano-biochar as a renewable substitute of carbon black for reinforcing styrene-butadiene rubber. *Waste Manag.* 102: 732–742.

John, E.M. and J.M. Shaike. 2015. Chlorpyrifos: Pollution and remediation. *Environ. Chem. Lett.* 13: 269–291.

Jung, C., J. Park, K.H. Lim, et al. 2013. Adsorption of selected endocrine disrupting compounds and pharmaceuticals on activated biochars. *J. Hazard. Mater.* 263: 702–710.

Kearns, J., E. Dickenson and D. Knappe. 2020. Enabling organic micropollutant removal from water by full-scale biochar and activated carbon adsorbers using predictions from bench-scale column data. *Environ. Eng. Sci.* 37: 459–471.

Keiluweit, M., M. Kleber, M.A. Sparrow, B.R.T. Simoneit and F.G. Prahl. 2012. Solvent-extractable polycyclic aromatic hydrocarbons in biochar: Influence of pyrolysis temperature and feedstock. *Environ. Sci. Technol.* 46: 9333–9341.

Kloss, S., F. Zehetner, A. Dellantonio, et al. 2012. Characterization of slow pyrolysis biochars: Effects of feedstocks and pyrolysis temperature on biochar properties. *J. Environ. Qual.* 41: 990–1000.

Kniss, A.R. 2017. Long-term trends in the intensity and relative toxicity of herbicide use. *Nat. Commun.* 8: 1–7.

Lacorte, S., P. Viana, M. Guillamon, R. Tauler, T. Vinhas and D. Barceló. 2001. Main findings and conclusions of the implementation of Directive 76/464/CEE concerning the monitoring of organic pollutants in surface waters (Portugal, April 1999 – May 2000). *J. Environ. Monit.* 3: 475–482.

Laine, J., S. Simoni and R. Calles. 1991. Preparation of activated carbon from coconut shell in a small scale cocurrent flow rotary kiln. *Chem. Eng. Commun.* 99: 15–23.

Lee, Y.G., J. Shin, J. Kwak et al. 2021. Effects of NaOH activation on adsorptive removal of herbicides by biochars prepared from ground coffee residues. *Energies* 14: 1297.

Lehmann, J. and S. Joseph. 2015. *Biochar for environmental management: Science, technology and implementation.* London: Routledge.

Lenart-Boroń, A., A. Wolanin, E. Jelonkiewicz and M. Żelazny. 2017. The effect of anthropogenic pressure shown by microbiological and chemical water quality indicators on the main rivers of Podhale, southern Poland. *Environ. Sci. Pollut. Res. Int.* 24: 12938.

Leovac, A., E. Vasyukova, I. Ivančev-Tumbas, et al. 2015. Sorption of atrazine, alachlor and trifluralin from water onto different geosorbents. *RSC Adv.* 5: 8122–8133.

Li, H., E.Y. Zeng and J. You. 2014. Mitigating pesticide pollution in China requires law enforcement, farmer training, and technological innovation. *Environ. Toxicol. Chem.* 33: 963–971.

Li, R., J.J. Wang, B. Zhou, et al. 2016. Enhancing phosphate adsorption by Mg/Al layered double hydroxide functionalized biochar with different Mg/Al ratios. *Sci. Total Environ.* 559: 121–129.

Liang, L., F. Xi, W. Tan, X. Meng, B. Hu and X. Wang. 2021. Review of organic and inorganic pollutants removal by biochar and biochar-based composites. *Biochar* 3: 255–281.

Lipczynska-Kochany, E. 2018. Effect of climate change on humic substances and associated impacts on the quality of surface water and groundwater: A review. *Sci. Total Environ.* 640: 1548–1565.

Liu, C.-H., W. Chu, H. Li, et al. 2019. Quantification and characterization of dissolved organic carbon from biochars. *Geoderma* 335: 161–169.

Liu, W.-J., H. Jiang and H.-Q. Yu. 2015. Development of biochar-based functional materials: Toward a sustainable platform carbon material. *Chem. Rev.* 115: 12251–12285.

Liu, Y., L. Lonappan, S.K. Brar and S. Yang. 2018. Impact of biochar amendment in agricultural soils on the sorption, desorption, and degradation of pesticides: A review. *Sci. Total Environ.* 645: 60–70.

Lua, A.C., T. Yang and J. Guo. 2004. Effects of pyrolysis conditions on the properties of activated carbons prepared from pistachio-nut shells. *J. Anal. Appl. Pyrolysis* 72: 279–287.

Mandal, A., N. Singh and T.J. Purakayastha. 2017. Characterization of pesticide sorption behaviour of slow pyrolysis biochars as low cost adsorbent for atrazine and imidacloprid removal. *Sci. Total Environ.* 577: 376–385.

Mandal, S., R. Thangarajan, N.S. Bolan, et al. 2016. Biochar-induced concomitant decrease in ammonia volatilization and increase in nitrogen use efficiency by wheat. *Chemosphere* 142: 120–127.

Manyà, J.J., M.A. Ortigosa, S. Laguarta and J.A. Manso. 2014. Experimental study on the effect of pyrolysis pressure, peak temperature, and particle size on the potential stability of vine shoots-derived biochar. *Fuel* 133: 163–172.

Mayakaduwa, S.S., P. Kumarathilaka, I. Herath, et al. 2016. Equilibrium and kinetic mechanisms of woody biochar on aqueous glyphosate removal. *Chemosphere* 144: 2516–2521.

McNamara, P.J., J.D. Koch, Z. Liu and D.H. Zitomer. 2016. Pyrolysis of dried wastewater biosolids can be energy positive. *Water Environ. Res*. 88: 804–810.
Melo, L.C.A., A.R. Coscione, C.A. Abreu, A.P. Puga and O.A. Camargo. 2013. Influence of pyrolysis temperature on cadmium and zinc sorption capacity of sugar cane straw – derived biochar. *BioResources* 8: 4992–5004.
Mendez, A., A. Gomez, J. Paz-Ferreiro and G. Gasco. 2012. Effects of sewage sludge biochar on plant metal availability after application to a Mediterranean soil. *Chemosphere* 89: 1354–1359.
Meyer, S., B. Glaser and P. Quicker. 2011. Technical, economical, and climate-related aspects of biochar production technologies: A literature review. *Environ. Sci. Technol*. 45: 9473–9483.
Mia, S., F.A. Dijkstra and B. Singh. 2017. Aging induced changes in biochar's functionality and adsorption behavior for phosphate and ammonium. *Environ. Sci. Technol*. 51: 8359–8367.
Mohan, D., A. Sarswat, Y.S. Ok and C.U. Pittman Jr. 2014. Organic and inorganic contaminants removal from water with biochar, a renewable, low cost and sustainable adsorbent – a critical review. *Bioresour. Technol*. 160: 191–202.
Morin-Crini, N., E. Lichtfouse, G. Torri and G. Crini. 2019. Applications of chitosan in food, pharmaceuticals, medicine, cosmetics, agriculture, textiles, pulp and paper, biotechnology, and environmental chemistry. *Environ. Chem. Lett*. 17: 1667–1692.
Mukherjee, A., A.R. Zimmerman and W. Harris. 2011. Surface chemistry variations among a series of laboratory-produced biochars. *Geoderma* 163: 247–255.
Netto, M.S., J. Georgin, D.S.P. Franco, et al. 2021. Effective adsorptive removal of atrazine herbicide in river waters by a novel hydrochar derived from Prunus serrulata bark. *Environ. Sci. Pollut. Res*. 1–14.
Oliveira Jr, R.S., W.C. Koskinen and F.A. Ferreira. 2001. Sorption and leaching potential of herbicides on Brazilian soils. *Weed Res*. 41: 97–110.
Palansooriya, K.N., Y. Yang, Y.F. Tsang, et al. 2020. Occurrence of contaminants in drinking water sources and the potential of biochar for water quality improvement: A review. Crit. Rev. *Environ. Sci. Technol*. 50: 549–611.
Panahi, H.K.S., M. Dehhaghi, Y.S. Ok, et al. 2020. A comprehensive review of engineered biochar: Production, characteristics, and environmental applications. *J. Clean. Prod*. 270: 122462.
Papadakis, E.N., Z. Vryzas, A. Kotopoulou, K. Kintzikoglou, K.C. Makris and E. Papadopoulou-Mourkidou. 2015. A pesticide monitoring survey in rivers and lakes of northern Greece and its human and ecotoxicological risk assessment. *Ecotoxicol. Environ. Saf*. 116: 1–9.
Pereira, R.C., J. Kaal, M.C. Arbestain, et al. 2011. Contribution to characterisation of biochar to estimate the labile fraction of carbon. *Org. Geochem*. 42: 1331–1342.
Premarathna, K.S.D., A.U. Rajapaksha, B. Sarkar, et al. 2019. Biochar-based engineered composites for sorptive decontamination of water: A review. *Chem. Eng. J*. 372: 536–550.
Qian, K., A. Kumar, D. Bellmer, W. Yuan, D. Wang and M.A. Eastman. 2015. Physical properties and reactivity of char obtained from downdraft gasification of sorghum and eastern red cedar. *Fuel* 143: 383–389.
Qin, Y., X. Zhu, Q. Su, et al. 2020. Enhanced removal of ammonium from water by ball-milled biochar. *Environ. Geochem. Health* 42: 1579–1587.
Qiu, Y., Z. Zheng, Z. Zhou and G.D. Sheng. 2009. Effectiveness and mechanisms of dye adsorption on a straw-based biochar. *Bioresour. Technol*. 100: 5348–5351.
Rangabhashiyam, S. and P. Balasubramanian. 2019. The potential of lignocellulosic biomass precursors for biochar production: Performance, mechanism and wastewater application-a review. *Ind. Crops Prod*. 128: 405–423.
Sarfaraz, Q., L.S. da Silva, G.L. Drescher, et al. 2020. Characterization and carbon mineralization of biochars produced from different animal manures and plant residues. *Sci. Rep*. 10: 1–9.

Schimmelpfennig, S. and B. Glaser. 2012. One step forward toward characterization: Some important material properties to distinguish biochars. *J. Environ. Qual.* 41: 1001–1013.

Shaheen, S.M., A. El-Naggar, J. Wang, et al. 2019. Biochar as an (Im)mobilizing agent for the potentially toxic elements in contaminated soils. *Biochar from biomass and waste*. Amsterdam: Elsevier, 255–274.

Shakoor, M.B., S. Ali, M. Rizwan, et al. 2020. A review of biochar-based sorbents for separation of heavy metals from water. *Int. J. Phytoremediation* 22: 111–126.

Shakoor, M.B., Z.-L. Ye and S. Chen. 2021. Engineered biochars for recovering phosphate and ammonium from wastewater: A review. *Sci. Total Environ*. 146240.

Shinogi, Y. and Y. Kanri. 2003. Pyrolysis of plant, animal and human waste: Physical and chemical characterization of the pyrolytic products. *Bioresour. Technol*. 90: 241–247.

Singh, B., B.P. Singh and A.L. Cowie. 2010. Characterisation and evaluation of biochars for their application as a soil amendment. *Soil Res*. 48: 516–525.

Spokas, K.A. 2010. Review of the stability of biochar in soils: Predictability of O: C molar ratios. *Carbon Manag*. 1: 289–303.

Sun, K., M. Kang, K.S. Ro, J.A. Libra, Y. Zhao and B. Xing. 2016. Variation in sorption of propiconazole with biochars: The effect of temperature, mineral, molecular structure, and nano-porosity. *Chemosphere* 142: 56–63.

Sun, K., M. Keiluweit, M. Kleber, Z. Pan and B. Xing. 2011. Sorption of fluorinated herbicides to plant biomass-derived biochars as a function of molecular structure. *Bioresour. Technol*. 102: 9897–9903.

Suo, F., X. You, Y. Ma and Y. Li. 2019. Rapid removal of triazine pesticides by P doped biochar and the adsorption mechanism. *Chemosphere* 235: 918–925.

Takaya, C.A., L.A. Fletcher, S. Singh, K.U. Anyikude and A.B. Ross. 2016. Phosphate and ammonium sorption capacity of biochar and hydrochar from different wastes. *Chemosphere* 145: 518–527.

Tan, X., Y. Liu, G. Zeng, et al. 2015. Application of biochar for the removal of pollutants from aqueous solutions. *Chemosphere* 125: 70–85.

Tang, J., H. Lv, Y. Gong and Y. Huang. 2015. Preparation and characterization of a novel graphene/biochar composite for aqueous phenanthrene and mercury removal. *Bioresour. Technol*. 196: 355–363.

Tao, Y., S. Hu, S. Han, et al. 2019. Efficient removal of atrazine by iron-modified biochar loaded Acinetobacter lwoffii DNS32. *Sci. Total Environ*. 682: 59–69.

Tomczyk, A., Z. Sokołowska and P. Boguta. 2020. Biochar physicochemical properties: Pyrolysis temperature and feedstock kind effects. Rev. *Environ. Sci. Bio/Technology* 19: 191–215.

Tripathi, M., J.N. Sahu and P. Ganesan. 2016. Effect of process parameters on production of biochar from biomass waste through pyrolysis: A review. Renew. *Sustain. Energy Rev*. 55: 467–481.

Vacca, A., L. Mais, M. Mascia, E.M. Usai and S. Palmas. 2020. Design of experiment for the optimization of pesticide removal from wastewater by photo-electrochemical oxidation with TiO_2 nanotubes. *Catalysts* 10: 512.

Vanraes, P., N. Wardenier, P. Surmont, et al. 2018. Removal of alachlor, diuron and isoproturon in water in a falling film dielectric barrier discharge (DBD) reactor combined with adsorption on activated carbon textile: Reaction mechanisms and oxidation by-products. *J. Hazard. Mater*. 354: 180–190.

Varjani, S., G. Kumar and E.R. Rene. 2019. Developments in biochar application for pesticide remediation: Current knowledge and future research directions. *J. Environ. Manage*. 232: 505–513.

Vieira, W.T., M.B. de Farias, M.P. Spaolonzi, M.G.C. da Silva and M.G.A. Vieira. 2020. Removal of endocrine disruptors in waters by adsorption, membrane filtration and biodegradation. A review. *Environ. Chem. Lett*. 18: 1113–1143.

Vonberg, D., D. Hofmann, J. Vanderborght, et al. 2014. Atrazine soil core residue analysis from an agricultural field 21 years after its ban. *J. Environ. Qual.* 43: 1450–1459.

Wagner, S., R. Jaffé and A. Stubbins. 2018. Dissolved black carbon in aquatic ecosystems. *Limnol. Oceanogr. Lett.* 3: 168–185.

Wang, B., B. Gao and J. Fang. 2017. Recent advances in engineered biochar productions and applications. *Crit. Rev. Environ. Sci. Technol.* 47: 2158–2207.

Wang, H., K. Lin, Z. Hou, B. Richardson and J. Gan. 2010. Sorption of the herbicide terbuthylazine in two New Zealand forest soils amended with biosolids and biochars. *J. Soils Sediments* 10: 283–289.

Wang, W., D. Xu, K. Chau and G. Lei. 2014. Assessment of river water quality based on theory of variable fuzzy sets and fuzzy binary comparison method. *Water Resour. Manag.* 28: 4183–4200.

Wei, D., B. Li, H. Huang, et al. 2018. Biochar-based functional materials in the purification of agricultural wastewater: Fabrication, application and future research needs. *Chemosphere* 197: 165–180.

Wei, L., Y. Huang, L. Huang, et al. 2020. The ratio of H/C is a useful parameter to predict adsorption of the herbicide metolachlor to biochars. *Environ. Res.* 184: 109324.

Wu, W., M. Yang, Q. Feng, et al. 2012. Chemical characterization of rice straw-derived biochar for soil amendment. *Biomass and bioenergy* 47: 268–276.

Xiao, R., H. Zhang, Z. Tu, et al. 2020. Enhanced removal of phosphate and ammonium by MgO-biochar composites with NH 3· H 2 O hydrolysis pretreatment. *Environ. Sci. Pollut. Res.* 27: 7493–7503.

Xiong, X., K.M. Iris, D.C.W. Tsang, et al. 2019. Value-added chemicals from food supply chain wastes: State-of-the-art review and future prospects. *Chem. Eng. J.* 375: 121983.

Xiong, Z., Z. Shihong, Y. Haiping, S. Tao, C. Yingquan and C. Hanping. 2013. Influence of NH 3/CO 2 modification on the characteristic of biochar and the CO 2 capture. *BioEnergy Res.* 6: 1147–1153.

Yang, G.-P., Q. Chen, X.-X. Li and X.-Y. Cao. 2010. Study on the sorption behaviors of Tween-80 on marine sediments. *Chemosphere* 79: 1019–1025.

Yavari, S., M. Abualqumboz, N. Sapari, H.A. Hata-Suhaimi, N.Z. Nik-Fuaad and S. Yavari. 2020. Sorption of imazapic and imazapyr herbicides on chitosan-modified biochars. *Int. J. Environ. Sci. Technol.* 17: 3341–3350.

Yoder, J., S. Galinato, D. Granatstein and M. Garcia-Pérez. 2011. Economic tradeoff between biochar and bio-oil production via pyrolysis. *Biomass and bioenergy* 35: 1851–1862.

Yu, J., H. Hu, X. Wu, et al. 2020. Coupling of biochar-mediated absorption and algal-bacterial system to enhance nutrients recovery from swine wastewater. *Sci. Total Environ.* 701: 134935.

Yuan, J.-H., R.-K. Xu and H. Zhang. 2011. The forms of alkalis in the biochar produced from crop residues at different temperatures. *Bioresour. Technol.* 102: 3488–3497.

Zabaniotou, A., G. Stavropoulos and V. Skoulou. 2008. Activated carbon from olive kernels in a two-stage process: Industrial improvement. *Bioresour. Technol.* 99: 320–326.

Zhang, P., H. Sun, L. Yu and T. Sun. 2013a. Adsorption and catalytic hydrolysis of carbaryl and atrazine on pig manure-derived biochars: Impact of structural properties of biochars. *J. Hazard. Mater.* 244: 217–224.

Zhang, P., H. Sun, L. Yu and T. Sun. 2013b. Adsorption and catalytic hydrolysis of carbaryl and atrazine on pig manure-derived biochars: Impact of structural properties of biochars. *J. Hazard. Mater.* 244: 217–224.

Zhang, Y.-J., Z.-J. Xing, Z.-K. Duan, M. Li and Y. Wang. 2014. Effects of steam activation on the pore structure and surface chemistry of activated carbon derived from bamboo waste. *Appl. Surf. Sci.* 315: 279–286.

Zhang, Y., B. Zhang, J. Yu, et al. 2021. Functional biochar for efficient residue treatment of sulfonylurea herbicides by weak molecular interaction. *Biochar*. 1–12.

Zhao, X., W. Ouyang, F. Hao, et al. 2013. Properties comparison of biochars from corn straw with different pretreatment and sorption behaviour of atrazine. *Bioresource technology*, 147: 338-344.

Zhelezova, A., H. Cederlund and J. Stenström. 2017. Effect of biochar amendment and ageing on adsorption and degradation of two herbicides. *Water, Air, Soil Pollut.* 228: 216.

Zheng, S., B. Chen, X. Qiu, M. Chen, Z. Ma and X. Yu. 2016. Distribution and risk assessment of 82 pesticides in Jiulong River and estuary in South China. *Chemosphere* 144: 1177–1192.

Zheng, W., M. Guo, T. Chow, D.N. Bennett and N. Rajagopalan. 2010. Sorption properties of greenwaste biochar for two triazine pesticides. *J. Hazard. Mater.* 181: 121–126.

Zheng, Y., B. Wang, A.E. Wester, et al. 2019. Reclaiming phosphorus from secondary treated municipal wastewater with engineered biochar. *Chem. Eng. J.* 362: 460–468.

Index

Note: Page numbers in *italics* indicate a figure and page numbers in **bold** indicate a table on the corresponding page.

A

B

M

N

O

P

Q

R

S

T

U

V

W

X

Z